# Deadly Metal Rain: The Legality of Flechette Weapons in International Law

# International Humanitarian Law Series

VOLUME 32



The International Humanitarian Law Series is a series of monographs and edited volumes which aims to promote scholarly analysis and discussion of both the theory and practice of the international legal regulation of armed conflict.
The series explores substantive issues of International Humanitarian Law including,

- protection for victims of armed conflict and regulation of the means and methods of warfare
- questions of application of the various legal regimes for the conduct of armed conflict
- issues relating to the implementation of International Humanitarian Law obligations
- national and international approaches to the enforcement of the law and
- the interactions between International Humanitarian Law and other related areas of international law such as Human Rights, Refugee Law, Arms Control and Disarmament Law, and International Criminal Law.

*The titles published in this series are listed at brill.nl/ihul*

# Deadly Metal Rain: The Legality of Flechette Weapons in International Law

## A Reappraisal Following Israel's Use of Flechettes in the Gaza Strip (2001-2009)

*Edited by*

Eitan Barak

Assistant Professor

The Hebrew University of Jerusalem

MARTINUS NIJHOFF PUBLISHERS

LEIDEN • BOSTON

2011

This book is printed on acid-free paper.

ISBN 9789004167193

# Table of Contents

# Illustrations and Maps

# Abbreviations

| | |
|---|---|
| ACR | *Advanced Combat Rifle* Program |
| AEF | American Expeditionary Forces |
| AFLR | Air Force Law Review |
| AJIL | American Journal of International Law |
| AMC | Army Materiel Command |
| AP-I | Additional Protocol I |
| APERS-T | Anti-Personnel (Round) with Tracer |
| ARMCOM | U.S. Army Armament Command |
| ASIL | American Society of International Law |
| AVM | Anti-Vehicle Mines |
| BRL | Ballistics Research Laboratory |
| CCD | Conference of the Committee on Disarmament |
| CMs | Cluster Munitions |
| CMC | Cluster Munition Coalition |
| CRDL | Chemistry Research Development Laboratory |
| CRS | Congressional Research Service |
| CYIL | Canadian Yearbook of International Law |
| Commentary on the Additional Protocols | *Commentary on the Additional Protocols of 8 June 1977 to the Geneva Conventions of 12 August 1949* (ICRC, Y. Sandoz et al., Eds., 1987) |
| ICRC Customary Law Study | *Customary International Humanitarian Law, 2 Vol.* (ICRC, Jean-Marie Henckaerts & Louise Doswald-Beck, Eds., 2005) |
| EJIL | European Journal of International Law |
| ENMOD | Environmental Modification |
| ERW | Explosive Remnants of War |
| FAS | Federation of American Scientists |
| FEAF | Far East Air Force |
| GGE | Group of Government Experts |
| GPF | General Purpose Flechette |
| GOC | General Officer Commanding |
| GWILR | George Washington International Law Review |
| HCJ | High Court of Justice |
| HRW | Human Rights Watch |
| HESH | High Explosive Squash Head |
| IAC | International Armed Conflict |

| | |
|---|---|
| ICBL | International Campaign to Ban Landmines |
| ICC | International Criminal Court |
| ICJ | International Court of Justice |
| ICJ Rep. | Reports of the International Court of Justice |
| ICRC | International Committee of the Red Cross |
| ICTY | International Criminal Tribunal for the Former Yugoslavia |
| IDF | Israeli Defense Force |
| IHL | International Humanitarian Law |
| ILM | International Legal Materials |
| ILR | International Law Reports |
| IMT | International Military Tribunal |
| IRRC | International Review of the Red Cross |
| J. OF C&S LAW | Journal of Conflict & Security Law |
| JCS | Joint Chiefs of Staff |
| LIAC | Law of International Armed Conflict |
| LPR | Laboratory Posture Report |
| *Laws of Armed Conflicts* | *The Laws of Armed Conflicts: A Collection of Conventions, Resolutions and Other Documents* (D. Schindler and J. Toman, Eds., 4th ed., 2004) |
| MAG | Military Advocate General |
| MOTAPM | Mines Other than Anti-Personnel Mines |
| NATO | North Atlantic Treaty Organization |
| New Rules | *New Rules for Victims of Armed Conflicts: Commentary on the Two 1977 Protocols Additional to the Geneva Conventions of 1949* (M. Bothe, K. J. Partsch and W. A. Solf eds., 1982) |
| NGO | Non-Governmental Organization |
| NYIL | Netherlands Yearbook of International Law |
| PCATI | Public Committee Against Torture in Israel |
| PCHR | Palestinian Centre for Human Rights |
| PHR-I | Physicians for Human Rights-Israel |
| R.F.C. | Royal Flying Corps |
| SCWS | Small Caliber Weapons Systems |
| Sess. | Session |
| SIPRI | Stockholm International Peace Research Institute |
| SIrUS | Superfluous Injury or Unnecessary Suffering (The ICRC Project) |
| SOAR | Special Operations Aviation Regiment |
| SPIW | Special Purpose Individual Weapon |
| Supp. | Supplement |
| UN | United Nations |
| UNESCO | United Nations Educational, Scientific and Cultural Organization |
| UNGA | United Nations General Assembly |
| UNIDIR | United Nations Institute for Disarmament Research |
| UNTS | United Nations Treaty Series |
| UNSC | United Nations Security Council |

| | |
|---|---|
| UNSG | United Nations Secretary General |
| USAF | United States Air Force |
| WMD | Weapons of Mass Destruction |
| WP | White Phosphorous |

# List of Cases



**Israeli Cases**

# List of Treaties

# Acknowledgments

During the years of research spent on this book, I have received invaluable assistance from numerous NGOs, foundations and individuals. In acknowledging their valuable help, I hope that I have not overlooked anyone.

I wish to thank to Roee Ariav, John Barrie, Guy Harpaz, Moshe Hirsch, Brian Rappert, Robbie Sabel, Yuval Shany, Moshe Tirosh and Shelomi Zecharia for their comments on earlier versions or chapters of this manuscript. I am indebted to Dominic Barker, Robin Coupland, Conrad Crane, Elizabeth Greenhalgh, Richard Guthrie, Scott Hamilton, Yossi Kaufman, Milton Leitenberg, Noam Ophir, Jullian Perry Robinson and Yossi Zur, who provided me with valuable sources. Special thanks go to Eric Prokosch for sharing his extensive personal files on flechette weapons as well as for his valuable comments on various chapters of this book.

Special thanks go to Physicians for Human Rights–Israel, B'Tselem–The Israeli Information Center for Human Rights in the Occupied Territories and the International Committee of the Red Cross (ICRC)–Israel and the Occupied and Autonomous Territories Delegation, for allowing me access to their resources and archives.

My sincere appreciation goes to The Minerva Center for Human Rights, The Hebrew University of Jerusalem and the Professor Nathan and Judith Feinberg Foundation for Promoting the Study of International Law for the support provided when writing of this book. A special word of gratitude goes to Martinus Nijhoff Publisher's Alan Stephens and Lindy Melman for their assistance, encouragement and patience throughout the long period this project has taken.

I also wish to mention Yonatan Fox, Efrat Hakak, Deena Pinson, Alex Porat, Yeela Porat, Gal Sever, Asaf Shamis and Daniel Shkedi, and especially Lior Avni, for their able research assistance and dedication. I am also indebted to Michele Manspeizev, and especially to Nina Reshef, my language editor throughout the years, for her consistently superb work and commitment.

I cannot close without expressing the debt to my family. My parents, Zila and Amos Berkovich, have persistently stood by me. My wife, Michal, has always given me her unfailing support and understanding while managing the responsibilities of a career and care of our three beloved children, Roee, Inbal and Eran. I dedicate this book to her.

Eitan Barak
The Hebrew university
Jerusalem

# Introduction

On December 27, 2008, Israel initiated a major 22-days long military operation in the Gaza Strip (Operation *Cast Lead*).[1] Similar to the case of the 2006 Second Lebanon War, [2] Israel was accused of using illegal weapons (cluster munitions in the former, white phosphorus [WP] in the latter). This claim appeared during a time of intense fighting[3] and became increasingly acerbic after the Operation's conclusion on January 17 (local time).[4] Ten days later, however, Amnesty International reported that "Apart from white phosphorus" its fact-finding team in Gaza found "hard evidence" of the use of flechettes – in the Israeli context armor rounds containing thousands of 'dart-like projectile[s that] stabilized [their] flight by small fins'[5] – during

1 For an overview on the events preceding the operation *see, e.g.*, Jim Zanotti *et al.*," Israel and Hamas: Conflict in Gaza (2008-2009)", *Congressional Research Service* (*CRS*) Rep. (Jan. 15, 2009).

2 *See*, for *example Israeli Cluster Munitions Hit Civilians in Lebanon*, Human Rights Watch (HRW) Press Release (July 24, 2006), available at <hrw.org/english/docs/2006/07/24/isrlpa13798_txt.htm>. Notably, in the context of the Law of Weaponry, the term 'weapons' includes 'ammunitions'. For the reader's benefit, we use the term "weapons" to cover both although the term "means of warfare" is used as well. The term "methods of warfare", however, refers specifically to the manner in which these means are used. For a discussion of these terms *see,* Hays, W. Parks, "Means and Methods of Warfare", *The George Washington International Law Review (GWILR)* (2006), 511, at 512, fn. 2.

3 *See e.g.,* Michael Evans & Sheera Frenkel, "Gaza Victims' Strange Burns Increase Concern over Phosphorus Shells", *The Times*, (Jan. 8, 2009) at 6; Peter Beaumont & Richard Norton-Taylor, "Airburst Shells 'are danger to civilians'", *The Guardian*, (Jan. 9, 2009), at 19; "UN Accuses Israel Over Phosphorus", *BBC News*, (Jan. 15, 2009), available at <http://news.bbc.co.uk/go/pr/fr/-/2/hi/middle_east/7831424.stm>.

4 *See*, for example, *Rain of Fire: Israel's Unlawful Use of White Phosphorus in Gaza* (HRW Rep., Mar. 25, 2009). For a critique of this report *see, HRW's "Rain of Fire": Neither Thorough Nor Impartial*", (Non-Governmental Organization [NGO] Monitor Rep., Apr. 2, 2009). For internal criticism *see*, for example, Amos Harel, "IDF Probes Improper Use of Phosphorus Shells in Gaza Strip", *Haaretz*, (Jan. 21, 2009), at A1; "Investigate Now" (Editorial), *Haaretz*, (Jan. 22, 2009), at 5. *Note: Haaretz* has published a concise English-language edition as of September 1997 (also available online). Citations were taken from this source whenever possible to avoid translation errors. Unless otherwise indicated, Hebrew sources were translated by the author.

5 For the definition of flechette as it appears in the professional literature *see, Jane's Ammunition Handbook 2005-2006* 14th ed. (London: Jane's Information Group 2005), 9 [Glossary section] [hereinafter, *Ammunition Handbook 2005-2006*].

the Gaza Operation.[6] This news release was followed the next month by a 45-page brief on the types of munitions used in the Operation, in which a few cases of civilian casualties resulting from flechette rounds were first mentioned.[7] Yet, the emphasis was placed on Israel's use of WP and this growing debate detracted attention from the use of the flechette weapons. Neither the media[8] nor Amnesty's reports on civilian casualties (i.e., collateral damage) due to the use of flechette weapons during the Operation were seemingly powerful enough to alter this state of affairs.

In September 2009, however, nine months after the Operation's conclusion, the legality of flechette weapons became the subject of interest following the release of the report prepared by the Gaza Fact-Finding Mission, commissioned by The United Nations (UN) Human Rights Council.[9] The "Goldstone Report," as it was called after the Mission's head, Justice Richard J. Goldstone, highlighted flechettes and a few other types of weapons (mainly WP) that were allegedly used by Israel during the Operation "in relation to the obligation to take all feasible precautions in the choice of the means and methods of warfare."[10] After finding that "in all cases those hit by these devices were civilians,"[11] the Mission stated that "(f)lechettes, as an area weapon, are particularly unsuitable for use in urban settings".[12] Hence, despite acknowledgment by the report's authors that the weapon is not "currently proscribed under international law,"[13] they called for an urgent UN General Assembly (UNGA) discussion on the "future legality of the use of certain munitions referred to in this report, and in particular white phosphorous, flechettes and heavy metal such as tungsten".[14]

Unlike the case of cluster munitions, Israel's use of flechettes, an area weapon (i.e., a means of warfare that is effective over a broad area, even if it is targeted against a specific point) by definition, within the Arab-Israeli conflict, has not been reserved

6 *Israeli army used flechettes against Gaza civilians*, Amnesty International Press Release (Jan. 27, 2009), <http://www.amnesty.org/en/news-and-updates/news/israeli-army-used-flechettes-against-gaza-civilians-20090127>.

7 *Israel-OPT – Fuelling conflict: Foreign arms supplies to Israel/Gaza* (Amnesty International MDE 15/012/2009, Feb. 23, 2009), 12-13.

8 *See*, for example, Amira Hass, "Penetrating the Laws of War", *Haaretz*, Apr. 17, 2009, at 3.

9 *See,* Human Rights Council Res. S-9/1, (Apr. 3, 2009).

10 *Human Rights in Palestine and other Occupied Arab Territories*, Report of the United Nations Fact-Finding Mission on the Gaza Conflict (A/HRC/12/48) Sept. 15, 2009, Chap. XII ["The Use of Certain Weapons"] [hereinafter, *The Goldstone Report*], at 194, para. 886. In addition to flechettes and WP, Dense Inert Metal Explosive (DIME) and "depleted and non- depleted uranium munitions" were mentioned. As to (the first) weapon (DIME), the Mission cited allegations regarding its use by the IDF in its "factual findings" section and noted that "further clarifications" were required (*see, Id.*, at 198, para. 909). As to the latter (munitions containing uranium) it stated that: "While it cannot be excluded that such weapons were used, on the basis of the information received the Mission decided not to investigate the matter further". *Id.*, at 198, para. 911.

11 *Id.*, at 197, para. 904.

12 *Id.*, at 414, para. 1924

13 *Id.*

14 *Id.*, at 425, para. 1971(d).

exclusively to large-scale military clashes.[15] Hence, future discussions of the weapons' legality which may eventually lead to some regulation of its use would likely address the increasing humanitarian concerns among some states, especially by civil society. After all, these weapons may in fact be as harmful to non-combatants as the Goldstone Report suggests. Evidence for this possibility was brought forth during the low-intensity conflict waged by Israel in Lebanon prior to its withdrawal in May 2000 and in the Gaza Strip since 2001. As such, the fact that the legality of flechette weapons has been broached only obliquely in the legal and the academic literature leaves many questions unanswered.

## The Legality of Flechette Weapons: Origin of the Lacuna

The scant literature addressing the flechettes' legality is glaring given the vast literature on related weapons, written in the 1970s. During that decade, the international community debated the possibility of applying prohibitions or restrictions on certain conventional weapons. In 1980, the *Convention on Prohibitions or Restrictions on Use of Certain Conventional Weapons which may be Deemed to be Excessively Injurious or to Have Indiscriminate Effects*[16] was finally adopted (the treaty is usually referred to as the *Inhumane Weapons Convention or the Convention on Certain Conventional Weapons* [CCW]).[17] As an "umbrella treaty", it contained procedural articles with specific prohibitions elaborated in the attached protocols, a format designed to facilitate the convenient addition of agreed upon weapons prohibitions or restrictions.[18]

In practice, however, very few weapons became the subjects of proposals for categorical prohibitions within the framework of this convention. Cluster munitions, incendiary weapons and flechettes were grouped together in the same narrow category of weapons whose use was considered by many in the international community to be inconsistent with the principles of International Humanitarian Law (IHL) governing the employment of weapons.

At first glance, the reason for this lacuna is simple: despite grave concerns over the use of these weapons in the 1970s, they were not used to any major degree in any of the subsequent armed conflicts that had attracted much international attention,

15 Israel has rarely used cluster munitions outside of large scale clashes (i.e., they were used in the 1973 War, the 1982 First Lebanon War, the 1978 Litani Operation and the 2006 Second Lebanon War). For a succinct history of Israel's use of cluster munitions *see,* Eitan Barak, "Doomed to be Violated ? The U.S.-Israeli Clandestine End-User Agreement and the Second Lebanon War: Lessons for the Convention on Cluster Munitions", *Denver Journal of International Law and Policy* 38 (2009) 1, at 6-13.

16 Dietrich Schindler and Jiri Toman, (eds.) *The Laws of Armed Conflicts* (Leiden: Martinus Nijhoff, 2004), 181 [hereinafter, *Laws of Armed Conflicts*]. Note: for the reader's convenience, conventions, resolution and other documents were cited from this source whenever possible.

17 *See*, for instance, Jozef Goldblat, *Arms Control: the New Guide to Negotiations and Agreements* 2nd ed., (Washington D.C.: Sage Publications, 2002), 287. The acronym CCW will be used to denote the Convention throughout this book.

18 For the term "umbrella treaty" *see,* for instance, Ashley J. Roach, "Certain Conventional Weapons Convention: Arms Control or Humanitarian Law ?", *Military Law Review* 105 (Summer 1984) 3, at 13.

such as the North Atlantic Treaty Organization (NATO) bombing of the Federal Republic of Yugoslavia (1999) or the two Gulf Wars (1991 and 2003). In fact, with the exception of the Israeli case, firearm flechettes have not been employed *en masse* since the end of the American intervention in Vietnam, with small-scale use of the weapon greatly in doubt. Flechette use thus differs sharply in this respect from the massive use of cluster munitions in various battlefields during the last two decades: the U.S. use against Iraqi forces and the NATO use against Serbian forces in the previously mentioned conflicts and, naturally, Israel's massive use of these weapons in the Second Lebanon War.

## Flechette Weapons: Their Use by Israel (1982-2009)

An explanation is not necessarily a justification. And so, given Israel's continued employment of flechettes in armed conflicts, the need for a literature on the weapon's legality has become acute. To emphasize this point, we briefly review the military campaigns in which Israel has been involved since the early 1980s: (a) The 1982 First Lebanon War and the subsequent protracted conflict between Israeli forces and Lebanese guerillas in South Lebanon up to Israel's withdrawal in May 2000. We should note that in the 2006 Second Lebanon War, tactical considerations are thought to have been the primary factor preventing widespread Israeli use of flechette rounds;[19] (b) The Israeli-Palestinian conflict waged in the Gaza Strip since 2001, with a two-year interval between Israel's September 2005 withdrawal and the Hamas achievement of full control of the Gaza Strip in June 2007. As the December 2008 *Operation Cast Lead* demonstrated, calls by human rights non-governmental organizations (NGOs) for an end to the use of "this prohibited weapon in the Gaza Strip",[20] repeated since 2001, have been of limited impact.[21]

---

19 For first-hand testimony on why flechette use was forestalled during the early stages of the Second Lebanon War, *see,* "When Permission is Given to Drive on a Civilian Road, Armored Forces Members Understand that a War has Broken Out", *Haaretz*, (July 23, 2006), at A7 (in Heb.). After the war, however, IDF soldiers published online accounts of the eventual small-scale use of the weapon. *See,* for example, <www.fresh.co.il/vBulletin/ showthread.php?t=252404#post1644400> (in Heb.).

20 *See, Firing of Flechette Shell that Killed Reuters Cameraman Violates Laws of War*, B'Tselem – The Israeli Information Center for Human Rights in the Occupied (hereinafter, B'Tselem) Press Release (Apr. 17, 2008) available at < http://www.btselem.org/english/Press_Releases/20080417.asp>. This specific call for an inquiry was motivated by the well-documented April 2008 case in which a flechette weapon was responsible for the killing of a Reuters TV cameraman, Fadel Shana, as well as three Palestinian bystanders. For the case *see, e.g.,* Rory McCarthy, "Palestinian Journalist Killed by Metal Darts from Israeli Shell", *The Guardian* (London), (Apr. 18, 2008), at 24. For similar calls against the use of flechettes in the Gaza Strip, as early as June 2001, *see,* for instance, HRW, *Israel: Dart Shells Pose Civilian Threat,* (Press Release, June 15, 2001), available at <http://www.hrw.org/en/news/2001/06/15/israel-dart-shells-pose-civilian-threat>.

21 Israel has recently acknowledged its own use of flechette weapons over the years, and mainly during *Operation Cast Lead,* in its official position paper regarding the legality of its military conduct during this Operation. *See,* Government of Israel, *The Operation*

In other words, a weapon accused of being inherently indiscriminate, and responsible for causing unnecessary suffering among combatants and non-combatants alike, became the standard anti-personnel weapon used by Israel's armor forces.[22] Moreover, it seems that this type of weapon was used in conflicts where its inherent characteristics were at odds with the conditions warranting its use. Especially troublesome is the weapon's continuous though sporadic use in the Gaza Strip. As opposed to the sparsely settled target zones in South Lebanon, the Gaza Strip, one of the most densely populated areas on earth, became the target of a weapon that, by its nature, cannot be employed in manner that distinguishes between its victims.

### Flechette Weapons: The Domestic Legal Debate

If any additional reason was needed to address the lacuna in the literature with in-depth research, it can readily be found. In October 2002, the legality of flechettes use in the Gaza Strip was raised before Israel's Supreme Court sitting as the High Court of Justice (HCJ).[23] In an unprecedented move given the recognized stance of worldwide domestic tribunals – all the more Israel's HCJ[24] – to refrain from involvement in issues related to the means of warfare employed by their governments,[25] the HCJ

---

*in Gaza: 27 December 2008 – 18 January 2009, Factual and Legal Aspects*, (July 2009), paras. 431-435.

22 Excluding the tank's secondary armament. The firing of two flechette rounds followed by a single round of a High Explosive Squash Head (HESH, an explosive warhead that creates a powerful explosion equivalent to that of approximately 40 grenades) at suspicious figures along the frontier of the Israeli-declared Security Zone in Southern Lebanon was a long-standing practice employed by Israel's armored forces. The HESH round was fired in response to the possibility that the targeted persons had taken shelter under rocks. *See,* Meron Rapoport, "They Walked in the Fields", *Yedioth Ahronoth*, (Mar. 8, 2002), (Friday Magazine) at 2 (in Heb.). In the Gaza Strip, although the aforementioned practice was not observed, the firing of flechette rounds as a first choice for targeting suspicious figures was repeatedly recorded. *See,* for instance, Avihai Beker, "The View from There" [Netzarim settlement in the Gaza Strip], *Haaretz*, (Friday Magazine) (Dec. 8, 2003), at 22 (in Heb.).

23 *Physicians for Human Rights & The Palestinian Centre for Human Rights v. O.C. Southern Command & The State of Israel* – Ministry of Defense, Petition for an Interim Decree. The Petition in its original text is available at <http://www.btselem.org/hebrew/Legal_Documents/8990_02_ Flechette_Appeal.doc>. An unofficial, translated version is available at <http://www.btselem.org/english/Legal_Documents/ HC8990_02_flechette_Appeal.rtf> [hereinafter, the *Flechette Petition*]. For the reader's convenience, Petitioners, Petition and State's Response, when related to the *Flechette Petition*, are capitalized throughout.

24 *See,* Ariel L. Bendor, "Justiciability of The Israeli Fight Against Terrorism" *GWILR* 39 (2007), 149-163.

25 As Prof. Falk argued in reference to the 1965 *Shimoda case* in which the legality of using nuclear weapons to bomb Hiroshima and Nagasaki was discussed in Tokyo District Court: "[i]t is the one and only attempt by a court to assess the legality of atomic and, by extension, nuclear weapons". *See,* Richard A. Falk, "The Shimoda Case: A Legal Appraisal of the Atomic Attacks upon Hiroshima and Nagasaki", *The American Journal of International Law* 59 (*AJIL*) (1965) 759, at 759. For the case itself *see, Ryuichi Shimoda et al. vs. The State*, Tokyo, 7 Dec. 1963. English translation is available at: *Japanese Annual*

agreed to review this Petition (hereinafter the *Flechette* Petition), which requested that the Court ban the use of flechette rounds in the Gaza Strip.

The Court's awareness of the highly problematic character of flechette use – as reflected in its troublesome *obiter dictum*: "[r]espondents do not dispute that the use of flechettes... entails increased danger towards civilians that inadvertently enter the zone of fire" – seems to have had a considerable impact on this surprising position.[26] The Petitioners, Physicians for Human Rights-Israel (PHR-I) and the Palestinian Centre for Human Rights (PCHR), argued that the use of rounds which explode in mid-air and scatter thousands of steel darts across a wide expanse in densely populated areas (i.e., the Gaza Strip) represents a breach of the Law of War.[27] The Court's discussion of the legality of flechette weapons *per se* was the only attempt made by any court, national or international, to assess this weapon's legal status, which part of the international community had either attempted to abolish in the 1970s or had labeled as 'dubious'.[28]

---

*of International Law* 8 (1964), 213 (An English summary is available at *"Existing Rules of international law Concerning the Prohibition or Restriction of Use of Specific Weapons," Survey Prepared by the Secretary-General of the United Nations*, UN Doc. A/9215 (Nov. 7, 1973), Vol. II, at 14-21). Although the legality of nuclear weapons was raised in various national courts after 1963, they generally refrained from serious discussion of the issue. In the 1989 *Little Affair*, for instance, a man known as Leonard John Lindon was charged with trespassing on a joint U.S.-Australia defense space research facility in central Australia. The defendant contended, *inter alia*, that the facility's purpose was to assist the U.S. strategy for waging wars, involving nuclear weapons, whose existence, deployment and use are illegal under international law since they "caused (and were designed to cause) terror and indiscriminate slaughter of non-combatants and the long-term poisoning and terminal devastation of the environment" (*See,* 88 International Legal Materials (ILM) (1992), 481-513, at 481-82). Therefore, he claimed that he was forced to take affirmative measures to prevent the continued commission of "that crime". The Northern Territory of Australia Court of Appeals nonetheless refrained from any discussion of the weapons' legality under IHL, while stressing that "we were not referred to any treaty or convention which expressly addresses the legality of the development and stockpiling of nuclear weapons". *Id.*, at 504.

26 HCJ 8990/02, *Physicians for Human Rights v. O.C. Southern Command*, P. D. 57(4) 193, at 194. An official English translation, to which reference has been made, is available at <http://elyon1.court.gov.il/Files_ENG/02/ 990/037/a32/02037990.a32.pdf> [hereinafter, the *Flechette Decision*]. However, as the official translation is completely inconsistent with the Hebrew original in several important parts where the Courts' statements were ascribed to the Respondents, there are instances in which the author's translation is preferable.

27 *The Flechette Petition*. In addition to relying on international law, the petitioners sought to review the legality of flechette use according to Israel's domestic law: *Basic Law: Human Dignity and Liberty* (1992, S.H. 1391, an official English translation is available at <http://www.knesset.gov.il/laws/special/eng/basic3_eng.htm>) and the exercise of "reasonableness" as demanded by Israeli administrative law. As the subject of this book is limited to the flechette's legality within international law, a concise summary of the petitioners' arguments, vis-à-vis Israeli domestic law (*infra* chapter 5A) will suffice here.

28 For the perception of the flechette as a dubious weapon *see,* Stockholm International Peace Research Institute (SIPRI), *The Law of War and Dubious Weapons*, (Stockholm: Almqvist & Wiksell, 1976) 70-72.

In April 2003, the Petition, which introduced evidence that included post mortem reports, was rejected unanimously by a three-justice panel after a short ruling by Justice Eliyahu Mazza (*hereinafter Flechette* Decision).[29] As expected, the rejection drew the Petitioners' ire[30] but severe criticism was also voiced by NGOs and human rights activists in Israel and abroad.[31] Human Rights Watch (HRW), the largest human rights organization in North America, which reiterated its call to the IDF to "immediately stop using U.S.-supplied flechette shells in the Gaza Strip", even promptly translated and publicized the Decision.[32] Shulamit Aloni, Israel's former Minister of Education (1992-1993) and recipient of Israel's highest civilian honor (Israel Prize) wrote: "In its decision of April 27, Israel's highest court has essentially issued a license to kill civilians by determining that the use of flechette rounds fired from tanks is not prohibited by international law".[33]

Nevertheless, despite the outcry following the HCJ ruling and its subsequent translation and publication in the July 2003 issue of *International Legal Materials* (ILM),[34] the international law community has continued to overlook this important subject.[35] Unlike other legal-moral issues that have become subjects of ener-

29 *The HCJ's Flechette Decision*, Justices Mishael Cheshin and Esther Hayut (a temporary nomination at the time) consenting. Both Mazza and Cheshin have since retired.

30 For the response of PHR-I Director-General, Dr. Tomer Fesser, ("It joins a long list of similar decisions ignoring the rights of the Palestinian residents of the territories and giving a free hand to the army to continue its policy of injuring innocent civilians") *see*, Margaret Neighbour, "Israeli Court Refuses to Outlaw Army's Use of Deadly Dart Shells", *The Scotsman*, (Apr. 28, 2003), at A13. For the PCHR's response ("a continuation of the Israeli judiciary's practice of legalizing Israeli war crimes against Palestinian civilians") *see*, *PCHR*, *Press Release*, 27 Apr. 2003, Ref: 52/ 2003, available at <http://www.pchrgaza.ps/files/PressR/English/2003/52-2003.htm>.

31 A leading Israeli NGO, B'Tselem, argued, for instance, that "the justices chose to favor the formalistic arguments and ignored basic rules of international law". *See*, *Flechette Shells: An Illegal Weapon*, available at <http://www.btselem.org/english/Firearms/Flechette.asp>.

32 As HRW declared, "the use of such antipersonnel weapons in densely populated areas makes the risk of civilian casualties intolerably high under international law." *See*, *'Israel: Stop Using Flechettes in Gaza'*, HWR Press Release, Apr. 29, 2003, available at <http://hrw.org/english/docs/2003/04/29/isrlpa5821_txt.htm>. For its translation *see* ,<hrw.org/press/2003/04/israelisupremecourt.htm (no longer available, on file with the author).

33 "The judges found grounds to permit – she added – or more precisely did not find grounds to forbid [use of this weapon in this manner], as if the impermissible could be made permissible". Shulamit Aloni, "A License to Kill Civilians", *Haaretz*, (May 4, 2003), at B2 (in Heb.). An unofficial English translation from which the citations were taken appeared at <http://www.countercurrents.org/pa-aloni240503.htm> [no longer available, on file with the author].

34 Supreme Court: *Physicians for Human Rights et al. v. Almoj* [*sic.*] *and the State of Israel*, *ILM* 42 (2003), 882. An English summary was made available ten months later. *See*, Fania Domb, "Judgments of the Supreme Court of Israel Relating to the Administered Territories" *Israel Yearbook on Human Rights* 34 (2004) 293, at 319-321.

35 For the scant references (up to two pages) to the HCJ Flechette Decision *see, e.g.*, Jeffery K. Walker, "Introductory Note", *ILM* 42 (2003) 880, 880; Yair Tirosh and Matan Goldblatt, "Judicial Review Over the IDF Response to Palestinian Violence since

getic public discussion during the Israeli-Palestinian conflict (e.g., Israel's "targeted killing" policy[36] and the IDF's "neighbor procedure"[37]) and have clearly affected the international law community, the flechette issue has remained untouched.

In fact, the scant attention given to this incident is surprising and lamentable regardless of the specific weapon involved. At the heart of the dispute is a crucial distinction in the Law of War, the distinction between *legal* and *illegal* weapons. A discussion of this distinction in national courts, especially in the highest courts, is a remarkably rare event, perhaps because the courts have chosen to avoid the issue or simply because no petitions have been submitted.[38] The wording of the ILM's introduction to the translated ruling, in itself an indication of the importance of the event, adequately makes this point: "It is rare for a domestic tribunal – in particular the highest appellate court – *to take up a pure law of war issue like this*".[39] In fact, in January 2002, only a year previous to the submission of the *Flechette Petition*, the HCJ stated in the context of the legality of targeted killing that: "[T]he choice of weapons used by the respondents in order to prevent murderous terror attacks is not an issue that

---

September 2000", *IDF Law Review* 1 (2003) 311, at 317; David Turns, "Weapons in the ICRC Study on Customary International Humanitarian Law" *Journal of Conflict and Security Law* (J. OF C&S LAW) 11 (2006) 201, at 215.

36 For the HCJ's decision regarding 'targeted killings' *see*, HCJ 769/02, *The Public Committee against Torture in Israel v. The Government of Israel*, 53(4) P.D. 817. For analyses of the legality or illegality of this practice prior to this decision *see*, for instance, Orna Ben-Naftali & Keren R. Michaeli, "We Must Not Make a Scarecrow of the Law: A Legal Analysis of the Israeli Policy of Targeted Killings" *Cornell International Law Journal* 36 (2003) 233; Gross Michael, "Fighting by Other Means in the Mideast: A Critical Analysis of Israel's Assassination Policy" *Political Studies* 51 (2003) 350; David Kretzmer, "Targeted Killings of Suspected Terrorists: Judicial Executions or Legitimate Means of Defence" *European Journal of International Law* (*EJIL*) 16 (2005) 212. For a comprehensive analysis of this subject in general *see*, Nils Melzer, *Targeted Killing in International Law* (Oxford: Oxford University Press, 2008).

37 For the HCJ's decision regarding "the neighbor procedure" (i.e., using local civilians as human shields) *see*, HCJ 3799/02, *Adalah et al., v. OC Cent. Command et al.*, 2003(1) 60(3) P.D. 67. For an English translation *see*, *ILM* 45 (2006), 491-503. For a legal analysis *see*, for instance, Michael N. Schmitt, "Human Shields in International Humanitarian Law" *Columbia Journal of Transnational Law* 47 (2009) 292, at 312-16. For a legal analysis on this subject prior to the HCJ's decision *see*, for instance, Roland Otto, "Neighbours as Human Shields? The Israel Defense Forces' "Early Warning Procedure" and International Humanitarian Law" *International Review of the Red Cross* 856 (*IRRC*) (2004), 771.

38 A pertinent case, concerning the use of cluster munitions (specifically, "Orkan" rockets [For details *see*, *Ammunition Handbook 2005-2006, at 660*] fitted with a cluster munitions warhead), was raised before the International Criminal Tribunal for the former-Yugoslavia (ICTY) in *The Prosecutor v Martić* [hereinafter, *the Martić case*]. While reviewing the indictment in that case, The Trial Chamber, avoided any serious discussion of the weapon's legality and merely noted that there was nothing in IHL formally forbidding the use of cluster bombs *per se*. *See*, The Prosecutor v. *Martić*, Trial Chamber I, Decision in Rule 61 Proceedings, Mar. 8, 1996, at 146, available at <http://www.icty.org/case/ martic/4>. For more on this case *see*, *infra* chapter 6(D).

39 Emphasis added. Walker, "Introductory Note", *supra* note 34, at 880.

this court would see fit to intervene in".[40] Three months later, during Operation "*Defensive Shield*," in which most West Bank cities were re-occupied by the IDF, a number of NGOs submitted a petition requesting, *inter alia*, that Israeli armed forces "refrain from shelling and from striking civilian targets in the Jenin refugee camp, by means of tanks, combat helicopters and/or all other types of shooting".[41] In rejecting the petition, the HCJ held that "to the extent it invites the courts to review operational measures during combat, the petition is non-justiciable".[42]

The dispute over the formulation of the prohibited weapons category in the Statute of the International Criminal Court (ICC)[43] testifies to the political sensitivity surrounding this ostensibly legal distinction. The opposition to the 1997 New Zealand–Switzerland proposal to make the ICC directly responsible for determining which weapons would be banned according to customary international law (i.e., "of such a nature to cause superfluous injury or unnecessary suffering or being inherently indiscriminate"), albeit motivated almost solely by possible reference to nuclear weapons, could serve as evidence of the general attitude.[44]

40 *See*, HCJ 5872/01, *Barakeh v. The Prime Minister*, 56(3) P. D. 1. Unsurprisingly, it was the only HCJ decision to be citied in the *Flechette* Decision and explained why the HCJ was barred from granting the Petition if "... the use of such artillery is not prohibited by international conventions". *See, Flechette* Decision, at 196A. For a critique of the *Barakeh* case *see*, Orna Ben-Naftali & Keren R. Michaeli, "Justice-Ability: A Critique of the Alleged Non-Justiciability of Israel's Policy of Targeted Killings" *Journal of International Criminal Justice* 1 (2003) 368.

41 HCJ 3022/02, Canon (LAW) – The Palestinian Organization for the Defence of Human Rights and the Environment v. The Commander of the IDF Forces in the West Bank, Petition for an Interim Decree, available at <http://www.adalah.org/features/opts/3022petition-heb.doc>. An unofficial translated version of this petition and the state's response are available at <http://www.adalah.org/features/opts/3022response-eng.doc>.

42 HCJ 3022/02, Canon (LAW) – *The Palestinian Organization for the Defence of Human Rights and the Environment v. The Commander of the IDF Forces in the West Bank*, 56(3) P. D. 9. For the legal issues raised before the HCJ during and immediately after this extensive operation (~ 30,000 Israeli soldiers participated) *see*, Yuval Shany, "Israeli Counter-Terrorism Measures: Are They 'Kosher' under International Law?", in: Michael N. Schmitt & Gian L. Beruto (eds.) *Terrorism and International Law: Challenges And Responses* (Ardsley, N.Y.: Transnational Publishers Inc, 2003) 96-113.

43 Rome Statute of the International Criminal Court, July 17, 1998, [hereinafter, *Rome Statute*], *Laws of Armed Conflicts*, 1309.

44 As McCormack and Robertson have noted, the declared nuclear powers "were so adamant about ensuring that the Court not deal with the question of the use of nuclear weapons without their consent that they also insisted that there be no reference to a violation of the general prohibition... in and of itself". Timothy McCormack & Sue Robertson, "Jurisdictional Aspects of the Rome Statute for the New Industrial Criminal Court", *Melbourne University Law Review* 23 (1999) 635, at 665. For a description of the various groups into which the participating states coalesced, *inter alia*, over the issue of the inclusion nuclear weapons' use in the provisions of the draft statute on prohibited weapons, *see*, Philippe Kirsch & John T. Holmes, "The Rome Conference on an International Criminal Court: The Negotiating Process" *AJIL* 93 (1999) 2, at 4. For a detailed account of the dispute *see*, Elaina Kalivretakis, "Are Nuclear Weapons above the Law–A Look at the International Criminal Court and the Prohibited Weapons Category" *Emory International Law Review* 15 (2001) 683-732, esp. 700-704. For the

It is important to note that the Petitioners in the *Flechette Petition* only sought to review the legality of flechette use in the Gaza Strip given the area's endemic features, mainly its dense population, Israel's overwhelming military superiority, and the weapon's known effects. The State, in adopting a prudent legal strategy, channeled the discussion to a relatively benign route from its perspective when opening its response: "The main petitioners' claim in this petition is that the use of flechette rounds is contrary to [the] fundamental principles of the Law of War in International Law." From this point on, the State presented an expansive presentation of the legality of flechettes under International Law – comprising nearly half of its response and concluding with the statement, "flechette ammunition [rounds] is not an illegal weapon in itself".[45]

Throughout the Petition's hearing, legal questions were blurred even further. The State dropped the words "in the Gaza Strip" when referring to "the legality of use" in its response, even when it referred to the Petition itself as creating – perhaps inadvertently but more likely intentionally – a platform from which to review the weapons' legality in theory. The Petitioners themselves also relied perhaps too heavily on the well-known 1996 *Advisory Opinion* of the International Court of Justice (ICJ) on the Legality of the Threat or Use of Nuclear Weapons (hereinafter, *Nuclear Weapons Advisory Opinion*).[46]

Whether it was due to the State's Response or because of the Petition's wording,[47] the HCJ engaged in a legal exchange requested neither by the State nor explicitly by the Petitioners. Its final conclusion, declaring that "the Petitioner's claim that the use of flechette shells is prohibited by the Law of War is incorrect and should be rejected"[48] simply did not address the relevant issues. A total ban on the weapons then in use by every Israeli tank in the Gaza Strip was not the Petitioners' goal.[49] Not

proposal and its rejection *see id.*, at 694-5. For the final wording of this pertinent subsection *see*, *Rome Statute*, Art. 8(2)(b).

45 *See*, HCJ 8990/02, *A Response by the State Attorney's Office*, para. 1. The State's *flechette response* in its original text is available at <www.btselem.org/hebrew/Legal_Documents/8990_02_Flechette_Appeal_State_Response.pdf> [hereinafter, *State's Response*]. No English translation is available. For the section's conclusion *see, Id.*, para. 22. Only in the second part of its response does the State address the question of "the use of Flechette in the Gaza Strip", where it finally focuses on the "specific cases which were mentioned in the petition" (*Id.*, para. 45).

46 For the ICJ's legal opinion, which the Petitioners extensively cited within the Petition, as well as the full summary of the opinion, attached as an appendix, *see*, *Legality of the Threat or Use of Nuclear Weapons, Advisory opinion* (International Court of Justice, July 8, 1996), *ILM* 35 (1996), 809 & 1343.

47 *See*, for instance, *the Flechette Petition*, para. 19, as follows: "The Appellants shall argue that the use of the Flechette shell is illegal...[and] constitutes a violation of accepted and/or customary international humanitarian law, since the shell causes indiscriminate damage and unnecessary suffering".

48 *See*, the *Flechette Decision*, at 195.

49 A petition would not have been submitted if the targets had been, for instance, Syrian infantry forces trying to retake the Golan Heights. Telephone Interview with the Petitioners' advocate, Adv. Yossi Zur, Carmeli-Arnon Law Office (May 5, 2003), additional interview (Dec. 13, 2006). Interestingly, the two advocates who represented the petitioning NGOs (Zur and Yosef Arnon) are former IDF officers. Zur served in the

surprisingly, judging from the ruling *per se*, it appears that the Petitioners had relied to some extent on the CCW. "The argument that flechettes are specifically banned by the CCW was a stretch at best", wrote a commentator shortly after the HCJ ruling.[50] The truth, however, was quite different; the CCW was not mentioned *even once* throughout the entire Petition. It was the State and only the State that resorted to this treaty.

Yet, as intriguing as this rare legal discussion and its moral overtones may be, it represents a point of departure from which to conduct a comprehensive, unique and up-to-date reappraisal of the legality of flechette weapons. The Goldstone Report's call for an urgent UNGA discussion on the weapons' future legality indicates the importance of this reappraisal. Moreover, given the IDF's continued use of flechettes in the Gaza Strip (especially after it obtained HCJ approval) on the one hand, and the Report's call for a moratorium on its use "in the light of the human suffering and damage they have caused"[51] on the other, this reappraisal is particularly timely.

## Outline of the Book

In the literature on the proper procedures for determining the legality of "new" weapons in accordance with Article 36 of the 1977 Protocol I to the 1949 Geneva Conventions [hereinafter, AP-I],[52] up to seven criteria were suggested (also labeled

armored corps during his compulsory military service whereas Arnon was a lieutenant colonel during his service in the reserves.

50 Walker, "Introductory Note", *supra* note 35, at 881. Walker elaborated on this point as follows: "In this case, two NGO petitioners..., petitioned for an order declaring the use of flechette artillery shells in the occupied territories *illegal as a matter of domestic law incident to Israel's membership in the 1980 United Nations Convention...*," [emphasis added]. *Id.*, at 880. As a translation of the Petition was not yet available, and only an unofficial translation of the ruling was at his disposal, Walker's flawed description could have been expected.

51 *The Goldstone Report*, at 425, para. 1971.

52 This key article, denoted as "new weapons", reads: "In the study, development, acquisition or adoption of a new weapon, means or method of warfare, a High Contracting Party is under an obligation to determine whether its employment would, in some or all circumstances, be prohibited by this Protocol or by any other rule of international law applicable to the High Contracting Party"; *Protocol Additional to the Geneva Conventions of 12 August 1949, and relating to the Protection of Victims of International Armed Conflicts*, 8 June 1977, (Protocol I), *Laws of Armed Conflicts*, 711, at 730. It should be noted that our use of the term "new" in this context is not restricted to new or innovative types of weapons. Rather, the term reflects the time of the weapon's acquisition by any given state party to AP-I. *See,* Claude Pilloud *et al.*, (eds.) *Commentary on the Additional Protocols of 8 June 1977 to the Geneva Conventions of 12 August 1949*, (Leiden: Martinus Nijhoff, 1987), at 424-5, para. 147 (hereinafter, *ICRC Commentary*). The ICRC guide for weapon review even stressed that "the purchasing State is under an obligation to conduct its own review of the weapon it is considering to acquire, and cannot simply rely on the vendor or manufacturer's position as to the legality of the weapon, nor on another State's evaluation". *See,* Lawand, Kathleen (principle author), *A Guide to the Legal Review of New Weapons, Means and Methods of Warfare: Measures to Implement Article 36 of Additional Protocol I of 1977* (Geneva: ICRC, revised version Nov. 2006), 24. Furthermore, some ICRC-affiliated experts claimed that "The provision

as "areas"[53]), among which the following four are most widely accepted:[54] (a) treaty obligations; (b) the prohibition against superfluous injury or unnecessary suffering which – as a leading scholar rightly noted – "is primarily intended to protect combatants rather than the civilian population";[55] (c) the weapon's capability of distinction; and (d) environmental considerations. Naturally, only the first three were perceived by practitioners "as of direct relevance in perhaps 99.9 percent of legal reviews"[56], and they are to be carefully explored here.

The book is divided into three parts: Part One (Chapters 1, 2 and 3) provides the legal and factual background necessary for a better understanding of the legality of the flechette weapon while Chapter 2 is devoted to *treaty law*. Part Two (Chapters 4, 5 and 6) focuses on the situation in Israel. Part Three (Chapters 7 and 8, which address areas B and C respectively), integrates analyses of the relevant literature (legal and military-medical), the 1970s prolonged debate over the weapon's legality, and up-to-date findings to provide an in-depth reappraisal of the weapon's legality.

Within Part One, Chapter 1 introduces the reader to the weapon *per se*, explaining the reasons for its developments and the technical aspects of its operation. It is a deliberately extended chapter given the legal framework. For a legal review of new weapons – as stated by a leading practitioner in the Law of Weaponry (Hays Parks) – two things are essential: first, "research and an understanding of diplomatic, legal and military history, including state practice in wartime"; and second, "an understanding of weapons and ammunition".[57] Regrettably, in hindsight, the lack of sufficient technical information on the weapon and its operation is one of the causes fellow researchers are reticent to explore the associated legal questions. To compensate for the void in contemporary research, the use of free-fall air-dropped flechettes during World War I (WWI) in addition to both air-dropped flechettes and firearm flech-

may also cover existing weapons, for example, those subsequently modified after an initial review". *See, Daoust* Isabelle *et al.*, "New Wars, New Weapons? The Obligation of States to Assess the Legality of Means, and Methods of Warfare", *IRRC* 846 (2002), 345, at 352.

53 *See,* Justin McClelland, "The Review of Weapons in Accordance With Article 36 of Additional Protocol I", 850 *IRRC,* (2003), 397, at 406-410.

54 *See,* for instance, Timothy McCormack *et al., Report on States Parties' Responses to the Questionnaire: International Humanitarian Law and Explosive Remnants of War,* (Melbourne, Asia Pacific Centre for Military Law, 2006), 35-36; *Daoust* Isabelle *et al., id.*, at 347-350; McClelland, *id.* The latter labeled the four criteria as "areas" and like McCormack, added a fifth, "Future legal issues", while arguing that the "acquisition process may take decades from the concept of a new weapon to its final disposal" (id., at 409). Hence, the argument goes, "it would be folly to omit consideration of possible future trends in the law. *Id.* In addition to these five criteria, McCormack has mentioned the following two: (f) *Harmful effects*: "will the weapon's harmful effects escape the control, in space and time, of those using them?"; (g) *Limitations on use*: do specific international laws or general principles of IHL "restrict or limit the circumstances in which the weapon may lawfully be used?". *Id.*

55 McCormack, *id.,* at 25.

56 W. Hays Parks, "Conventional Weapons and Weapons Reviews", *Yearbook of International Humanitarian Law 8* (2005) 55, at 131.

57 *Id.*, at 66-7.

ettes' use in Vietnam are reviewed. The later case is in fact the only case of free-fall air-dropped anti-personnel flechette use since WWI and the only case of massive firearm flechette use other than the Israeli case. Such marginal use of the projectiles may explain why a weapon that was at the heart of harsh political-legal debate more than 30 years ago was forgotten until 2001.

Chapter 2 focuses on historical disputes. These provide a background for the legality issue through a survey of the prolonged legal efforts to ban certain "conventional"weapons in general and flechette weapons in particular.[58] The efforts – eventually leading to the drafting of the CCW – initiated in principle since 1968, are described in this chapter in considerable detail. Following the impact of the Vietnam War – the US 'battlefield laboratory' for its post-Korean War weapons, such as the flechettes[59] – the 1968 UN International Conference on Human Rights in Tehran requested the UNGA to invite the UN Secretary General (UNSG) to study, amongst other issues, "the need for additional humanitarian international conventions... to *ensure the prohibition and limitation of the use* of certain methods and *means of warfare*".[60]As noted by Professor Frits Kalshoven, more than anything else, the Vietnam War prompted the renewed interest of the international community in "an almost forgotten topic": a ban on the use of specified weapons.[61] The last part of Chapter 2 reviews the flechette issue as it evolved during the three CCW review conferences, a review requisite for any evaluation of the State's Response in the *Flechette Petition* that partly relied on the discussion – or, more precisely, the lack of discussion – during those conferences. Most importantly, this review demonstrates the rollercoaster-like twists that the flechette legality question has taken in the legal community's agenda: from a hot and disputed issue to oblivion. Five reflections (Chapter 3) close this section. The first explores the disappearance of free-fall air-dropped flechettes to be followed by assessing the legality of their use in WWI. Then, the effectiveness of firearm flechettes is assessed while the involvement of U.S. political authorities in their introduction into the Vietnam battlefield is explored in the fourth reflection. In examining the reasons behind the failed attempts to ban flechette weapons, the underlying political context of the various frameworks are assessed and their respective impacts are explored. These frameworks are the 1974-1977 Diplomatic Conference and the 1978-1980 UN Conference on the Use of Certain Weapons and its Preparatory Sessions. Finally, the reasons behind the non-renewal of these attempts during the review Conferences in general, and why

58 In any case, unlike the review of treaty law under Art. 36 of AP-I which requires reviewing only "international treaty law to which the reviewing State is a party" (*see,* Daoust Isabelle *et al., supra* note 52, at 349), once the legality of flechette weapon in the abstract is determined, all specific rules became relevant.

59 This is especially true regarding anti-personnel weapons as the introduction of the M79 40mm grenade launcher illustrated (it employed the then new technique of controlled fragmentation). *See,* Eric Prokosch, *The Technology of Killing: A Military and Political History of Antipersonnel Weapons.* (London: Zed Books Ltd, 1995), 56-57.

60 Emphasis added. For the request *see, Res. XXIII, Final Act of the International Conference on Human Rights,* UN Doc. A/conf. 32/41, at 18 (1968).

61 Frits Kalshoven, "Arms, Armaments and International Law", *Recueil Des Cours* 19 (1985[II]) 181, at 225.

the international outcry (mainly in 2002-2003) following the Israeli use in the Gaza Strip was not translated into a concrete proposal in particular, are addressed.

Chapter 4 begins the second part of the book, and first focuses on the Israeli use of flechette weapons. Similar in format to Part One, it provides the factual material. Following a brief survey of Israel's procurement of flechette weapons and local improvements, the chapter reviews the use of flechettes in Lebanon (1987-2000). An extensive assessment of the weapon's use in the Gaza Strip during the period preceding the Petition (2001-2003) then follows. Particular attention is given to the three incidents in which Palestinian civilians were victims, irrespective of the IDF's recognition of the victims' identity as non-combatants in only two of them. These incidents served as part of the Petition's factual grounds (hereinafter *the three incidents*). This chapter, together with the general background provided in the first part of the book, supports the Petition's analysis, the State's Response and, most importantly, the HCJ ruling, found in Chapter 5. Our discussion of the internal legal dispute begins with a succinct survey of the parties' positions during the legal debate and concludes with a review of the HCJ's arguments and its final ruling. Five reflections (Chapter 6) close this part. The first explores the reasons for the IDF's continuous efforts to maintain a low profile regarding the use of flechettes, although no further legal intervention appeared forthcoming after the 2003 HCJ Decision. Then, a survey of this use, mainly during *Operation Cast Lead* and the renewed debate the Operation generated over the weapon's legality is provided. Israel's pertinent position, as stated in a document prepared in advance for the Goldstone Report, and the Report itself, are analyzed as well within this reflection. The following third reflection discusses and explores the reasons for the disturbing gap between the Court's seemingly superficial approach to the Petition and in broader context, which should have encouraged a distinctly profound discussion. Finally, the principles of proportionality and precautions in reference to the local use of flechettes in the Gaza Strip are discussed, with reference to two notable cases in which the principle of proportionality was applied to the issue of weapons selection: the *Martic case* [62] and the review of NATO's use of cluster munitions and depleted uranium projectiles by the Committee Established to Review the NATO Bombing Campaign against former Yugoslavia.[63] A discussion on the ability of the military's legal advisers to ensure a proper balance between the armed forces' inherent tendency to employ all available means at their disposal and the need to protect civilians in conflict zones, seals this chapter.

The third and main part of the book joins the two previous parts conceptually. The relevant literature (legal and military-medical), the prolonged debate conducted on the legality of the flechette weapon during the 1970s, and up-to-date findings (e.g., interviews and Israeli post-mortem reports[64]) provide a solid basis for an in-depth reappraisal of the weapon's legal status.

---

62 *See, supra* note 38.

63 Final Rep. to the Prosecutor by the Committee Established to Review the NATO Bombing Campaign Against the Federal Republic of Yugoslavia, *ILM* 39 (2000), 1257, at 1271.

64 Medical reports constitute important evidence when conducting weapon reviews. As McClelland noted: "The difficulty in conducting the legal review when a new method

Following a short preface on the historical-legal sources of the principle prohibiting weapons causing "unnecessary suffering" or "superfluous injury"[65] (the English and French wording of Article 23(e) of the 1907 Regulations of the *Hague Convention (IV) Respecting the Laws and Customs of War on Land,* [hereinafter, 1907 Hague Regulations]),[66] Chapter 7 assesses flechette weapons according to measures traditionally used to quantify human suffering, such as the degree of suffering caused at the time of injury or the mortality rate from the total number of injuries. The chapter continues with an analogy to an incident during WWI when Germany disputed the legality of the American Expeditionary Force's use of shotguns in trenches. Similar to the flechette weapon, a shotgun disperses a large number of buckshot pellets (or flechettes) with each shot, thereby creating a large number of near-simultaneous wound trajectories. Following is a review of the argument, often repeated in expert discussions during the 1970s, that high velocity flechette darts tend to bend or break on impact, thereby causing superfluous injury to human tissue. The discussion concludes with an assessment of the weapon's legality according to this principle, based on relevant criteria from the seven criteria proposed in 1997 by the International Committee of the Red Cross (ICRC). These were in fact meant to "facilitate the review of the legality of weapons" in an important ICRC-sponsored SIrUS (an ICRC acronym for Superfluous Injury or Unnecessary Suffering) Project, crowned by its initiators as "an attempt to bring objectivity to the legal notion of superfluous injury or unnecessary suffering".[67]

Nevertheless, as Dr. Hans Blix, the energetic member of the Swedish delegation to the 1970s discussions on banning or restricting conventional weapons noted, "All weapons cause suffering and weapons can be used indiscriminately".[68] Jurists have also argued that "...when the effects of any weapon are described in detail a case could probably be made for its prohibition". Given these normative conundrums, the more crucial questions remain: *"how and in what circumstances are these weapons used*; what are the military advantages sought by their use; could other weapons, or the

of injury is being assessed will be the absence of any medical evidence upon which to make an assessment of the balance between the military utility and the nature and amount of suffering". *See,* McClelland, "The Review of Weapons...", *supra* note 53, at 412.

65 The principle prohibiting unnecessary suffering is referred as the 'Principle of Humanity' in various sources; *see*, for example, UN Doc. A/9215 (Vol. I), (Nov. 7, 1973), at 208, para. 203(1).

66 Convention (IV) respecting the Laws and Customs of War on Land and its annex: Regulations concerning the Laws and Customs of War on Land. The Hague, 18 October 1907, *Laws of Armed Conflicts*, 55, at 73.

67 Robin Coupland & Peter Herby, "The SIrUS Project and Reviewing the Legality of New Weapons" *IRRC* 835 (1999) 583, at 586.

68 Hans Blix, "Current Efforts to Prohibit the Use of Certain Conventional Weapons", *Instant Research on Peace and Violence* 4(1) (1974) 21, at 25. Blix – praised as the "driving force of the Swedish effort for antipersonnel weapons bans..." – uses the example of a knife that "can cause terrible wounds and can be used blindly". It can also be used to stab the victim's body time and again, thereby ensuring unnecessary suffering. For the example *see, id*; for the praise showered on Blix *see*, Prokosch, *The Technology of Killing*, *supra* note 59, at 170.

same weapons with less-potency, accomplish these same ends?".[69] These core questions are addressed in the final section of Chapter 7.

Chapter 8 is devoted to a review of flechette weapons according to the principle of distinction, the second cardinal principle governing the use of conventional weapons.[70] In the ICJ's wording, this principle demands that states never "use weapons that are incapable of distinguishing between civilian and military targets", whereas in our context (anti-personnel weapons), the distinction lies between combatants and non-combatants.[71] After a short introduction, the chapter discusses the feasibility of defining any type of weapon as inherently discriminatory, especially after the ICJ *Nuclear Weapons Advisory Opinion*. A detailed discussion as to whether flechette weapons have a strong tendency to harm combatant and non-combatants alike, as argued during the 1970s, follows.

The book closes with a summary and conclusions. First, the findings regarding the flechette weapon's legality under the three relevant review criteria (i.e., treaty obligations, the prohibition against unnecessary suffering or superfluous injury, and the weapon's capacity to distinguish between targets) are discussed. Second, we explore whether the mechanism for generating *specific rules* to regulate certain types of (conventional) weapons, together with the relevant *generic rules*, adequately address associated humanitarian concerns. Some thoughts on how civil society might counter or compensate for the inability of the aforementioned mechanism

69 Emphasis added. *See,* Andrew M. Stone, "Conventional Weapons, Their Deployment and Effects from a Humanitarian Aspect: Recommendations for the Modernization of International Law" *AJIL* 68 (1974) 149-150 (book review).

70 For the ICJ's position that this principle, along with the principle prohibiting weapons causing 'unnecessary suffering' or 'superfluous injury' are 'the cardinal principles' of IHL governing the distinction between legal and illegal weapons *see, Nuclear Weapons Advisory Opinion*, para. 78. Some legal experts have distinguished an additional ground for prohibition of treacherous or perfidious weapons, referring to the effects of mines, booby traps, and delayed-action bombs on non-combatants. This concept is an elaboration of the general principle found in Article 23(b) of the 1907 Hague Regulations: "In addition to the prohibitions provided by special Conventions, it is especially forbidden...(b) To kill or wound treacherously individuals belonging to the hostile nation or army"; For details of this concept *see,* for example, *'Existing Rules of international law Concerning the Prohibition or Restriction of Use of Specific Weapons', Survey Prepared by the Secretary-General of the United Nations*, UN Doc. A/9215, (Nov. 7, 1973), Vol. I, at 19, para. 6; R. R. Baxter, "Conventional Weapons under Legal Prohibitions" *International Security* 1(3) (Winter 1977) 42, at 47. Nevertheless, as Kalshoven remarked with respect to the 1974 Lucerne Conference of Governmental Experts on the Use of Certain Conventional Weapons [hereinafter, *1974 Lucerne Conference*], "the prevalent feeling was that the concept did not merit a separate place among the legal criteria governing use of conventional weapons". *See,* Kalshoven, "Arms, Armaments..." *supra* note 61, at 237.

71 *Nuclear Weapons Advisory Opinion*, para. 78. In addition to these two 'cardinal principles', it was argued that "the Court identified a third fundamental principle: the principle of neutrality, whereby *(inter alia)* the effects of weapons must be contained within the territories of the belligerent States". Yoram Dinstein, *The Conduct of Hostilities under the Law of International Armed Conflict* (Cambridge: Cambridge University Press 2004), 57. It is hard to conceive, however, of weapons other than nuclear or biological ones, to which this principle is more relevant.

and generic rules to respond to these concerns, as reflected in the 1997 *Anti Personnel Mine Ban Convention*[72] and the 2008 *Convention on Cluster Munitions* (CCM)[73], closes this book.

72 Convention on the Prohibition of the Use, Stockpiling, Production and Transfer of Anti-Personnel Mines and on their Destruction, 18 September 1997, *Laws of Armed Conflicts*, 285.

73 The Convention's text in all official UN languages is available at <http://treaties.un.org/doc/ Publication/CTC/26-6.pdf>.

# Part I.

## Flechette Weapons: Factual and Legal Background

# Chapter 1

## The Weapon's Features and History

### A. Introduction

This chapter is the first of two designed to provide the factual and legal background necessary for a better understanding of the flechettes' legality, to be discussed in Part III. Such a discussion has been hindered so far by lacunae in the legal, military, professional and historical literature, due to the virtual unavailability of secondary sources regarding the weapon's history and development. This caveat applies particularly to documentation of the weapon's technical and operational features.

Some practitioners and legal scholars of the Law of Weaponry may insist that introducing the reader to what is known about the weapon *per se*, its development and technical aspects, is justified. In their view, an extended survey – including "an understanding of weapons and ammunition"[1] – is required for an in-depth analysis of any new weapon's legality, let alone one as controversial as the flechette. This chapter provides such a survey. It starts by tracing the etymology of the word flechette, reviews the weapon's early development in the pre-WWI era and discusses the use of air-dropped flechettes during WWI. After reviewing France's use of free-fall air-dropped flechette in the First Indochina War's famous Battle of Dien Bien Phu (March-May 1954) as well as the similar use made by U.S. forces in the Second Indochina War (the Vietnam War), the chapter shifts focus to the type of flechette weapon relevant to our main discussion: firearm flechettes.

This section therefore opens with a description of the first type of munitions in which modern firearm flechettes were introduced: small arms munitions. Despite the prolonged effort and the heavy resources invested, the project has not yielded a satisfactory weapon (thus far). It next traces the origin and development of flechette armor rounds as employed by Israel in Lebanon, but especially in the Gaza Strip. A review of the first use of these rounds and similar kinds of projectiles, known as Beehive munitions, by U.S. and U.S.-supplied forces in the Vietnam War follows. The Vietnam theatre was, in fact, the only case of free-fall air-dropped anti-personnel flechette use since WWI and the only case of massive firearm flechette use outside the Israeli case. A discussion of the proliferation of flechettes in the Post Vietnam Era closes the chapter. As this section indicates, while the deployment of anti-personnel flechette projectiles for either armored or artillery forces has been limited, flechette-filled air-to-ground rocket warheads have undergone a dramatic

1 W. Hays Parks, "Conventional Weapons and Weapons Review" *Yearbook of International Humanitarian Law* 8 (2005) 55, at 66.

change since their first use in the Vietnam War. Not only has the U.S. used these warheads in military operations across the globe, a few states have opted to manufacture their own flechette-filled warheads and have even competed with the U.S. in the international weapons market, ensuring the weapon's proliferation. Although these warheads are not the book's main focus, having constant and widespread use in contemporary war zones such as Afghanistan, they are discussed because of their potential to inflict a considerable collateral damage given the multitude of darts released and wide blanket coverage.

The etymology of the word flechette is *flèche,* French for a small dart, in reference to the "steel arrow" introduced by the French military on the eve of WWI.[2] There is apparently nothing new to this means of warfare. By the Middle Ages, metal arrows or darts with a finned or fluted shaft were already in use as projectiles.[3] Crossbows from that period employed quarrels (short, heavy, square-headed arrows) or bolts of iron with integrated stabilized fins.[4] Nor is there anything new about *firearm* flechettes. Fourteenth century cannons, for instance, could shower up to 50 metal darts with each volley.[5]

Nevertheless, the context in which it entered French military jargon on the eve of WWI was quite new, utilizing the ancient method of combat with darts in the third theatre of warfare – aerial warfare. The contemporary term "flechettes" (up to the Vietnam War) consequently signified the appearance of a new weapon: large steel darts (approximately 12cm-15cm long, on average) dropped in large quantities from attack aircrafts as a method of attacking troop concentrations or marching

2 *See,* for instance, entry in J. A. Simpson and E. S. C. Weine (eds.*), The Oxford English Dictionary* 2nd ed. (London: Clarendon Press 1989), 1035. *See also* entry *at Merriam-Webster's Online Dictionary*, available at <www.britannica.com/dictionary?book=Dictionary&va=Flechette %20&query=Flechette>. In addition, the word as it entered French army slang meant *débrouillard* (a resourceful person; a survivor). *See,* Milton Garver, "French Army Slang" *Modern Language Notes*, 32 (1917), 151, at 154. However, with the reintroduction of flechette weapons by the U.S. army in the 1970s, *flechette* or *fleshette,* although defined as "aircraft missiles resembling a dart" in 1933, came to mean "one of a spray of darts exploded as an antipersonnel weapon". *See,* Garland Cannon & Beatrice Mendez Egle, "New Borrowings in English" *American Speech* 54, (1979), 23, at 27-8.

3 In August 1346, during the Battle of Crécy, one of the most important battles of the Hundred Years' War, 6000 Welsh archers shot 36,000 – 72,000 arrows with metal darts per minute at the French cavalry. Given the fact that one dart weighed about 100 grams, this was more than seven tons of metal per minute. *See,* Andrew Ayton & Sir Phillip Preston, "Topography and Archary: Further Reflection on the Battle of Crécy", in: Andrew Ayton *et al.* (eds.), *The Battle of Crécy 1346* (Rochester, NY: Boydell Press 2007), 351, at 361.

4 *See,* Ralph Payne-Gallwey, *The Book of the Crossbow* (New York: Dover Publications, 1995) (first published: 1903, Longmans, Green, and Co.), 126-28.

5 In fact, the first known metal cannon, *Pot-de-fer* (iron pot in French) fired a flechette projectile (probably a single arrow) as early as 1326 A.D (the time it was first depicted in a manuscript, *De officiis regum* by Walter de Millimete); it was subsequently used by the French in the Hundred Years' War (1337- 1453). *See,* Konstantin Nossov, *Ancient and Medieval Siege Weapons: A Fully Illustrated Guide to Siege Weapons and Tactics*, (Guilford, CT: The Lyons Press 2005), 207-9.

troops.[6] As an author writing on aviation in 1915 claimed: "In effect they are the modern form of arrow".[7]

## B. Air-dropped Flechettes

From the scant attention given to flechette use in the WWI literature one may imagine that the phenomenon was short-lived. References to the weapon are virtually absent from general WWI history books,[8] while studies focusing on WWI air warfare include only short, if any, references to the weapon.[9] Nonetheless, sources written in French as well as WWI memoirs, literature and newspapers suggest that the weapon was employed – albeit on a smaller scale after 1915 – by the nascent air forces of various belligerents' almost until the War's end.[10] Furthermore, contemporary images of the flechette as a dreadful anti-personnel weapon – rightly or not – have an old pedigree, as outlined in the following section.

6 *See*, for instance, *Illustrated War News*, Jan. 27, 1915, at 18; Rene Martel, *French Strategic and Tactical Bombardment Forces of World War I*, (Lanham: Scarecrow Press 2006) [first published as L'Aviation Francaise de Bombardement (Des Origines au 11 Novembre 1918), 1939] (translated by Allen Suddaby; edited by Steven Suddaby), 14. It has been argued that flechettes were also used as the projectile of choice by the French against the Germans in the first electromagnetic rail guns (*see*, "Flechettes Shotgun Ammunitions – Historical Background", Sabot Designs LLC, available at <http://www.antipersonnel.net/sdllc/001.html>). However, electromagnetic rail guns were not actually employed in the battlefield and retained, in fact, their experimental status during that war. See Ian R. McNab, "Early Electric Gun Research" *IEEE Transactions on Magnetics* 35, (1999), 250, at 251-2.

7 Dommett William Erskine, *Aeroplanes And Airships*, (New York: Whittaker & Co, 1915), 80.

8 For instance, Cyril Falls' classic book on WWI warfare, while mentioning a variety of weapons such as the torpedo, chemical agents and machine guns, does not include any reference to flechette weapons. Moreover, despite the documented use of air-dropped flechettes during the early stages of the War, Falls stated as follows: "The air forces were very small...It is the naked truth, though hard to realize, that the armament of an 'aeroplane' or 'machine' – never an 'aircraft' in those days – was a rifle, perhaps even a revolver". Cyril B. Falls, *The First World War*, (London: Longmans 1960), 17.

9 A good example is the single reference to be found in Lee Kennett's 288-page work. *See*, Lee B. Kennett, *The First Air War: 1914-1918*, (New-York: Free Press 1999), 21. Even John H. Morrow's 450-page comprehensive study of the significance of airpower during WWI has no more than three references (up to two sentences each). *See*, John H. Morrow, *The Great War in the Air: Military Aviation from 1909 to 1921* (Washington D.C.: Smithsonian Institution Press 1993), 16, 62, 84.

10 There is, for instance, a recorded use of flechettes by the Ottoman Air Service against the forces of the Arab Revolt in southern Jordan as late as February 4, 1918. The plane, however, was shot down by Arab ground fire near Uneiza. For the British report of the crew's interrogation, dated March 1, 1918, *see*, *The British National Archives*, Air 1/2360/226/6/3. The author thanks Dr. Dominic Barker, University of Southampton, an expert for WWI Military archaeology, for bringing this affair to his attention.

### BI. The Balles Bon Weapons: Air-dropped Flechettes in the Pre WWI Era

Initially called *balles Bon*[11] (after their inventor, Lieutenant Colonel Bon, who took the idea from the famous French engineer and inventor Clément Ader[12]), these fluted metal darts (pointed and grooved to remain vertical), weighing approximately 20 grams, were experimented with by the French as early as 1910.[13] With prewar France expressing considerable enthusiasm for using the offensive capabilities of aircrafts, aero clubs, lobbyists and prominent ministers (e.g., Paul Painlevé and Louis Barthou) joined the call for exploring their capacity for bombardment in addition to reconnaissance missions.[14] Hence, in 1911, bombing exercises with flechettes and shells were ordered by General Pierre Auguste Roques, Inspecteur général de l'Aéronautique in the French war ministry.[15] However, after recognizing these weapons' poor accuracy, it was suggested that they be used *over retreating columns* (i.e., badly protected infantry) or *convoys.*[16] Nevertheless – as pilots were to learn in WWI – any concentration of troops or artillery batteries constituted targets that did not require much precision weaponry.[17] The French high command's strong support for developing incendiary and explosive flechettes suitable for attacking airships in response to the menace inherent in Germany's impressive aerial military arsenal, indirectly prompted work on anti-personnel flechettes in April 1912.[18] By 1913, pow-

11 Additional names that were later used included *Bon fleche ball*, *fleche Bon*, *fleche Ader*, *Bon ball*, and *Bon incendiary ball. See,* Jean-Pierre Leroy, *Les Fléchettes D'avion De La Grande Guerre À Nos Jours* [The Aerial Flechettes of the Great War of Our Time], (Paris: Emotion Primitive 2007), 54. This well-illustrated small book is the most comprehensive source to date on free-fall air-dropped flechettes. Bon himself retired before the last experiments conducted at the Chalon Camp in 1913. *Id.*, at 17.

12 On Ader's view, expressed in his 1909 book (Clement Ader, *Military Aviation* [first published as *L'Aviation militaire,* 1909] [Lee Kennett Editor &Translator], Montgomery, AL: Air University Press 2003), *see,* Lee Kennett, *A History of Strategic Bombing* (New-York: C. Scribner's Sons 1982), 44.

13 On the suggestion that Ader was influenced by a traditional Japanese weapon called the *suriken* (lit. "hand-hidden blade"), a set of which were given to the French government during a visit by the Japanese delegation to Paris in 1880, *see,* Leroy, *Les Fléchettes D'avion…, supra* note 11, at 11. See also the sub-section "German Air Darts and French flechettes, World War I era", in Jason Wotherspoon, *Introduction & Short History of Shuriken*, available at <http://www.secrets-of-shuriken.com.au /history.htm#5>.

14 *See, e.g.,* Martel, *French Strategic and… supra* note 6, at 14; Email correspondence with Dr. Elizabeth Greenhalgh, a Research Fellow at the University of New South Wales, July 31, 2008. Dr. Greenhalgh is currently writing a book on General Ferdinand Foch, the French supreme allied commander in 1918.

15 Morrow, *The Great War in the Air, supra* note 9, at 16.

16 For a note by the French Ministry of War, dated Nov. 4, 1911, *see,* Leroy, *Les Fléchettes D'avion… supra* note 11, at 80.

17 In fact, France and Germany even used flechettes when targeting crews on deck at sea. For such attacks during 1915 *see, Id.*, at 93.

18 Germany possessed ten military aircraft of which the Zeppelin type, based only a short air distance from Paris, were able to load tons of bombs. *See,* Morrow, *The Great War in the Air, supra* note 9, at 35. For France's incentive in developing flechettes *see,* Kennett, *A History of Strategic Bombing, supra* note 12, at 11.

erful flechettes known as "the Bon ball" or "the Bon flechette" were ready and pilots had begun highly classified training in preparation for launching these missiles from aircrafts.[19]

In early 1913, darts weighing 300kg were shipped to Morocco, a French protectorate since March 1912. On May 11, 1914, a few months before the outbreak of WWI, these antipersonnel flechettes were first used against insurgent tribes (the Riata and Tsoul) during the French occupation of Taza in northern Morocco. The darts, which were dropped not only on troop concentrations but also on civilians targets such as marketplaces, left a profound impression on both sides. Using small aerial bombs (named "Aasen bombs" for their famous Norwegian inventor, Nils Waltersen Aasen) and incendiary flechettes, which were found to be 'useful in setting fire to grain fields', this was the first aerial operation in colonial areas.[20] In fact, as Rene Martel stated in his 1939 classic book on WWI French bombing forces, the two airmen who died during the campaign "were the first aviators killed in the course of [a] war operation".[21]

### *BII. Air-dropped Flechettes in WWI*

As a French bombardier during WWI recalled, when the War broke out in August 1914, "[t]he only projectiles existing in quantity were the antipersonnel steel darts".[22] Air weapons were, after all, unique. At the War's onset, bombs were therefore frequently makeshift and the outcome of individual initiatives. Lee Kennett has noted that, "[o]f all the 'new' weapons of the Great War, it alone has no predecessor and no precedent".[23] Mass-produced by the Renault factory at Billancourt, the low-cost,

19 Even a dirigible (*Sella de Beauchamps*) was tested for its fitness for employing various armaments, including flechettes. *See*, Leroy, *Les Fléchettes D'avion…*, *supra* note 11, at 19; Martel, *French Strategic and..*, *supra* note 6, at 6-7. However, as the small fleet of French airships had been equipped with bombs during WWI, including incendiaries, flechettes were not employed extensively. For a French airship's ("Conte") with flechettes night attack on German forces next to Sarrebourg on December 17, 1914 *see*, Leroy, *Les Fléchettes D'avion…*, *Id.*, at 50.

20 The operation was carried out by two squadrons, established by Morocco's Military Governor, Hubert Lyautey, in 1912, that were based in Casablanca and Oujda. See Martel, *French Strategic and… supra* note 6, at 8-9; Kennett, *A History of Strategic Bombing*, *supra* note 12, at 15; Leroy, *Les Fléchettes D'avion…*, *supra* note 11, at 99-100.

21 Martel, *French Strategic* and… *supra* note 6, at 9.

22 *Id.*, at 14. The famous British WWI pilot, James McCudden, recalled that upon his arrival on August 13, 1914 at an aerodrome a few miles from Amiens where the entire *Royal Flying Corps* (R.F.C.) had convened in France, besides "small hand grenades which were to be used as bombs", they also received "some flechettes, or steel darts, for dropping". James Thomas Byford McCudden, *Flying Fury: Five Years in the Royal Flying Corps*, (Barnsly: Greenhill Books, reprinted edition 2000) [first published in 1919 as *Five Years in the Royal Flying Corps*], 25.

23 Kennett, *The First Air War*, *supra* note 9, at 21; Kennett, *A History of Strategic Bombing*, *supra* note 12, at 18. The tank, he explained, had been preceded by the chariot and the submarine by the dreadnought. However innovative these new models, they were nonetheless direct continuations of their forerunners. *See*, Kennett, *The First Air War*, *id.*, at 21.

easy-to-manufacture flechettes were carried in boxes of 500, to be released in batches of 20 while fastened to the sides or under the fuselages of airplanes, with movable bottoms that opened upon release. With a weight of only 20 grams each, an ordinary military aircraft, capable of carrying 250 kilos, could carry no fewer than 12,500 darts.[24]

Once released (from an altitude of 800m during the War's early stages, and later above 2km, following the initial results), the flechettes formed an elliptical shower, gushing out while maintaining their "formation" all the way down [as the rare 1915 illustration shows, see p. 27], hitting with a ground speed of 150 to 200 m/s (when released at 2km).

As to the dispersal area, we have varied estimates, beginning at 91m and ranging up to a rectangle of 500m by 200m. French calculations suggested that a person standing out in the open, within the area of dispersion of two boxes (1000 darts) of flechettes dropped from an altitude of 1 km had a 1 in 80 chance of being hit, a chance that reached 1 in 20 once the person is lying on the ground. As to horses, their chances of being hit are 1 in 10.[25]

Like its successor, firearm flechette, free-fall air-dropped flechette enjoyed outstanding success psychologically. As a rain of steel silently dropped from the sky – earning another of its nicknames, "the silent death" – it caused widespread panic among soldiers on the ground.[26] In fact, it had such grave consequences on the forces' morale that the German and the French press alike attempted to minimize its dreadful effects by providing contradictory reports suggesting that the physical injures could be easily addressed.[27] The weapon and its damage to the human body, which certainly contributed to the demoralization, were described in a British journal, *The Field*, in late November 1914:

24 *See*, Leroy, *Les Fléchettes D'avion…, supra* note 11, at 29-30. Moreover, most French aircraft were capable of carrying a considerably superior bomb load; every additional kilogram represented approximately 50 more darts. For visibility reasons, however, some aircraft, especially bi-planes with propulsive motors, were better suited than others to carry out these bombardment missions and thus were widely adapted. *Id.*, at 48.

25 Dommett, *Aeroplanes and Airships, supra* note 7, at 80; Martel, *French Strategic* and…, *supra* note 6, at 14; Morrow, *The Great War in the Air, supra* note 9, at 61-2; *British Illustrated War News*, Jan. 27, 1915, at 18; "Aircraft and the War" *Flight*, (Nov. 27, 1914), 1163, at 1164; Leroy, *Les Fléchettes D'avion…, supra* note 11, at 21, 23; "Aeroplane Darts and Fire Darts" *Scientific American Supplement* (Feb. 20, 1915) 2042 [a translated report by a German war correspondent taken from *Kriegstechnische Zeitschrift*].

26 As Davilla & Soltan noted in their comprehensive study of French military aircraft during WWI, "[a]lthough crude this method of attack was an effective psychological weapon", James J. Davilla & Arthur M. Soltan, *French Aircraft Of The First World War*, (Boulder, CO: Flying Machines Press, 2002), 315; For French pilots' accounts *see, e.g.*, Leroy, *Les Fléchettes D'avion…, supra* note 11, at 81.

27 For the effect on morale as reflected in various soldiers' testimonies *see*, Leroy, *Les Fléchettes D'avion…, supra* note 11, at 73-78. For an attempt to mitigate this dreadful image *see*, *Id.*, 75. Interestingly, soldiers used to collect flechettes and send them to their families as a proof of the War's cruelty. *Id.*, at 78.

> Many of the French avions carry into the air a large supply of darts (flechettes), which have proved to be deadly weapons. ..., the darts are of great utility to [*sic*] spread panic by decimating corps of troops advancing to the attack or retreating. There can be no reason to refrain from speaking of them because the Germans have already copied the French dart and used it against the Allies. ... It is needless to describe the means by which they are dropped. It suffices to say that falling on a man from the low altitude of only 200 meters (656 ft.), the dart will as surely kill or wound him as a bullet fired from a rifle... A weekly medical periodical published at Munich contains a note by Dr. Volkmann concerning the character of the wounds inflicted by the French darts. A soldier hit on the head was killed on the spot. The dart which struck another man on the shoulder traversed the whole of his body, &c[sic]. Dr. Volkmann and one of his German colleagues, Dr. Gunberg, affirm that the darts dropped from aeroplanes are extremely dangerous, and that the wounds inflicted by them are almost always mortal.[28]

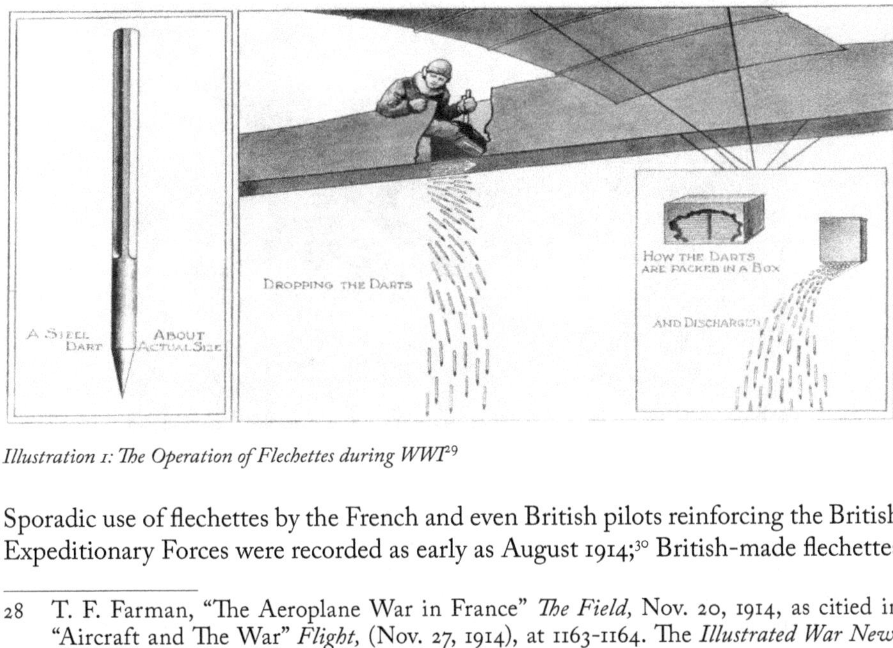


*Illustration 1: The Operation of Flechettes during WWI*[29]

Sporadic use of flechettes by the French and even British pilots reinforcing the British Expeditionary Forces were recorded as early as August 1914;[30] British-made flechettes

28 T. F. Farman, "The Aeroplane War in France" *The Field,* Nov. 20, 1914, as citied in "Aircraft and The War" *Flight,* (Nov. 27, 1914), at 1163-1164. The *Illustrated War News* reported in Jan. 1915 that "the Bavarian General Meyer was so killed not long ago, the darts piercing his helmet into the brain", The *Illustrated War News, Jan. 27, 1915*, at 18. A later German report, from Jan. 1915, was less conclusive regarding the weapons' effects after reviewing the results of two attacks. *See, Aeroplane Darts and Fire Darts, supra* note 25.

29 *Illustrated War News, Jan. 27, 1915*, at 18. This British magazine was known for its publication of quality photographs and drawings during WWI.

30 On August 23, 1914, for instance, a British airplane "harassed advancing enemy troops and transport columns with grenades, flechettes (steel arrow) and improvised petrol bombs". *See,* Ralph Barker, A *Brief History of the Royal Flying Corps in World War I*, (rep. London: Robinson, 2002) [first published as *The Royal Flying Corps in France* (2 Vols., 1995], 41.

(the first of many developed during the War) were probably used during one of the initial dogfights conducted in history.[31] Nevertheless, logbooks kept by French squadrons indicate that systematic, organized use of the weapon began only in September 1914; within one month, a daily average of 50,000 flechettes were dropped, mainly on German troops in the Argonne and the Champagne regions.[32] To overcome the poor results anticipated once the attacked forces had retreated into the trenches, some pilots would release leaflets motivating the soldiers to leave the trenches prior to dropping the flechettes. A more common practice involved dropping anti-personnel flechettes after releasing an exlosive bomb. The explosion forced the soldiers to lie down to avoid the bomb's fragments, which substantially increased their vulnerability to the darts.[33]

Animals, however, were found to be especially helpless and easy targets. All the belligerents possessed a cavalry corps; however, unlike mounted soldiers, horses entangled in their harnesses while carrying heavy loads could not easily flee or seek refuge.[34]

*llustration 2: Flechettes in Action, WWI*[35]

31 The dogfight took place on August 26 between a Sopwith Tabloid armed "with a handful of flechettes" and a German biplane which had just finished an attack on a French aerodrome in the area of Compiègne. However, it is unclear from the British report as to whether the weapon was actually used. *See,* McCudden, *Flying Fury, supra* note 22, at 35-36. However, the appearance of machine guns and the possibility of shooting through propellers quickly turned flechettes into obsolete weapons for such purposes. Leroy, *Les Fléchettes D'avion…, supra* note 11, at 91.

32 Martel, *French Strategic and…, supra* note 6, at 16-17. For specific reports see also Leroy, *Les Fléchettes D'avion…, supra* note 11, at 81-82. One pilot was awarded a medal for dropping 18,000 flechettes (i.e., 36 *Bon* boxes) in the course of eight bombardment missions conducted within a single day (Apr. 15, 1915). *Id.*, at 82.

33 Leroy, *Les Fléchettes D'avion…, supra* note 11, at 82.

34 In 1914, the French army had 175,000 horses and mules while in 1917, the British used 530,000 horses and 230,000 mules. *See, id.*, at 84.

35 This illustration was posted on May, 18, 2007 at < http://www.militaryimages.net>[no longer available].

Due to their efficiency, the darts were quickly copied by most belligerents (Britain, Belgium, Germany, Austria-Hungary, Russia, Turkey and Italy), with the majority developing their own versions (see Illustration No. 3).[36] German airplanes were equipped with flechettes *(fliegerpfeil)* by September 1914 and within one month flechettes, ironically engraved with the words "French Invention, German Made", were dropped on French towns (Betheny and Bar le Duc).[37] Even the small Austro-Hungarian air force dropped flechettes on the Serbian front that same month (October 1914);[38] the British did the same when they attacked German-occupied towns the following month (November 1915).[39]

The Russians, with half their imperial fleet comprised of French-made airplanes, received French flechettes as well. Yet, the absense of launching boxes and proper user manuals hindered extensive use of the weapon at the War's onset, while the introduction of increasing numbers of powerful locally produced bombers (e.g., 20 Ilya Muromets by summer 1916), carrying up to 800 kg. of bombs, ensured the continuation of this limited use given the attractiveness of using "ordinary" bombs.[40] In contrast, Belgium, with its young fleet made up of many French-made airplanes and flechettes, faced fewer technical difficulties; they were thus able to drop flechettes on German soldiers, who viewed such attacks as "aerial terrorism", early in the War.[41] Turkish forces, which were subjected to British flechette attacks once Turkey joined the War in November 1914, retaliated with German-made flechettes beginning in

36 Each type had different features. The British flechette version, for instance, the *fliegerpfeil* (lit. flying arrows), could be dropped from airplanes and zeppelins alike from a height of 500m-1,500m, reaching speed of 200m and spreading over a 200m radius. Furthermore, due to the need to counter the German Zeppelin raids, an anti-Zeppelin incendiary flechette was developed in 1916. It had a steel weighted point with a brass hollow tube tail filled with an incendiary charge for igniting the pierced gas bags. *See,* Leroy, *Les Fléchettes D'avion…, supra* note 11, at 124-131.

37 *Id.*, at 75. Interestingly, German flechette attacks were highly effective once they were carried out from the obsolete Rumpler Taube (German, dove) monoplane as ground-based observers were unable to detect it at in an altitude above 400m. *Id.*, 62. The rapid advance of aviation during the war nonetheless pushed "the Invisible Aircraft" – as the French called it – aside by the end of 1914; hence, only few attacks were recorded, *Id.* In fact, unlike the French, the German air force found flechettes "hardly suitable" when dropped from the air. *See,* Wolfgang Fleischer, *German Air-Dropped Weapons to 1945*, (London: Midland Publishing, 2003), 13. On the German air force *see,* also Morrow, *The Great War in the Air, supra* note 9, at 35-40.

38 Leroy, *Les Fléchettes D'avion…, supra* note 11, at 31; Morrow, *The Great War in the Air, supra* note 9, at 84.

39 *See,* report by Philip Gibbs, British General Headquarters, *The Daily Telegraph*, Nov. 15, 1915 as citied in "The Work of The R.F.C" *Flight* (Nov. 19, 1915), at 903. Mass production of British flechettes began in the Rolls Royce Derby factory. *See,* "Sir Henry Royce, Bart" *Flight*, Oct. 5, 1956, at 575.

40 Sporadic use of flechettes continued up to 1917. *See,* Leroy, *Les Fléchettes D'avion…, supra* note 11, at 63-4; on the Russian air fleet *see,* also Morrow, *The Great War in the Air, supra* note 9, at 45-6.

41 Leroy, *Les Fléchettes D'avion…, supra* note 11, at 64-5. For the German expression, which became fashionable, *see, Id.*, 73.

September 1915. Shortly thereafter, Turkey introduced locally made flechettes, which were dropped during the final stages of the Gallipoli Campaign .[42]

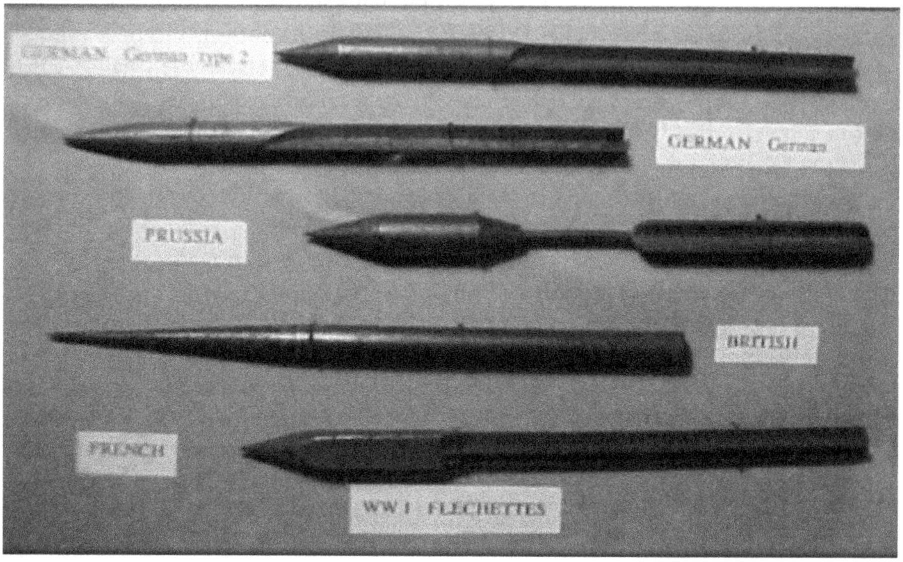


*Illustration 3: Flechettes Employed by the Various Belligerents, WWI*[43]

An interesting case regarding flechettes is Italy, which joined the conflict only in May 1915 by declaring war on the Austro-Hungarian Empire. As a state that possessed a relatively modern air force and as the first to carry out an aerial bombardment, in November 1911 (the Italo-Turkish War)[44], no use of flechettes by its 150 frontline aircraft was expected.[45] Yet, flechettes of different types were used during the 1915 battles of the Isonzo and the Trentin.[46] Only a year later, in 1916, did bombs begin to push flechettes into the margins of Italian weaponry.[47]

42 British use of flechettes resulted mainly from the lack of modern ammunition at the British air base in Lemnos island. *See*, Leroy, *Les Fléchettes D'avion…*, *supra* note 11, at 100-1. For data on the original Turkish flechette *see*, *Id.*, 63.

43 Darts offered for sale by a private flechettes collector in an auction site (no longer available). Most types of darts are displayed in the WWI "Air Artillery" section of the Imperial War Museum, London.

44 *See*, Kennett, *A History of Strategic Bombing*, *supra* note 11, at 12-14. Furthermore, Italy carried out the first ever night bombing raid (May 8, 1912), with an electric lamp attached to pilot's helmet due to the absence of any lights on the instrument panel. *Id.*, at 13.

45 On Italian aviation in 1915 *see*, Morrow, *The Great War in the Air*, *supra* note 9, at 127-9; Kennett, *A History of Strategic Bombing*, *supra* note 12, at 29-31.

46 For a reproduction of a painting by Guido Zanoni showing an Italian plane showering an Austro-Hungarian trench with flechettes, *see*, Ezra Bowen, *Knights of the Air*, (New-York: Time-Life Books, 1980), 53. The original is available at the WWI exhibit at the Heeresgeschichtliches Museum [Museum of Military History Arsenal], Vienna.

47 Leroy, *Les Fléchettes D'avion…*, *supra* note 11, at 65.

Even the Middle Eastern arena witnessed flechette attacks. As flechettes were easily dropped from most types of planes, they were readily used against Turkish troops (February 1915) by French naval seaplanes (the Nieuports model) based at Port Said. As later occurred with firearm flechettes in the Vietnam War, the darts' ability to penetrate heavy vegetation resulted in their being dropping on areas such as oases, inflicting much pain on soldiers and camels.[48] In contrast, besides dropping German-made flechettes on Allied forces during the Suez Canal campaign in July 1916, the small Turkish air force used them against rebel Arab forces in Saudi Arabia and Jordan.[49]

Flechettes continued to play an important role in WWI throughout 1915 but afterward their use became increasingly sporadic.[50] Like all other means of warfare, these weapons soon gave way to more sophisticated and far more deadly weapons.[51] Despite their terrifying image,[52] more efficient ways to employ high explosive and incendiary gravity bombs – effective against hard targets whilst flechettes use even against airships proved to be a failure[53] – soon became available.

### *BIII. The "Lazy Dog" Weapons: Air-dropped "Finned Bullets" in Vietnam*

More than 40 years elapsed before air-dropped anti-personnel darts reappeared on the battlefield. During late WWII, the U.S. army initiated a project designated to develop effective anti-personnel flechette weapons for the U.S. Air Force ("The

48 The French squadron was replaced by a larger English seaplane squadron in 1916. *See,* Tonnele Jean, "Une Escadrille D'hydravions a Port-Said" (1914-1916) [A Seaplane Squadron at Port Said, 1914-16] *Revue Maritime* 228 (1966) 6, at 11-12. On the adaption issue see also Martel, *French Strategic and…, supra* note 6, at 20. The naval seaplanes' attacks hastened the Turkish retreat following the failed attempt to conquer Egypt or to undermine British control over the Suez Canal.

49 *See,* Leroy, *Les Fléchettes D'avion…, supra* note 11, at 101-2. For a record of 1918 flechette attacks in Jordan *see, supra* note 10.

50 In addition to the 1918 Turkish attacks mentioned earlier, German flechettes attacks were continued until as late as 1918. *See,* for instance, a personal record of a heavy German flechette night attack on May 14, 1917 in Loos-en-Gohelle. Martin Fox, *Corporals All: With the Special Brigade R E, 1915-1919,* (New-York: C.H.Harman, 1967), 71.

51 In WWI literature, the flechette is therefore mentioned – if at all – in the context of 1914 and is treated as an obsolete weapon, to be replaced by more modern ones. *See, e.g.,* Anthony Clayton, *Paths of Glory: The French Army,* 1914-1918 (London: Cassell, 2003), 119.

52 According to one account, "there is a case on record of one [flechette] piercing the steel helmet of a soldier, passing through his body and the body of his horse, and burying itself deep in the ground". *See,* H. F. King, "Armamen*t*", *Flight* (June 28, 1934), 638–640. For a reflection of this image in the world press *see, e.g.,* "New Deadly Weapon Used by French Aviators", *The Newark Advocate,* (Newark, Ohio), Jan. 19, 1915, at 1.

53 Even in case of a successful attack– a hard task in and of itself–the multitude of holes created allows for the gas to escape slowly without exploding, which enables emergency landings as well as quick repairs. For such a failed attack on January 20, 1915, *see,* Leroy, *Les Fléchettes D'avion…, supra* note 11, at 88. To solve the problem, incendiary flechettes designed to ignite hydrogen were introduced but were quickly replaced by very effective Le Prieur fuses. *Id.,* at 90.

MX[Material eXperimental]-1600"),[54] while the UK came close to dropping poisoned darts on German forces.[55] The Germans themselves experimented with multiple fléchette loadings in artillery shells.[56] Both projects were, however, either terminated or abandoned by the end of the War.

Nevertheless, the Chinese military's practice of employing waves of soldiers to attack (mainly) U.S. forces in the Korean War (1950-1953) prompted the U.S. army to develop more effective anti-personnel weapons for use by air and ground forces alike. Following an analysis indicating the necessity of a new, deadlier anti-personnel aerial weapon given the inefficiency of certain types of aerial bombardments against ground soldiers[57], the U.S. Air Force (USAF) secretly developed free-fall missiles and corresponding dispensing units, known as "HAIL missiles".[58] This name was quickly revised to an intentionally homespun alternative: *Lazy Dog* missiles.[59] As the authoritative history of the USAF's Aeronautical Systems Center (ASC) notes that "LAZY DOG [*sic*] anti-personnel missiles were designed to spray enemy troops with small projectiles with three times the force of standard air-burst bombs".[60] However, in contrast to WWI flechettes that were composed of bare steel, these flechettes had rounded tips, no lathe marks, and a Parkerized finish (see Illustration 4). As such, they indeed resembled missiles rather than darts.[61]

54 The MX 1600's formal title was *"Ballistic Bombs Dispenser for Antipersonnel Projectile"*. *See*, Leroy, *Les Fléchettes D'avion…*, *supra* note 11, at 142. The project's main incentive was the anticipated need to repel masses of Japanese troops (a large portion were to be transferred from China) during the planned U.S. large-scale invasion of Japan's main islands. Email correspondence with Mr. Paul Sutton, Technology Resources Specialist at Saint Petersburg College and expert for WWII weapons, Aug. 18, 2008 (on file with the author).

55 For this affair which was exposed only in mid 2009 *see*, Deborah Haynes, "Britain Came Close to Dropping Poisoned Darts on German Troops" *The Times*, June 26, 2009, available at <http://www.timesonline.co.uk /tol/news/uk/article6578583.ece>. It should be noted that Canadian scientists participated in this project as well.

56 *See*, Daniel Watters, *Special Purpose Individual Weapons: A Brief History of Fléchette and the SPIW Project*, available at <http://www.thegunzone.com/spiw.html>.

57 For the well-researched study, conducted by Lieutenant Colonel Haile *see*, Leroy, *Les Fléchettes D'avion…*, *supra* note 11, at 141.

58 *See*, Conrad C. Crane, *American Airpower Strategy in Korea 1950-1953*, (Kansas City: University Press of Kansas, 2000), 135.

59 For the intention behind the choice of "attractive or homely names" *see*, Edward Nell & Onora Nell, "War Words" *College English 28* (May 1967), 603, at 606. Though he has no formal reference, Prof. Crane assumes that the name HAIL missiles was supposed to resemble the weather effects of hail (rather than reflecting the figure behind the project's initiation: Colonel Haile) and was replaced to avoid hinting at the weapon's purpose. Email correspondence with Prof. Conrad C. Crane, Sept. 30, 2008 (on file with the author).

60 *See*, Lori S. Hawthorne-Tagg, *On The Front Line of R&D: Wright-Patterson Air Force Base in the Korean War, 1950-1953* (Wright-Patterson AFB, OH: Wright-Patterson AFB, 2001), 34.

61 Indeed, they are often treated as "technically small, unguided kinetic missiles" rather than flechettes or darts. *See*, for instance, the entry "Lazy Dog" at <http://en.wikipedia.org/wiki/Lazy_Dog_(bomb)>.

*Illustration 4: Lazy Dog Missiles*[62]

In late 1951, various shapes and sizes of "[t]he strangest air projectile" were tested.[63] The most effective shape chosen (4.45 cm in length and approximately 19 grams in weight) enjoyed high penetration capabilities.[64] By mid 1952, while the project was yet to be completed, 16,000 bombs (226kg each) were ordered by the Far East Air Force (FEAF), while crew members began their training in Japan. Designated by the FEAF's Target Analysis Devision to "decimate large offensive concentrations", suppress flak and destroy construction projects, expectations for the weapon were quite high.[65] After all, one Superfortress (B-29) could carry up to 80,000 "finned bullets" – as they were once called – and cover a wide area.[66]

The higher the expectations, the greater the potential disappointment. As the War progressed, Communist forces remained so dispersed or dug in that utilization of the weapon "was never considered practical". Leaving aside a single instance of its use, once the FEAF Bomber Command's request to employ the new weapon against enemy searchlights was denied, the weapon's deployment was doomed. By July 1953 the War had ended, with no further cases of use recorded.[67]

62 Lazy Dog Missiles offered for sale at a popular U.S. auction site (no longer available, on file with the author).

63 For the quotation *see*, Crane, *American Airpower Strategy…*, *supra* note 59, at 134.

64 As they develop an impressive amount of kinetic energy (terminal velocity of 213 m/s), the darts can penetrate some 30cm of sand. *See*, Hawthorne-Tagg, *On The Front Line..*, *supra* note 61, at 34; "Resume of Technical Data on Conventional Munitions Used in Special Air Warfare Activities" *USAF Manual*, (Ap. 3, 1969), at 6-2, available online at <http://www.texastradingpost.com/militaria/lazydog.html>; *see*, also "Lazy Dog Missiles", available at <http://vietnamgear.com/kit.aspx?kit=512>.

65 Crane, *American Airpower Strategy…*, *supra* note 59, at 135.

66 *Id.*, at 134.

67 *Id.*, at 134-5. For the single case of flechette use *see*, Leroy, *Les Fléchettes D'avion…*, *supra* note 12, at 145. In fact, FEAF's policy held that the weapon would only be used against major concentrations of enemy troops in conjunction with friendly, or to counter-hostile, ground offensives. However, no record exists of such a use. E-mail correspondence with Prof. Crane, *supra* note 59; Crane, *American Airpower Strategy…*, *supra* note 58.

However, in March 1954, the existence of this weapon was formally disclosed to the public.[68] Concurrently, it had its first large-scale use in the Battle of Dien Bien Phu (March 13-May 8, the First Indochina War). In numerous sorties, heavy French bombers (PB 4Y Privateers) and, very likely, U.S. bombers as well, hailed huge quantities of newly arrived U.S. Lazy Dogs ("LD's" or "Hail Leaflets"as they were called at that time) down on Communist supply lines and positions.[69] However, despite the weapon's large-scale use (each bomber was capable of carrying up to four tons of flechettes), the desperate situation in which the besieged French forces found themselves remained unchanged. Excellent Communist camouflage tactics and unloading of the flechettes from improper heights due to a strong anti-aircraft fire seemed to render the weapons ineffective.[70] As claimed by the authors of the most comprehensive study on this battle:

> [w]e shall never know how many collies and mountain tribesmen ... died under the strafing, the napalm, the delayed-fuse bombs and the hail of "Lazy Dogs," but Giap's siege force never ran out of ammunition.[71]

A decade later, when it was the U.S.'s turn to wage war in Vietnam, the weapons were brought back into the arena (see Illustration 5) under new nickames: *Devil Dogs* or *Red Dod*.[72] These darts could be hurled from buckets, dropped by hand, or thrown into small paper bags, and were usable with almost any type of airplane.[73] However, a new dispenser releasing up to 10,000 darts was eventually introduced, the MK-4 Cluster Adapter.[74] Designed to shower the target with flechettes, the adapter was basically a simple hinged casing with bins inserted to hold the projectiles. The bins were mechanically opened by a time fuse (shown in Illustration 6), which allowed dispersion during the aircraft's flight. Once released, the darts, designed to penetrate

68 *See*, "Here and There" *Flight*, (26 Mar. 1954), at 358.

69 For more on the codename 'hail leaflet' *see, e.g.*, Martin Windrow, *The Last Valley: Dien Bien Phu and the French Defeat in Vietnam* (Cambridge: Da Capo Press, 2005), 577. According to Windrow, on May 1, 1954, a US B-26 dropped napalm and Lazy Dogs. *Id.*

70 Indeed, on April 12, 1954 a bomber which plastered communist positions with LD's was shot down. Bernard Fall & Bernard B. Fall, *Hell in a Very Small Place: The Siege of Dien Bien Phu*, (Cambridge: Da Capo Press, 2002), 240. Release of the weapons was carried out from a height of 17,000ft-20,000ft instead of the required 15,000ft. Leroy, *Les Fléchettes D'avion…*, *supra* note 11, at 145. According to Leroy, however, experimental French flechettes were dropped. *Id.*

71 Fall and Fall, *id.*, at 133. This lengthy book is considered the definitive account of the battle.

72 *See*, Leroy, *Les Fléchettes D'avion…*, *supra* note 11, at 147.

73 For the airplanes (*e.g.*, Skyraider A-IE) and helicopters (*e.g.*, the famous Bell UH-1 "Huey") available for such use *see*, *id.*, at 146-7. *See*, also 'Lazy Dog Missiles', available at <http://vietnamgear.com/kit.aspx?kit=512>.

74 *See*, 'Free-Falling Missiles', *Aviation Ordnanceman* 3 & 2 (Bureau of Naval Personnel, Rate Training Manual, NAVPERS 10345-C) (1970) at 264.

"soft targets", fell freely on the target area.[75] With no effect on hard targets, it was crowned "an anti-personnel weapons from [*sic*] the first order".[76]

*Illustration 5: Mk-44s at Bien Hoa Air Base, South Vietnam, 4 July 1965*[77]

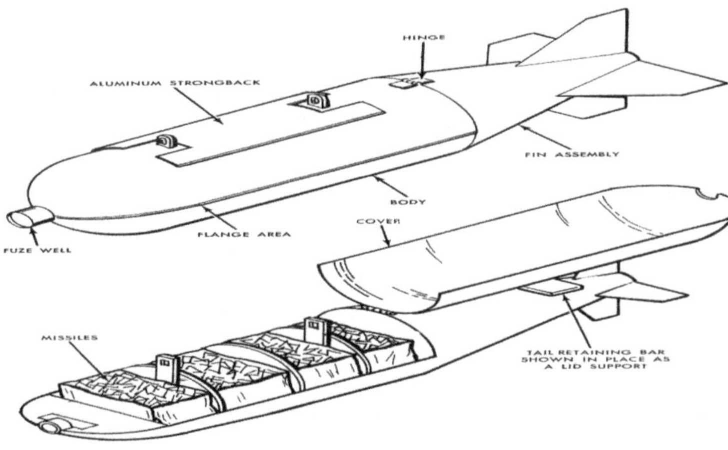


*Illustration 6: The Mk-44 Cluster Adapter*[78]

75 *See,* H. F. King, "Missiles and Projectiles" *Flight*, Jan. 28, 1955, at 114-122, at 129; Paul Dickson, *War Slang: American Fighting Words and Phrases Since the Civil War*, (London: Brassey's, 2003), 353; *Lazy Dog Missiles and Flechette Darts*, available at <http://www.texastradingpost.com/militaria/lazydog.html>.

76 *See,* "Here and There", *supra* note 69, at 358.

77 The Calvin Chapman Collection, The Vietnam Center & Archive, Texas Tech University, available at <http://www.virtual.vietnam.ttu.edu/cgi-bin/starfetch.exe?> (Item #: VAS004750).

78 *See, Bombs and Bomb Components*, TM 9-1325-200 to 11-1-28 (Washington D.C.: Departments of The Army, The Navy and The Air Force, Apr. 1966) at 2-65-6. A similar diagram is available online at <http://en.wikipedia.org/wiki/Lazy_Dog_(bomb)>.

Nonetheless, no records of large-scale use of this weapon, which was perceived from the onset as having "a very restricted application', appear to exist".[79] By 1969, the idea of using free-fall air-dropped darts as an effective modern anti-personnel weapon was abandoned.[80]

When the idea of free-fall air-dropped darts re-emerged in the 1990s, the darts in this new generation were designated as anti-material rather than anti-personnel weapons. In fact, some of these new weapons (e.g., the U.S. precision-guided cluster bomb, CBU-107, with its 3,750 tungsten and steel flechettes) were developed with the clear intention of minimizing collateral damage with reference to non-combatants.[81]

With that in mind, this book focuses on the anti-personnel flechette weapons still in use, the firearm flechettes.

## C. Firearm Flechettes: Development and Operation

### *CI. Flechette Rifles: The Rise and Demise of the "Salvo Project"*

In February 1951, Irwin R. Barr, inventor and co-founder of Aircraft Armaments Inc. (later the AAI Corp.) in Maryland, formulated the concept of high-velocity, low-drag, finned projectiles for increased range and reduced flight-time in small-caliber ammunition. Seven months later, AAI was awarded a contract by the US Army to study the application of flechette ammunition to a varierty of weapons – a contract that 'initiated the development of modern flechettes'.[82]

By 1953, although the Korean War had just ended, the first flechette shotguns (32 flechettes) were successfully produced for the US army standard combat shotguns.[83] By the next major U.S. military conflict, the Vietnam War, flechette shotgun shells (12-gauge) were in use; although poor media coverage prevented any public

79 *See*, H.F. King. "Missiles and Projectiles", *supra* note 76, at 121.

80 *See*, Leroy, *Les Fléchettes D'avion…*, *supra* note 12, at 148. Reliable information on the reasons for the idea's eventual abandonment is still missing. A leading anti-personnel weapons expert has suggested, however, that the appearance of cluster munitions – perceived as more efficient and deadly anti-personnel weapons – when the war intensified (1965) might have made the Lazy Dogs obsolete. Even so, the question is still open regarding the period prior to 1965. Email correspondence with Eric Prokosch, Oct. 8, 2008; Nov. 10, 2009 (on file with the author).

81 For information on a classified 2000 program to develop air-dropped flechettes, albeit with a guided dispenser, designated to puncture storage containers holding chemical or biological warfare agents, *see*, Ramon Lopez/Eglin AFB, "US Air Force Tests Latest 'Agent Defeat' Weapons" *Flight International*, Dec. 5-11, 2004, at 6; for more on the CBU-107 Passive Attack Weapon (PAW) see David Hambling, "Blast Reduction; New Airborne Bombs Localize Destruction and Reduce Casualties", *Defense Technology International* 1(7), (Sep. 1, 2007) at 54.

82 Thomas F. Swearengen, *The World's Fighting Shotguns*, (New-York: T.B.N. Enterprises, 1978), 464. Swearengen was the Chief Warrant Officer of the US Marine Corps and an explosive ordnance expert in the Combined Material Exploitation Intelligence Center in Vietnam during the 1960s.

83 For a detailed account of the use of flechette with small arms see *Id.*, at 464-471; see also Daniel Watters, *A 5.56 X 45mm Timeline – Small Arms History*, available at <http://www.thegunzone.com/556dw.html>, as well as, *56mm Background–Special Purpose Individual*

debate over the weapon's attributes and strategic value.[84] Barr's inventive genius went on to influence two U.S. weapons development programs involving flechette-loaded munitions, the "Salvo Project" and the "Beehive Project".

The Salvo Project was a prolonged (1952-1960) but unsuccessful multi-agency effort aimed at producing small arms capable of firing several projectiles *simultaneously*, in a "salvo" (as opposed to ordinary automatic fire), for the purpose of maximizing human hit probabilities.[85] In the late 1950s, this project focused on the flechette. The offshot of this project was the *Special Purpose Individual Weapon* (SPIW – a bland and somewhat deceptive name for the flechette rifles being developed[86]), three 40 mm grenades set in piggyback launchers, a product that could provide what ordnance officers sought: the addition of an area fire capability to designated semi-automatic rather than single-shot flechette rifles.[87] The SPIW's – much lighter and faster than rifle bullets – flechettes were planned to be enclosed in a fiberglass or nylon sleeve (*sabot*) that separated from the flechette upon the dart's emergence from the gun barrel.[88] However, the firing trials of three SPIW prototypes, conducted during 1964, demonstrated that none of them were particularly reliable or durable, and that a host of problems remained to be fixed. The contracting companies were instructed to submit 'second generation' prototypes but none were ready before August 1966. Given the general feeling that the rifles' flaws were unlikely to be solved in the near future, the Army's Chief of Staff, General Harold K. Johnson, recommended in late 1966 that the SPIW be reoriented to the status of exploratory development and that the XM16E1 (known as M16) should be adopted instead.[89] In December 1973, despite

*Weapons: A Brief History of Fléchette and the SPIW Project*, available at <http://www.thegunzone.com/spiw.html>.

84 These shotguns were used "particularly in river boat operations by the US navy". *See*, SIPRI, *Anti-Personnel Weapons*, (New-York: Taylor and Francis, 1978), 108; On the use of shotguns during the Vietnam War, see also Parks, "Conventional Weapons...", *supra* note 1 , at 81, fn.79; Swearengen, *The World's Fighting Shotguns, supra* note 82, at 18.

85 Eric Prokosch, *The Technology of Killing: A Military and Political History of Antipersonnel Weapons*, (London: Zed Books Ltd 1995), at 43; For an overview of the entire project *see*, Blake Stevens and Edward .C. Ezell, *The SPIW: The Deadliest Weapon that Never Was*, (Toronto, Collector Grade Publications, 1985), chap. 1.

86 However, the company that won the first contract, the AAI Corp., had conducted researched on flechette prior to 1962. *See*, Eric Prokosch, *The Simple Art of Murder: Antipersonnel Weapons and Their Developers*, (Philadelphia: narmic, 1972), 69.

87 Stevens and Ezell, *The SPIW, supra* note 85, at 1. Watters, *Special Purpose Individual Weapons, supra* note 56. For test results of a five-barrel machine gun firing flechettes conducted during 1961 *see*, S. S. Lentz, "An Investigation of the Characteristics of Flechette Rounds When Fired from A Multi-Barreled Test Gun", Ballistics Research Laboratory (BRL) Tec. Note 1379, AD 355191, (Feb. 1961).

88 Prokosch , *The Technology of Killing* , *supra* note 85, at 47.

89 Watters, *Special Purpose Individual Weapon*, supra note 56. For the results of the 'second generation" SPIW trails *see*, Stevens and Ezell, *The SPIW, supra* note 85, at 85. In addition, a micro-rocket powered flechette was researched during the 1960s for use as a covert weapon (*see, e.g.*, US patents 3521564 and 3344711). Email correspondence with Julian Perry Robinson, Feb. 21, 2010 (on file with the author).

the anticipated future of the flechette-firing rifle,[90] the program was officially cancelled after some $30 million had been spent.[91] It is important to note that the program ended not due to harsh internal crticism regarding the "inhumane" damage it would inflict upon its victims[92] but, rather, on the basis of the rifles' technical flaws: insufficient accuracy and difficulties with the sabot's separation from the dart.[93]

Yet, in 1969, the *Future Rifle Program* was initiated, under whose cover rifles with either multiple flechettes or bullets ("micro-caliber") were tested in 1974.[94] Perceived as a waste of funds by Congress, the program's activities were scaled back. Yet, the idea of a flechette-firing rifle refused to die. In 1986, the U.S. Army inaugurated the *Advanced Combat Rifle* (ACR) Program, mandated to design a replacement for the M16A2 assault rifle, specifically, a new "combat rifle for the twenty-first century" that would represent a major leap in firearms technology.[95] In response, two companies, AAI and the Austrian firm *Steyr-Mannlicher*, offered experimental flechette rifles. Both rifles could fire flechettes at a muzzle velocity of more than 1371m/s (4500f/s), apparently fast enough to ensure that the hooking effect would take place when the darts struck flesh.[96] Yet, although most of the ACR trial results are still unavailabe, it is known that not only were the offered fléchette cartridges considered to be less accurate than their conventional projectile rivals, none of the rifles offered a sufficiently high hit probability (i.e., exceeding that of the standard M16) to warrant further development. After years of research and testing, some $300 million invested and legal review of all weapons systems offered by the U.S. Military

90 John Wood, "Future Infantry Weapons", *The Infantry*, (Sept.-Oct. 1969). Wood was chief of the US Weapons Research Division.

91 *See*, U.S. Army Armament Command (ARMCOM) Laboratory Posture Report (LPR), FY74, at 3-60. For cost estimates see Prokosch, *The Simple Art of Murder*, *supra* note 86, at 68.

92 *See*, for instance, "Ottinger Says Army Developed Inhuman Dart Gun in Secret", *New York Times*, (May 24, 1970), at A50. Such a projectile (i.e., the dart), claimed then New York senatorial candidate (D) Richard Ottinger, "is much more devastating than the dumdum bullet". Ottinger, who charged that the rifle has been developed without Congress' knowledge, described the weapon as "diabolical and inhumane". In response, the Pentagon announced that "legal considerations were reviewed and it was determined that the weapons would not violate the rules of warfare". *Id.*

93 ARMCOM, LPR, FY74, at 3-60. For an overview of the entire project see also Prokosch, *The Technology of Killing*, *supra* note 85, at 44-47. Yet, as Stevens and Ezell concluded "It seems that whatever blame there is to be placed for the failure of the SPIW should be laid chiefly at the feet of the program's managers, whose gigantic case of overweening optimism created an impossible series of interdependent stumbling blocks". Stevens and Ezell, *The SPIW*, *supra* note 85, at 131.

94 U.S. Army research on the feasibility of flechette ammunition for rifles had been going on for years. *See, e.g.*, L.C. MAcAllister, *Drag and Stability Properties of The Xm144 Flechette with Various Head Shapes* (BRL Memo. Rep. 1981, AD 854724), May 1968. *See*, also Stevens and Ezell, *The SPIW*, *supra* note 85, at 101.

95 Kevin Dockery, *Future Weapons*, (Berkley: Berkley Books, 2007), 77. According to Parks, however, the flechette concept was reconsidered as part of the ACR Program already in 1982. *See*, Parks, "Conventional Weapons and…", *supra* note 1, at 81, fn. 97.

96 Dockery, *Id.*, at 78-80. However, as Parks noted "none possessed the lethality of standard 5.56x45mm ammunition". *Id.*

Advocate General (MAG),[97] the idea of fléchette rifles – together with the belief in the viability of the entire project – was abandoned.[98]

The notion of operative flechette rifles thus remained dormant – if not dead. However, as the following section details, development of flechette-loaded artillery shells or armor rounds within the framework of the Beehive Project had a different fate.

### *CII. "Beehive Projectiles": Origin and Development*

Initial interest in developing effective anti-personnel projectiles for artillery and armored forces emerged from the experience of the Korean War.[99] As mentioned with regard to the Lazy Dog Program, the Chinese tactic of human wave attacks against U.S. forces motivated weapons designers to produce a range of new anti-personnel weapons effective against large concentrations of soldiers.[100] New canister rounds filled with steel balls or slugs (little steel rods) were subsequently introduced to replace ordinary "shrapnel shells",[101] which were absent from the U.S. Army's arsenal due to their limited effectiveness.[102] The breakthrough in flechette projectiles came, however, only in the war's aftermath, when the potential of filling canister-type projectiles with small dart-like objects (flechettes) was demonstrated with shotgun ammunition.[103]

97 The flechette rifles were submitted for legal review on March 20, 1990. *See*, Hays Parks, "Memorandum of Law: Review of Weapons in the Advanced Combat Rifle Program", reprinted in *The Army Lawyer*, DA PM 27-50-211 (July 1990), 18.

98 The U.S. Army decided that "existing technology does not offer a significant improvement over the standard M16A2 assault rifle". *See*, Ramon Lopez, "Improved M16 May Emerge from US Army Program" *International Defense Review* 25 (Oct. 1992), 977, at 977. For more on the ACR program *see, e.g.*, Ian V. Hogg & John S. Weeks, *Military Small Arms of the 20th Century*, 7th Ed., (Iola, WIS: Krause Publications, 2000), 293; Sybil P. Parker, *McGraw-Hill Encyclopedia of Science & Technology* 8th Ed. (New-York: McGraw-Hill, 1997), 107.

99 For the Korean War's contribution to the interest in flechette projectiles see also information posted by The Federation of American Scientists (FAS), available at <www.fas.org/man/dod-101/sys/land/m546.htm>.

100 *See*, SIPRI, *Anti-Personnel Weapons*, *supra* note 84, at 25.

101 A common type of controlled fragmentation shell designed to break open over the heads of enemy troops, dispersing its deadly contents. The popular use of the term shrapnel to describe air burst artillery shells used in the Korean War is, however, technically incorrect as modern shells produce fragments or splinters, not shrapnel. For a concise overview of explosive air burst shells and their importance *see*, William Weir, *50 Weapons that Changed Warfare*, (New-Jersey: New Page Books, 2005), 89-92.

102 For the official U.S. Army history on the inefficiency of shrapnel *see*, Prokosch, *The Technology of Killing*, *supra* note 85, at 68.

103 *Id.*, at 79, fn. 63. The attractiveness of flechettes also reflects one of the important trends observed by experts regarding the development of anti-personnel weapons since WWII: a decrease in the average size of fragments, with a concomitant increase in their number, as "the more fragments, the greater the chances of hitting someone." Even a small fragment (a pre-controlled one in our case) can cause a bad wound if it has sufficient velocity. *See*, Eric Prokosch, "Antipersonnel Weapons", *International Social Science Journal*, 28 (1976) 341, at 344.

In November 1955, the International Harvester Co. began to develop flechette-loaded projectiles under the technical supervision of the Picatinny Arsenal (Dover, New Jersey), the "Cradle of American ammunition".[104] Following the company's March 1957 decision to terminate its ordnance activities, the contract was passed on to the Whirlpool Company (April 1957). Under the framework of the so-called "Beehive Program"[105] – a name appropriately chosen given the way that flechettes are compartmentalized in the various projectiles – Whirlpool continued its work throughout the 1960s.[106] Given the strong correlation between flechette lethality (i.e., the projectile's efficiency) and their density within the canister, efforts were invested on developing projectiles with the largest number of flechettes possible;[107] one contract even called for the manufacture of projectiles filled with some 9900 flechettes.[108]

Compression of such a large number of darts into standard projectiles was achieved by tightly packing the flechettes in alternating directions (i.e., a flechette tip nestled between the fins of its neighbor). Upon release from the cartridge, the four stabilized steel fins used to balance the projectile's flight caused the flechettes lying opposite to the projectiles' trajectory to reverse direction and travel forward.[109] Given the flechettes' high density and low velocity decay, such projectiles had greater area coverage and were therefore much more effective (e.g., superior wounding capability, effective range) than coventional cansiter muntions.

But technical problems plagued development. The first flechette-filled round to be developed (M377 90mm, 1958) suffered from so many production problems that it was never mass produced.[110] The initial contracts were followed by others in the early 1960s, designated to improve the weapons together with the production process; two

104 Whirlpool Corp., Final Progress Rep. No. 8, Contract No. DA33-008-ORD-1561, 90mm Canister Ammunition (Apr. 1957-Dec. 1957), (doc. date is illegible) at 1 [hereinafter, *Whirlpool Corp. 1957 Progress Rep.*]. For the exact quotation *see,* Harry W. Painter, "Picatinny Arsenal in Research and Development", *Army Research and Development News Magazine* (Jan.-Feb. 1974), 16, at 16.

105 *Whirlpool Corp. 1957 Progress Rep* , at 2.

106 Painter, "Picatinny Arsenal in…" *supra* note 105, at 17. The common assumption among users, however, is that the name reflects the noise made by flechettes flying through the air, a sound that reminds one of beehives.

107 *See,* Ian V. Hogg, *The Illustrated Encyclopedia of Ammunition*, (Berkeley: Quantum Books, 2003), 156; Prockosh, *The Technology of Killing, supra* note 85, at 68.

108 This projectile (106-mm caliber M581) was highly effective in Vietnam. For its details *see,* Anthony G. Williams & Leland Ness, (eds.) *Jane's Ammunition Handbook 2005-2006*, (London: Jane's Information Group, 14th ed., 2005), [hereinafter, *Ammunition Handbook 2005-2006*], at 316.

109 In addition, the company succeeded in overcoming the technical problem of darts fanning out in a cone, leaving the center of the target area undamaged. *See,* Prockosh, *The Technology of Killing, supra* note 85, at 69.

110 US Army Materiel Command- Historical Office, *Arsenal for the Brave-A History of the United States Army Material Command 1962-68*, (Washington D.C.: US Army Materiel Command, 1969), at 155. For a detailed overview on this projectile *see, Whirlpool Corp. 1957 Progress Rep.*, at 2-5.

105mm beehive rounds were thus developed.[111] At the same time, classified research was conducted by the army for the purpose of estimating the flechette's "wounding potential".[112]

### *CIII. Flechette Projectiles: Method of Operation*

In principle, all flechette projectiles are similarly operated (they differ, of course, in technical features such as the amount of darts they contain). Yet, in detailing their method of operation we have chosen the beehive model most commonly used in recent years (albeit with an advanced Israeli-made fuse): the 1968 U.S.-made anti-personnel tank with tracer (APERS-T) 105mm M494 round, which Israel has consistently deployed since the mid 90s in Southern Lebanon and later in the Gaza Strip.

This round was developed following an urgent request for a limited quantity of beehive rounds for the 105mm M68 rifled tank gun in February 1968.[113] It weighed some 14kg. (31lbs), contained 5000 darts, a tracer (M13) to enable detection of the round trajectory during its five-second burning time and a yellow dye that bursts into a yellow ball when the round splits. In the absence of a substantive explosive component, this ball was a simple and visible indication of a successful operation.[114] More importantly, it enabled the gunner to adjust subsequent fire. The round came with a hand-settable (i.e., no need for a fuse wrench) mechanical time fuse (the M571) having a *muzzle action* option as a default mechanism to enable use from even a very short range (3m.). This feature turned out to be highly important in Vietnam as it provided tank crews with an effective emergency defense against short-range assaults. The tank's gunner is also able to quickly set the fuse to operate along the direct firing trajectory at 100m increments starting at 200m, up to a maximun range of 4.4 km (imposed by the maximum fuse time marking), normally at a height of six

111 For more contracts *see, e.g.*, "Design, Development, Fabrication and Test 105mm. Beehive Ammunition", 1959; "Design and Development of New and Improved Flechettes and Applicable Weapon Systems", 1961. Whirlpool held at least nine army contracts for producing at least five different types of beehive projectiles from 1957 to 1966. *See*, Prokosach, *The Simple Art of Murder*, *supra* note 86, at 29.

112 The research, by the Army BRL and its Chemistry Research Development Laboratory (CRDL), was conducted on various types of flechettes up to 1965. *See*, Joseph Blair & Joseph Sperrazza, *Evaluation of Wound Data and Munitions Effectiveness as Based On Battlefield Surgeries in Vietnam* (WDMET), N.D. AD511909; William Kokinakis & Joseph Sperrazza, *Criteria for Incapacitating Soldiers with Fragments and Flechettes*, (BRL Rep. No. 1269), Jan. 1965.

113 *See*, *Advanced Production Engineering for the 105 MM XM494E3*, Picatinny Technical Rep. 4283 (Nov. 1971). For an earlier version of this round *see*, *Advanced Production Engineering Program for the 105 MM XM494* APERS Cartridge, Picatinny Technical Rep. 3407 (July 1966) [hereinafter, 1966 Technical Rep. 3407]. It should be noted that the "X" refers to an experimental munition, with the designation dropped once mass production began.

114 The yellow cloud's high visibility was the last footage caught by the camera of Fedel Shana, a Reuters TV cameraman accidently killed by a flechette round on April 17, 2008. The footage is available at <http://www.reuters.com/news/video?videoId=80475>.

meters. To ensure ultimate flechettes dispersion, however, the time fuse should be set to operate 60m short of the designated target.

A detonator that rests on an axial relay flashes through a central flash tube and ignites four radially positioned detonators to expel a charge located in the base. The detonators tear away and remove the forward portion of the aluminum projectile's ogive-shaped skin. Centrifugal forces then add to the darts' high velocity (more than 822m/s [2,700fps]) in their 20-degree cone-shaped dispersion.[115] The three rear tiers of flechettes are ejected using a piston plate propelled by the base charge, which imparts additional velocitiy to these flechettes while centrifugal forces disperse them in a way similar to the first four tiers.[116] In the end, 5,000 steel darts, each weighing less than one gram (the so-called 13-grain standard) and arranged in seven tiers, are dispersed by centrifugal forces in a conical pattern.[117] Once the target lies at least 300m away from the gun muzzle, the flechettes spread out in an arc form, 90m wide, with a maximum effective range of 300m from the point at which the fuse functioned (see Illustration 7).[118]

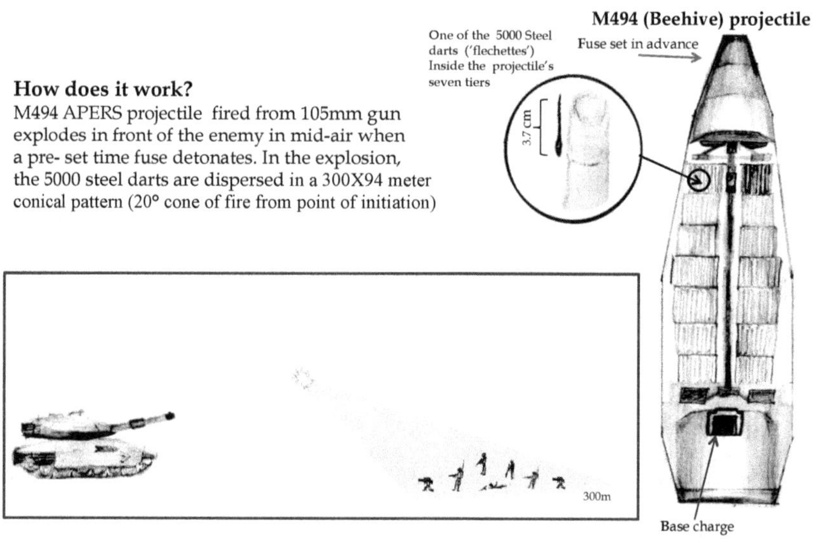


*Illustration 7: Operating the Flechette Round*[119]

115 The subprojectiles (flechettes) derive their effectiveness from the squaring of the velocity (MV2). *See*, 1966 Technical Rep. 3407, at 3.

116 Advanced Production Engineering for the 105 MM XM494E3, , *supra* note 114, at 9.

117 *Id.* at 9. For the XM494 dart's specification see Dr Ricardo Nachman, Autopsy Rep. No. 3/2293/2001, Jan. 16, 2002, at 11 (on file with the author). Those darts were removed from Palestinian flechette victims who had been transferred to The Israel National Center of Forensic Medicine (The Israeli Pathological Institute) by the IDF.

118 *See*, for instance, Terry Gander and Ian V. Hogg, *Jane's Ammunition Handbook 1993-94* (1992), 134. For an updated version *see*, *Ammunition Handbook 2005-2006*, at 295.

119 Courtesy of Gur Bakai, Jerusalem, Israel, June 2010. For a similar illustration of a different type of flechette round (M546 APERS-T 105mm) appearing following an

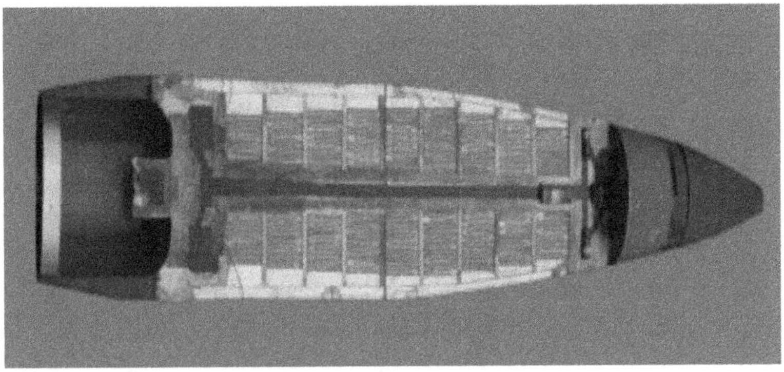

*Illustration 8: A Cutaway of an M546 Flechette Round*[120]

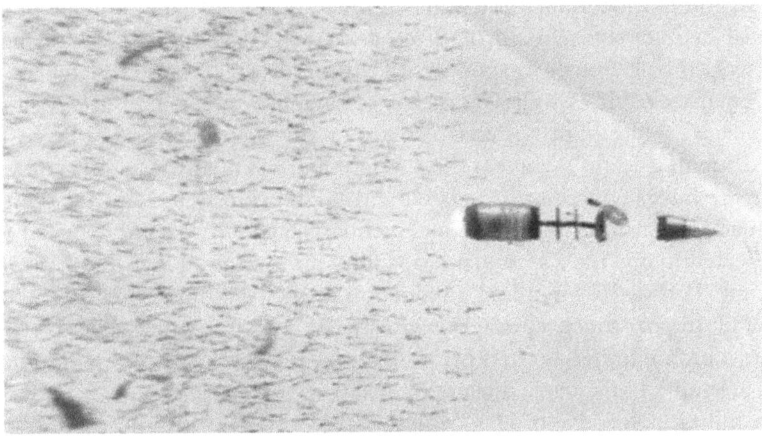

*Illustration 9: Scattering after the Round Splits*[121]

August 2002 incident involving non-combatant fatalities by flechette weapons in the Gaza Strip *see*, "Dart Bombs 'Killed Four Palestinians'", *BBC World News*, Aug. 30, 2002, available at <http://news.bbc.co.uk/2/hi/middle_east/2225706.stm>. As the flechette round employed by Israel is different (M494 APERS-T, instead of the M546 APERS-T, which appears in the BBC illustration and is loaded with 8,000 8-grain steel flechettes), the illustration's specifications are inaccurate. Nonetheless, its very appearance demonstrates the world's interest in these weapons, especially in comparison to the striking absence of worldwide interest in the use of other "conventional weapons" used in the same fighting zone.

120 For the cutaway illustration *see*, FAS, *supra* note 101.

121 Illustration used by the IDF when explaining the weapon's method of operation. Courtesy of D.B., a former IDF NCO.

*Illustration 10: Flechettes' Dimensions*[122]

The escalation in Indochina in the wake of the 1964 Gulf of Tonkin Incident (which intensified U.S. involvement in the Vietnam War), created an urgent demand for an effective anti-personnel weapon designed to provide "a close-in defence against massed enemy assault as well as a means of assault on personnel in deep grass or dense foilage".[123] The high number of darts in any round addresses the first need ("massed enemy assault") whereas their weight and mainly their aerodynamic shape – unlike the bulky fragments associated with high explosives – address the second need. In addtion to this anti-personnel "primary capability", some types of rounds were expected to provide a secondary capability for "defeating low-flying slow-moving aircraft", a goal that required a reduction in the weight of some flechette projectiles and, in consequence, adaption of smaller darts (i.e., the 13-grain standard) to ensure a high number of darts placed in every projectile.[124] In 1965, the Picatinny Arsenal attempted to address this urgent need.

The military objective of introducing a large number of flechettes in each projectile, multiplied by the projectiles production rate, required the manufacture and assembly of an astronomical number of flechettes. Dramatic improvements in design had to be developed in order to meet the required production rates. A unique mass production process was thereupon developed and within four months, 200 flechette rounds (M590, 90mm) were ready to be sent straight to the Vietnamese theatre for real-time field evaluation in combat.[125] As the United States Army Materiel Command (AMC) formal history reads: "The Beehive munitions round, developed and fielded on a crash basis…".[126] Ultimately, almost ten types of flechette rounds were manufactured, ranging from 105mm tank rounds or artillery shells (XM546

122 B'Tselem Photo Archive, available at <http://www.btselem.org/English/Photo_Archive/One_Photo.asp?start=17>.

123 Painter, "Picatinny Arsenal in…" *supra* note 105, at 17.

124 1966 Technical Rep. 3407, at 3.

125 Painter, "Picatinny Arsenal in…" *supra* note 106, at 17.

126 *See, A Brief Chronology of Events and Achievements 1962–1972* (U.S. Army Materiel Command, 1972), at 8 [hereinafter, AMC's Brief Chronology].

filled with over 8,000 flechettes)[127] to 155mm (XM396) howitzers (i.e., short guns for firing shells on high trajectories at low velocities).[128] As already mentioned, all had the ability to function at the muzzle or at any desired point downrange.[129]

In December 1965 the first shipment of beehive muntions reached Vietnam to await proof of their devastating effectiveness.[130]

### *CIV. 'The Most Efficient Killer': Use of Beehive Munitions in the Vietnam War*

Although flechette projectiles had become fully operational by mid-late 1966, they were exposed to their first instance of combat firing only in November 1966, when a single round (105mm XM546) killed nine North Vietnamese troops and repelled an attack on a U.S. outward base.[131] While this incident did not attract much attention, this situation was soon to change.

On 26 December 1966, two companies of the 22nd Regiment of the North Vietnamese Army, were close to overunning an artillery base in central South Vietnam (Landing Zone BIRD at Kim Son Valley) when "[a]lmost in desperation", the commander of the trapped 105mm battery (having one last gun intact), Capt. Leonard L. Schlenker, ordered the firing of two beehive rounds. In the wake of the rounds' firing, the North Vietnamese retreated, leaving behind 266 bodies – the majority of fatalities ascribed to the darts fired from the two rounds.[132] It was this battle, according to the formal history of the U.S. Artillery Corps, 'that really woke up field artillerymen and infantrymen to the effectiveness of this new round'.[133] As few writers on the Vietnam War made note of the event:

> News of the beehive's awesome battlefield debut traveled quickly through American bases in Vietnam. Before long there were beehive adaptations for both light and

127 For the XM546 specifications *see*, *Ammunition Handbook 2005-2006*, at 502; Edgar C Doleman *et al.*, *Tools of War (Vietnam Experience)*, (Boston: Boston Publishing, 1985), 46.

128 The prominent munitions were the 90mm (XM580), the 105 mm (XM494) and the 152mm tank rounds (the latter were designated for Sheridan Light Tanks). *See, Id.* For data on the XM580 *see*, *Ammunition Handbook 2005-2006*, at 280.

129 *See*, Painter, "Picatinny Arsenal in…" *supra* note 105, at 17.

130 AMC's Brief Chronology, at 8.

131 David Ewing Ott, *Vietnam Studies: Field Artillery, 1954-73* (Washington D.C.: Department of the Army, 1975), at 61. Major General Ott served in Vietnam as executive officer of the II Field Force Artillery during 1966 and as commander of the 25th Infantry Division Artillery during 1967. For more on that use (2d Battalion, 320th Field Artillery) see also Prokosch, *The Technology of Killing*, *supra* note 85, at 70.

132 Ott, *Vietnam Studies*, *Id.*, at 108; Prokosch, *id.*, at 70.

133 Jonathan B. A. Bailey, *Field Artillery and Firepower* , 2nd ed., (Annapolis, MD: Naval Institute Press, 2nd ed., 1993), at 51, fn. 7; Prokosch, *id*; Ott, *Id.*, at 108-09 (according to whom only one beehive round was fired).

> medium howitzers, for the 2.75-inch rockets carried by helicopter gunships, for 90 and 106MM recoilless rifles, and for tanks.[134]

In parallel, the media learned of this innovative means of warfare; news reports began being issued praising the weapon's effectiveness in the dense forests and tangled vegetation of Vietnam. Military reports of "enemy soldiers actually being nailed to trees" were cited[135], expressing the darts' relatively high penetration capability in soft targets (e.g., human tissue) as opposed to their limited penetration capability in harder substances (as shown in Illustration 11). As expected, flechette rounds, crowned by the media as "the most feared weapons" with which US tanks were equipped,[136] quickly earned an abhorrent image in the victims' eyes (in this case, the Vietcong or the North Vietnamese Army).[137]

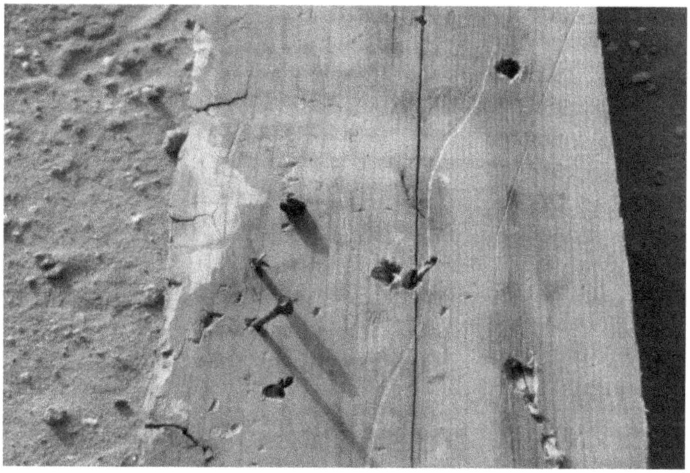

*Illustration 11: Anti-personnel Flechette's Low Penetration Capability against Hard Surfaces*[138]

134 Doleman *et al.*, *Tools of War, supra* note 128, at 48.

135 Bob Horton (AP) Times, 14 Dec. 1967 as reprinted in Prokosch, *The Simple Art of Murder, supra* note 86, at 28.

136 *See,* AP, "Defective' Tanks Help GIs Crush Red Ambush", *Los Angeles Times*, (Mar. 29, 1969) at A2. The article refers to the 152mm round delivered to Vietnam in January 1969 for the newly arrived Sheridan tanks. In March 1966 its development had been authorized; in addition to its anti-personnel capability, it was designated for use against "unarmored and lightly armored vehicles". *See,* Picatinny, 152MM XM617 APERS-T Cartridge Technical Characteristics, Formal In-Process Review (IPR), Aug. 16, 1967. On Jan, 1967, the Whirlpool Corp. was awarded a contract to facilitate the round's design, fabrication and testing. *See,* Contract No. DAAA25-67-C-0441 Whirlpool Monthly Cover Reports (Mar. -Oct. 1967).

137 In an 1971 documentary produced by the Vietnamese People's Army, medical experts argued that the weapon "is even more atrocious" than cluster munitions. *See,* 'U.S. War Technique and Genocide in Vietnam', (16mm, 1971, 35 min.) Film Transcript, at 4 (on file with the author).

138 This picture, showing flechettes on an old 105mm munitions box, is available at <http://www.fresh.co.il/ vBulletin/showthread.php?t=251734> (courtesy of Mr. Yossi

Three months later, on 21 March 1967, another artillery base (Fire Support Base GOLD near Tay Ninh) came under such a large-scale Vietcong assault that its defense perimeter was breached. At that point, the besieged infantry commander hurriedly requested permission to fire beehive rounds (M546). After obtaining permission, the battery's entire stock of 30 beehive rounds was used in direct fire; in the eyes of one eyewitness, every gun subsequently functioned as "a huge shotgun blast" [in reference to the image of multiple small pellets released upon firing].[139] The fact that despite being a small portion of the 1,000 rounds(!) directly fired that day, the 30 flechette rounds were "the largest number of these rounds fired in a single engagement to date",[140] attests to this weapon's singularity. The attacking forces eventually retreated, leaving at least 635 casualties[141]; the two artillery officers who carried out the beehive rounds' firing were awarded the Distinguished Service Cross several months later.[142]

Permission to fire the flechette rounds was required not for humanitarian considerations but, rather, because beehive rounds were then "in critical[ly short] supply throughout Vietnam".[143] Yet, after such a successful introduction, the amount of beehive rounds shipped to Vietnam increased dramatically, up to six times the original amount.[144]

Following the aforementioned success in the field, beehive rounds became mandatory items in tank and artillery arsenals throughout Vietnam; their use becoming more frequent up to the War's end.[145] Contrary to the fragments associated with high explosives, the flechettes' ability to penetrate dense foliage – a common feature of the Vietnamese terrain – while still maintaining wounding velocity provided these projectiles with a considerable military advantage over other weapons in that theater. Beehive rounds' extended killing zones, in conjunction with the flechette's high velocity, made these means of warfare "devastatingly effective against exposed infantry" by neutralizing a high number of infantrymen.[146]

---

Kaufman, a former IDF Armor Corps serviceman). Note that some flechettes passed through the box, previously undamaged, while others were stuck in the wood at different depths.

139 Ralph R. Steinke, "Redleg Heroism at Suoi Tre", *Field Artillery Journal* (Aug. 1993), 25, at 26.

140 Ott, *Vietnam Studies supra* note 132, at 113; Bailey, *Field Artillery, supra* note 134. According to Steinke, 40 rounds of beehives were fired that day. Steinke, *id.*, at 27.

141 Steinke, *id.*, at 27. Bailey, *id.* Note that according to Steinke, there were 647 Vietcong confirmed killed. U.S. losses were 31 soldiers killed and 109 wounded.

142 Lieutenant Colonel John William Vessey, Jr. and Sergeant First Class Raymond C. Childress, Steinke, *id.*, at 27.

143 *Id.*

144 63,000 M546 rounds (the type used in this famous battle), for instance, were procured in 1967, in comparison to 10,000 in 1966. For U.S. Department of Defense procurement data *see*, SIPRI, *Anti-Personnel Weapons*, *supra* note 101, at 36-7.

145 For additional cases of beehive use, for instance, Ott, *Vietnam Studies*, *supra* note 132, at Chap. IV.

146 *See*, FAS, *supra* note 100.

Not surprisingly, recalled Major General William C. Westmoreland, Commander of the U.S. forces in Vietnam during the war's peak (1964-1968), "[b] eehive artillery rounds were markedly effective at close range, sometimes making the difference between life and death when isolated artillery units came under attack".[147] When comparing the high number of casualties inflicted in any given engagement involving flechettes to the low number of flechette rounds fired during this war, a leading conventional weapons expert concluded that: "...flechette ammunition must have been one of the most efficient killers of the Vietnam War".[148]

### *CV. Flechette Weapons in the Post Vietnam Era: Proliferation and Use*

Surprisingly, after the initial enthusiasm – as a series of contracts indicated–production of beehive rounds "abruptly and mysteriously" ceased.[149] In fact, after 1968, no sizable production contracts were awarded for either beehive cartridge components or fuzes.[150] Furthermore, unlike other types of anti-personnel weapons (e.g., cluster muntions), which many states have either developed or acquired, very few states are known to have developed and produced flechette-filled anti-personnel *projectiles* (flechette-filled air-to-surface rockets will be discussed later) in the more than four decades following their successful employment in the Vietnam War.[151] Substantive allegations regarding their manufacture exist only in regard to the former Soviet Union.[152] Ironically, following reports on Soviet use of new flechette-

147 *See,* William C. Westmoreland, *A Soldier Reports*, (Cambridge: Da Capo Pr, 1976), 281.

148 Prokosch, *The Technology of Killing*, *supra* note 85, at 71. As a mark of their efficiency, the highest number of flechette rounds fired in a single engagement in the Vietnam War stood at 30 rounds.

149 Prokosach, *The Simple Art of Murder*, *supra* note 86, at 29. Prokosch suggests three reasons for the sudden cessation: (a) The U.S. army became aware of the danger this munitions posed to "friendly forces"; (b) Inherent technical problems; (c) Simple cost-benefit analysis given their high production cost in comparison to the "usual"/"ordinary" high explosive ammunition. *Id.*

150 One small contract for 105mm flechette tank rounds was awarded in mid 1970. *See, Commerce Business Daily*, July 10, 1970 as cited in Eric Prokosch, *Notes on the Development of Canister and Beehive Rounds for Artillery*, (unpublished draft, on file with author), July 1974, at 16, fn.25.

151 In 2002, experts counted China, Israel and Russia under this category. *See,* Erich Prokosch & Ernst Jan Hogendoorn, Antipersonnel Weapons, in: Ilkka Taipale *et al.*, (eds.) *War or Health? A Reader* (New-York: Zed Books 2002) 65, at 72. Israel has neither developed nor produced flechette projectiles though it did produce advanced fuses for its U.S.-made 105mm rounds. Since the early 1990s, however, it has made sure to adjust them for use in its new main battle tanks (Merkava Mark III & IV) that are equipped with a 120mm smoothbore main gun.

152 *See,* Hogg, *The Illustrated Encyclopedia*, *supra* note 108. While Hogg referred to Russian manufacturing of 122mm and 152mm flechette projectiles, Williams & Ness, suggested that 125mm projectiles were produced as well. *See, Ammunition Handbook 2005-2006*, at 338.

loaded Kalashnikov assault rifles(AKS) in Afghanistan, it was the U.S. that accused the Soviets of using a "vicious weapon" (July 1980).[153]

However, the above accusations required authentication and, until the mid 1990s, when Israel's use of flechette in South Lebanon became known – as Jane's researchers referred to the common flechette tank rounds – the "U.S. army was apparently the only armed force to utilize this round".[154] Furthermore, unlike Israel, the U.S. had refrained from adjusting its own arsenal of flechette tank rounds (i.e., 152mm, 90mm, and 105mm rounds) to the modern 120mm gun, installed on its principal combat tank, the M1 Abrams, since 1985.[155] As the U.S. seems to have renounced use of flechette tank rounds (although it retains the flechette artillery shells option), Israel's arsenal of flechette tank rounds is one of the last being employed, with all that implies for the internal legal debate over its legality.[156]

Despite the limited proliferation of anti-personnel flechette projectiles (tank rounds and artillery shells), deployment of flechette-filled air-to-ground rocket warheads has undergone a dramatic change. The 2.75in (70mm) Folding Fin Aircraft Rocket came to be one of the most widely used munitions in Vietnam; its "ever-growing success" in that theatre was to spur the U.S. Army's search for more effective warheads.[157] In Spring 1970, anti-personnel flechette-filled warheads (WDU-4) saw their initial operation.[158] As reflected in their production contracts, these warheads were the most widely used flechette weapons in Vietnam.[159] Given the rockets' high velocity, they posed a risk of being "tumbling flechettes" – an effect characterizing flechettes hitting the body above a certain velocity.[160]

153 This reference was made by the then State Department Spokesman, John Trattner. See (XINHUA News Agency), Soviet Troops Test New Assault Rifle in Afghanistan, July 11, 1980; for the reports *see, e.g.*, Drew Middleton, "Soviet Troops Said to Test New Weapons in Afghanistan", *The New York Times*, (July 10, 1980), at A8.

154 *See, Jane's Ammunition Handbook 1993-94*, *supra* note 119, at 134.

155 Until 2005, no dedicated antipersonnel rounds for the 120mm main gun were available in the U.S. In January 2005, a contract was awarded to two companies (General Dynamics Ordnance and Tactical Systems) to produce such a round, containing more than 1,000 tungsten balls (M1028). Unlike flechette rounds, however, this round is far more precise than an area weapon as it can only hurl out of a muzzle in a shotgun-like pattern, similar to the older canister ammunition. *See,* Michael Green & Greg Stewart, *M1 Abrams at War*, (New-York: Zenith Imprint, 2005), 72.

156 Unlike the 1993-1994 edition, the 2005-2006 edition of Jane's prestigious *Handbook* mentioned that: 'The US army was at one time the only armed force to use this round, although the M494 APERS-T was held by the ...(IDF) for operations involving prepared ambushed positions'. *See, Ammunition Handbook 2005-2006*, at 294.

157 Prokosch, *The Simple Art of Murder*, *supra* note 84, at 37.

158 *See,* Jonathan Bernstein, *US Army AH-1 Cobra Units in Vietnam*, (Oxford: Osprey Publishing, 2003), 43.

159 Prokosch, *The Simple Art of Murder*, *supra* note 86, at 37. Their main producer was The Northrop Corp., *Id.*, at 39. *See,* also Prokosch, *Antipersonnel Weapons*, *supra* note 103, at 348-9.

160 For this concern *see,* Prokosch & Hogendoorn, *War or Health?*, *supra* note 151, at 73.

Unlike flechette projectiles, a modern version of the HYDRA 70 Rocket System is still in widespread use.[161] In the U.S., for instance, this family of weapons has a rich history of providing close air support to ground forces from some 20 different firing platforms (fixed-wing and armed helicopters alike). The 1993 versions of anti-personnel flechette warheads contain 1,180 (3.8g.) or 2500 (1.8g.) steel flechettes and release a red marker during flight.[162] Following the Vietnam War, for instance, flechette-filled air-to-ground rocket warheads were used by the U.S. in various military operations across the globe.[163] Furthermore, a few states have opted to manufacture their own flechette-filled air-to-ground rocket warheads and therefore compete with the U.S. in the world weaponry market.

There is, however, one weighty difference between U.S. and non U.S. weapons systems, with the latter *designated for anti-material use rather than an anti-personnel use*. Most popular non-U.S. flechette-filled rockets, such as the Canadian HYDRA 70 (known as CRV-7 rockets) – which became the *de facto* standard for forces aligned with the West – have two types of flechette-filled warheads: an anti-tank warhead (WDU-5002/B containing five tungsten flechettes) and a "general purpose flechette" (GPF) warhead, containing 80 (20g.) tungsten alloy flechettes.[164] As such, these warheads are in constant and widespread use in war zones such as Afghanistan, where the UK's attack helicopter force (Apaches) extensively uses the GPF warhead.[165]

---

161 A comprehensive overview of this rocket system is available at <http://www.globalsecurity.org /military/systems/ munitions/hydra-70.htm>.

162 For more details on flechette warheads (M255 & M255E1) *see,* Chris Bishop and Jim Laurier, *Apache AH-64 Boeing (McDonnell Douglas) 1976-2005* (Oxford: Osprey Publishing, 2005), 14; Leroy, *Les Fléchettes D'avion…, supra* note 11, at 149. It should be noted that this weapon is considered "the primary area weapon system" both the widely used Apache Longbow gunship and the AH-1F 'modernized' Cobra. *See,* Bishop and Laurier, *Id.*

163 For cases of use by the Special Operations Aviation Regiment (SOAR) gunships (AH-6 model) during its various missions from Grenada in 1983 to contemporary Afghanistan and Iraq, *see,* Michal J. Durant and Steven Hartov, *The Night Stalkers*, (New-York: G.P. Putnam's Sons 2006). For use against Iranian naval forces during the Iran-Iraq War *see,* Harold Lee Wise, *Inside the Danger Zone – the U.S. Military in the Persian Gulf 1987-1988*, (Annapolis, MD: Naval Institute Press, 2007), 100-1, 128. In the 1991 Gulf War (Operation Desert Storm), gunships carried such rockets even during the first surprise attack. *See,* Thomas Taylor, *Lightning in the Storm – The 101st Air Assault Division in the Gulf War,* (New-York: Hippocrene Books 1994) 128, 187.

164 The first type is designed to defeat NATO-standard triple heavy armour while the second is marketed by Magellan Aerospace Corp. as an area weapon for attack helicopters against light to medium armour. *See,* the company's brochure, available at <http://www.magellanaerospace.com/_bin/rockets/rocket.cfm>; e-mail correspondence with Mr. Ken Kohut, Team Leader, Magellan Aerospace Marketing, Sept. 14 & 18, 2008. As to other states, since 1987, a French company (Thomson Brandt Armaments – TBA), for instance, is marketing 68mm AMV (*anti-mate'riel el ve'hicule*) flechette-type warheads (36 darts capable of penetrating 8mm steel plate at an impact velocity of 500m/s). For these and other types of flechette warheads *see,* Gunston Bill and Spick Mike, *Modern Fighting Helicopters*, (Little Rock, AR: Tiger books, 1991), 55-56; Braybrook Roy, *Aircraft Weaponry of Today*, (Sparkford: Haynes Publishing Group, 1988), 39-40.

165 Each Apache is armed with 16 CRV- flechette rockets (alongside two AGM-114 Hellfire missiles). *See,* Hoyle Craig, "Apaches Pass Trial by Fire", *Flight International,* (Jan. 9,

However, once used in salvos by attacking helicopters, these warheads do pose a high danger of collateral damage given the multitude of darts carried on every mission and their wide blanket coverage. For instance, the 16 GPE warheads with which the UK's Apaches are armed create conditions where a ripple against a ground target could release 1280 darts (i.e., 16 x 80 darts per warhead). For a typical helicopter, armed with French-made flechette warheads, 22 rounds of fire would result in the blanket coverage of a 1,000m by 70m area.[166] Furthermore, some states have adapted these air-to-ground rockets for use by ground forces.[167]

As will be discussed in the following chapter, the international community's attempts to outlaw this weapon in the wake of the Vietman War have dissipated and shifted toward other weapons, such as landmines, which may be said to be equivalent in their potential for human suffering. The flechette anti-personnel weapon remained in the shadows until its reemergence in the Middle East in the mid 1990s. The 2009 Goldstone Report has focused its attention on Israel's most recent use of the weapon (in Gaza), a development to be elaborated in Part II.

---

2007) available at <http://www.flightglobal.com/ articles/2007/01/09/211458/apaches-pass-trial-by-fire.htm>. As of November 2007, 800 GPF's (FL-7 rockets as they are referred to by British army) have been fired from the eight British Apaches deployed on combat operations since April 2006. *See,* Joris Janssen Lok, "Longbow Radar Key to British Apache Ops in Afghanistan", *Aviation Week*, (Nov. 22, 2007) available at <http://www.aviationweek.com/aw/blogs/defense/index.jsp>. For an account of their use *see,* Damien Lewis, "Britain's £4 Billion Killing Machine", *The Daily Mail Online*, (Sept. 5, 2008) available at <http://www.dailymail.co.uk/moslive/article-1044410/Britains-4-billion-Killing-Machine.html>.

166 Gunston & Mike, *Modern Fighting Helicopters, supra* note 165; Roy, *Aircraft Weaponry of Today, supra* note 165, at 39-40.

167 The U.S. army, for example, has adapted Avenger HMMWV 4x4 vehicles to the Hydra-70 family of 2.75 inch pods in place of Stinger SAM pods, while South Korea (Hanwha Corp.) has marketed 70 mm (40-round) MLRS using this rockets with flechette warheads containing 1200 darts. For the U.S. system, *see, e.g.*, <http://www.geocities.com/Pentagon/ 7963/avenger.htm>. For the South Korean system's specification *see,* the Hanwha Corp.'s brochure (on file with the author).

# Chapter 2

## From Pointed Controversy to Oblivion: The Non-banning of Flechette Weapons

### A. Introduction: Restricting Weapons of War throughout the Centuries

To date, no conventional rule in international law explicitly restricts or prohibits the use of flechette weapons per se. The reasons for this state of affairs (some might say failure) are to be found in the weapon's intrinsic features, especially the context of its use, and the interests of the parties taking one or another position regarding its legality. This chapter surveys the prolonged efforts to prohibit or restrict the use of certain conventional weapons in general and flechettes in particular. Efforts to ban flechettes were initiated in the late 1960s – shortly after the advent of the firearm flechette weapon during the Vietnam War – and eventually led to the conclusion of the *1980 Convention on Certain Conventional Weapons* (CCW).[1] The attention given to flechette weapons during the three CCW Review Conferences is traced chronologically in order to evaluate the State of Israel's Response in the Flechette Petition, which relied in part on the discussions held – or, to be more precise, the lack of discussion – regarding the weapon's legality during those conferences. The review describes the rollercoaster-like twists that the question of flechette legality took within the legal community's agenda: from a disputed issue to oblivion. However, since the law of weaponry is a lesser-known branch of international law, we begin this chapter with a brief overview of the efforts expended to regulate the means of warfare prior to the period of interest.

Throughout history, humanity has sought to ameliorate the horror of war by, *inter alia*, imposing legal restrictions on the permitted means of waging war. The Laws of Manu in India (circa 500 BC), for instance, prohibited the use of poisonous and other inhumane weapons,[2] while Innocent II and the second Lateran Council (1139) banned the crossbow as a military weapon hateful to God and unfit

1 Convention on Prohibitions or Restrictions on Use of Certain Conventional Weapons which may be Deemed to be Excessively Injurious or to Have Indiscriminate Effects, Geneva, 10 Oct. 1980 (entry into force. 2 Dec. 1983). Dietrich Schindler and Jiri Toman, (eds.), *The Laws of Armed Conflicts*, (Leiden: Martinus Nijhoff, 2004), 181 [hereinafter, *Laws of Armed Conflicts*].

2 A Swedish Working Group Study, *Conventional Weapons, Their Deployment and Effects from Humanitarian Aspects, Recommendations for the Modernization of International Law*, (Stockholm: Almqvist & Wiksell, 1973), 11.

for Christians.[3] The task of regulating weapons of war has thus remained a challenge throughout the centuries. In the interim, however, technological developments brought about a dramatic increase in the destructive potential of new types of weapons while the Industrial Revolution guaranteed their large-scale availability. Hence, restrictions or prohibitions on the use of selected means of warfare became almost permanent items on the international agenda in the mid-nineteenth century; the issue consequently achieved a central position in the formulation of the Law of War (alternatively, the Laws of Armed Conflict [LOAC] or, as currently referred to, the IHL).

Within this framework, the first modern multilateral treaty addressing weapons was the St. Petersburg Declaration of 1868 which, prohibited the use of "Explosive Projectiles under 400 Grams Weight" [hereinafter, *1868 St. Petersburg Declaration*].[4] The date of this declaration, almost a century and a half ago, indicates that various endeavors to regulate or prohibit weapons or weapon systems did constitute an integral part of the emerging modern Law of War.[5]

Additional instruments governing specific weapons were adopted in both the 1899 International Hague Peace Conference (e.g., the *Declaration (IV, 2) Concerning Asphyxiating Gases,* which prohibited employing "projectiles" with "asphyxiating or deleterious gases"[6]) and the 1907 Second Conference (e.g., *Convention (VIII) Relative*

3 The ban, which included longbows as well, was based on the perception that the wounds caused by the crossbow's quarrels (bolts with a square armor-piercing head) were 'barbarous'. This prohibition did not, however, include combat against infidels and was widely violated even within the Christian world. *See,* Sir R. Payne-Gallwey, *The Book of The Crossbow*, (New-York: Dover Publications, Inc. 1995) [originally published: Longman, 1903], 3. In any case, it seems that the political interest of preserving the knights' status – to be distinguished from the humanitarian concern – played an important role as well. *See, e.g.*, Preston Richard & Wise. Sydney, *Men in Arms: A History of Warfare and It's Interrelationships with Western Societies*, (New-York: Praeger, 1979), 75. For a modernist explanation according to which the ban resulted from Pope Innocent II's interests in winning the war he waged that same year against Roger II, the King of Sicily, given the latter's heavy reliance on Muslim knights *see*, M.S. Turner, *The Not So Diabolical Crossbow: A Re-Examination of Innocent II's Supposed Ban of the Crossbow at the Second Lateran Council*, (London: Lulu, 2004).

4 Declaration Renouncing the Use, in Time of War, of Explosive Projectiles under 400 Grammes Weight, 1868, *Laws of Armed Conflicts*, at 91.

5 Under the rubric "weapons systems" we include, for instance, efforts to regulate military aircraft or submarines in the inter-World War period. On these efforts regarding military aircraft *see,* for instance, W. Hays Parks, "Air War and the Law of War", *Air Force Law Review* 32 (1990) 1, at 23-36. On the efforts to ban submarines *see,* W. Hays Parks, *"Making Law of War Treaties: Lessons from Submarine Warfare Regulation"*, in: Michael N. Schmitt, (ed.), *International Law across the Conflict Spectrum: Essays in Honor of Professor L.C. Green* (Newport, RI: United States Naval War College 2001), 229-385.

6 Declaration (IV, 2) concerning Asphyxiating Gases. The Hague, 29 July 1899, *Laws of Armed Conflicts*, 95. For a detailed overview of the origin and negotiations of this Declaration *see, e.g.*, Richard Price, *The Chemical Weapons Taboo*, (New-York: Cornell University Press, 1997), 30-38.

*to the Laying of Automatic Submarine Contact Mines*[7]). Yet, the number of these instruments was few and their effect on the military arsenals of the participants marginal at best. As Barbara Tuchman notes with respect to efforts to restrict weapons in the first 1899 Conference, which was held in a much more convenient atmosphere than was its 1907 successor:

> For arbitration some hope sprouted but for arms limitation... or new weapons, there was none. Despite the desperate efforts of the Russians and the warm support of the small states and many civilian delegates, every proposal for restriction or moratorium was shown to be 'impractical' by the military delegates of the major powers.[8]

Furthermore, other than the 1899 *Hague Declaration Concerning Expanding Bullets*,[9] the only specific instrument that could have affected military planning in WWI – the *Asphyxiating Gases Declaration* – was easily violated.[10] In fact, even this exception was questioned as to its efficacy or relevance.[11] However, *the 1899 Hague Convention (II) Respecting the Laws and Customs of War on Land* and its 1907 *Convention (IV) Revised Version* [hereinafter, *1907 Hague Convention*], provided a solid source for the *generic rules* that outlawed weapons according to the characteristics of their employment. First and foremost, it elaborated a principle – "[T]he right of belligerents to adopt means of injuring the enemy is not unlimited"[12] – that the ICRC traced back to the

7 The Hague, 18 Oct. 1907, *Laws of Armed Conflicts*, 1071. For a detailed discussion on this Convention *see, infra* chapter 8, II.

8 *See*, Barbara W. Tuchman, *The Proud Tower: A Portrait of the World Before the War, 1890-1914* (New-York: Bantam Books, 1967), 301.

9 Declaration (IV, 3) concerning expanding bullets. The Hague, 29 July 1899, *Laws of Armed Conflicts*, 99. As chapter 3 details the 1907 Hague Declaration (XIV) "Prohibiting the Discharge of Projectiles and Explosives from Balloons" The Hague, 18 Oct. 1907 (*see, Laws of Armed Conflicts*, 309), could not have been an exception *a priori* given the fact that of the then major powers only Great Britain and the U.S. ratified the declaration. For a list of signatures, ratifications and accessions *see, Id.*, 313.

10 For the German willingness to violate the prohibition and its formal explanations of those breaches, according to which they were not the first to use lethal gas shells, *see*, Price, *The Chemical Weapons Taboo, supra* note 6, at 48-49. Yet, as Price notes, the "repeated invocations by German authorities of the doctrine of *kriegsraison* [i.e., the necessities of war take precedence over the rules of war]" cast doubts on this explanation. Furthermore, the German response to the 1916 alleged French first use was so quick that Price speculates whether the decision to use them had been made "regardless of the French actions". *Id*, at 49.

11 Parks argued, for instance, that the forces of all nations used, for the most part, only full metal-jacketed bullets both prior to, and following, the Declaration's adoption. As such, it is unclear if compliance with the Declaration resulted from submission to such a law of war or simply from firearms reliability issues. *See*, Parks Hays, "Conventional Weapons and Weapons Reviews" *Yearbook of International Humanitarian Law* 8 (2005), 55, at 69.

12 The 1907 Hague Regulations, Art. 22. For the ICRC commentary on this article *see*, Claude Pilloud *et al.*, (eds.) *Commentary on the Additional Protocols of 8 June 1977 to the Geneva Conventions of 12 August 1949* (Leiden: Martinus Nijhoff 1987), 389-399 [hereinafter, *ICRC Commentary*].

writings of Hugo Grotius.[13] This principle is the cornerstone of *any discussion on the legality of weapons;* in the 1940s, the International Military Tribunal at Nuremberg passed it – like all Hague provisions regarding weapons – into customary international law.[14] Since then, the principle has been reaffirmed in various legal instruments, such as Additional Protocol I (AP-I) (art. 35(1))[15] and the CCW Preamble.[16] Second, Article 23(e) of the authentic French text of the Hague Regulations (both 1899 and 1907) states that: "It is especially prohibited… [t]o employ arms, projectiles, or material of a nature to cause superfluous injury".[17] This principle was reaffirmed in various recent legal instruments, including the AP-I (art. 35(2), the CCW Preamble,[18] the International Criminal Tribunal for the former-Yugoslavia (ICTY) Statute,[19] and the Statute of the International Criminal Court (ICC).[20] Finally, as Professor Baxter stated: "Over-arching these stipulations and on a higher level of abstraction"[21] is the famous Martens Clause (para. 8), found in the *Preamble* of both *Conventions*, to wit:

> Until a more complete code of the laws of war has been issued, the High Contracting Parties deem it expedient to declare that, in cases not included in the Regulations adopted by them, the inhabitants and the belligerents remain under the protection and the rule of the principles of the law of nations, as they result from the usages established among civilized peoples, from the laws of humanity, and the dictates of the public conscience.[22]

13 Hugo Grotius, *De Jure Belli Ac Pacis Libri Tres* [On the Law of War and Peace: Three books], 1625. *See, ICRC Commentary*, at 389, para.1383.

14 Office of United States Chief of Counsel for the Prosecution of Axis Criminals, *The Nazi Conspiracy and Aggression: Opinion and Judgment* (Washington D.C: GPO, 1947), at 83.

15 *Laws of Armed Conflicts*, 711, at 730.

16 CCW, *Laws of Armed Conflicts*, 181, at 184.

17 For the 1899 French text, *see,* Pearce Higgins, *The Hague Peace Conferences and Other International Conferences Concerning the Laws and Usages of War*, (Cambridge: Cambridge University Press, 1909), 206, at 234. For the 1907 text *see, Id.* For the difficulties owing to linguistic differences between the official French text and the English translation *see, infra*, chapter 5 B.

18 *Laws of Armed Conflicts*, at 181.

19 *Statute of the International Tribunal for the Prosecution of Persons Responsible for Serious Violations of International Humanitarian Law Committed in the Territory of the Former Yugoslavia since 1991, 25 May 1993*, Article 3(a), *Id.*, 1285, at 1288.

20 *Rome Statute of the International Criminal Court, July 17, 1998*, Art. 8(2)(b)(xx). *Id.*, 1309 at 1319. See also *San Remo Manual on International Law Applicable to Armed Conflicts at Sea, 12 June 1994*, Para. 42(a), Id., 1153, at 1161. For the principle's applicably on non-IAC as well, *see,* Jean-Marie Henckaerts and Louise Doswald-Beck, *Customary International Humanitarian Law*, Vol. I., (Cambridge: Cambridge University Press, 2005) at 239-40 [hereinafter, *ICRC Study Report*].

21 *See,* Baxter, R. R. "Criteria of the Prohibition of Weapons in International Law", in: Horst Ehmke *et al.*, *Festschrift fur Ulrich Scheuner – Zum 70. Geburtstag* (Berlin: Duncker & Humblot, 1973), 41, at 41.

22 *Laws of Armed Conflicts*, 55, at 61. This clause remained essentially intact in both versions.

This clause, "the brainchild of M. de Martens",[23] although not considered a separate criterion for the assessment of weapons' legality,[24] proved – as the ICJ stated in its 1996 *Nuclear Weapons Advisory Opinion* – "to be an effective means of addressing the rapid evolution of military technology".[25]

However, the Third International Peace Conference, recommended by the 1907 Second Conference in its Final Act,[26] never convened due to the outbreak of WWI. The latter saw the first large-scale use of new weapons such as chemical weapons, tanks, and military aircraft,[27] events that provided an impetus for several endeavors to regulate controversial weapons throughout that war, such as chemical weapons.[28] Even so, besides the 1925 *Geneva Protocol* ban on the use of biological and chemical weapons in war [hereinafter, the 1925 Geneva Protocol],[29] post World War I endeav-

---

23 *See*, Yoram Dinstein, *The Conduct of Hostilities under the Law of International Armed Conflict* (Cambridge University Press, 2004), at 56. Prof. Friedrich Fromhold Martens was a Russian delegate to the 1899 and 1907 Hague Peace Conferences.

24 A few jurists have ventured to suggest that the element of "*dictates of the public conscience*" – to be iterated in AP-I, Article 1(2) – may provide an additional legal instrument for prohibiting specific means of warfare. For such an attempt in the context of nuclear weapons *see, e.g.*, Elaina Kalivretakis, "Are Nuclear Weapons above the Law–A Look at the International Criminal Court and the Prohibited Weapons Category", *Emory International Law Review* (*Emory Int'l l. Rev* 15 (2001), 683, at 717-721; Detlev F. Vagts, "The Hague Conventions and Arms Control", *The American Journal of International Law* (*AJIL*) 94 (2000), 31, at 36. While the ICJ evaded this issue in its 1996 "*Advisory Opinion on the Legality of the Threat or Use of Nuclear Weapons*" (International Court of Justice July 8, 1996), ILM 35 (1996), 809 & 1343 [hereinafter, *Nuclear Weapons Advisory Opinion*], para. 84), Judge Weeramantry assigned great importance to this element in his dissenting opinion (*see*, *Dissenting Opinion* of Judge Weeramantry, Part III, para. 4-5). Note: all Declarations, Separate and Dissenting Opinions of the Judges are available at <http://www.icj-cij.org/docket/ index.php?p1=3&p2=4&k=e1&p3=4&case=95>. Nonetheless, as Theodor Meron stressed, "...the Martens clause does not allow one to build castles of sand. Except in extreme cases, its references to principles of humanity and dictates of the public conscience cannot, alone, delegitimize weapons and methods of war, especially in contested cases". *See*, Theodor Meron, "The Martens Clause, Principles of Humanity, and Dictates of Public Conscience", *AJIL* 94 (2000) 78, at 88. For the origins of this clause as a "diplomatic gimmick" to bridge the gap between opposing positions in the First Hague Conference, the article's various interpretations and the criticism it attracted, *see*, Antonio Cassese, "The Martens Clause: Half a Loaf or Simply Pie in the Sky?" *European Journal of International Law* 11 (*EJIL*) (2000) 187-216.

25 *See*, *Nuclear Weapons Advisory Opinion*, at para 78.

26 *See*, Final Act of the Second Peace Conference. The Hague, 18 October 1907, *Laws of Armed Conflicts*, 45, at 47.

27 On the use of chemical weapons during WWI *see, e.g.*, Fredrich Brown, *Chemical Warfare: A Study in Restraints*, (Princeton: Princeton University Press, 1968), 1-48. On the importance of tanks in this war *see*, Orgill's classic book, Douglas Orgill, *The Tank*, (London: William Heinemann Ltd, 1970), chaps. 2-6. As to the realm of military aircrafts *see, e.g.*, Lee Kennett, *The First Air War: 1914-1918*, (New-York: Free Press, 1999).

28 For these efforts in respect to chemical weapons *see, e.g.*, Price, *The Chemical Weapons Taboo*, *supra* note 6, at chap. 4.

29 *The Geneva Protocol for the Prohibition of the Use in War of Asphyxiating, Poisonous or Other Gases and of Bacteriological Methods of Warfare*, 17 June 1925, (entry into force, Feb.

ors to ban other types of weapon such as incendiary weapons proved unsuccessful.[30] Nonetheless, dozens of states did include reservations in their ratification or accession to the 1925 *Geneva Protocol* making the *Protoco*l a *de facto* no-first-use agreement. [31] In sum, these attempts at introducing restraint in weaponry can be considered to have resulted in some limited success.

Having seen the horrors of Hiroshima and Nagasaki, the international community's attention in the aftermath of WWII naturally shifted to nuclear weapons. This may explain why the first resolution passed by the new UNGA on January 24, 1946 referred to this type of warfare. It established the UN Atomic Energy Commission, which was mandated, *inter alia*, with the urgent task of formulating a proposal "for the elimination from national armaments of atomic weapons and of all other major weapons adaptable to mass destruction."[32] Nonetheless, the tension between the U.S. and the Soviet Union, each of which submitted contradictory proposals, prevented any significant progress on this matter.[33]

Concurrently, on February 13, 1947, the UN Security Council (UNSC) established the UN Commission for Conventional Armament.[34] The previous failure repeated itself, with added disagreement between the superpowers regarding the overlap of the mission assigned to the Atomic Energy Commission. Yet, by adopting a US suggestion to define its jurisdictions through a process of elimination,[35] the Commission for Conventional Armament left an important, albeit inadvert-

8, 1928). [Hereinafter, *1925 Geneva Protocol*] *Laws of Armed Conflicts*, 105. On the negotiations of this protocol, *see*, R.R. Baxter and Thomas Buergenthal, "Legal Aspects of the Geneva Protocols of 1925", *AJIL* 63 (1970), 853, at 857-865; SIPRI, *The Problem of Chemical and Biological Warfare: CBW and the Law of War* (Stockholm: Almqvist & Wiksell, 1973), Vol. III, chap. 3.

30 For a brief overview of the use of incendiary weapons during WWI *see, e.g.*, John Wyndham Mountcastle, *Flame On: U.S. Incendiary Weapons, 1918-1945*, (Shippensburg, PA: White Mane Books, 1999), at 9-19. For the U.K.'s unsuccessful attempt to ban incendiary weapons during the 1932-1934 Geneva Disarmament Conference *see, e.g.*, SIPRI, *Incendiary Weapons* (Stockholm: Almqvist & Wiksell, 1975), 24-25; Pertti Joenniemi & Allan Rosas, *International Law and the Use of Conventional Weapons* (Tampere: Tampere Peace Research Institute, 1975), at 46.

31 For a list of these reservations *see*, *Laws of Armed Conflicts*, at 116-123.

32 *See*, UNGA Res. I(I) Jan. 24, 1946.

33 For an overview of the opposing positions taken by the U.S. and the Soviet Union *see*, United Nations, *The United Nations and Disarmament 1945-1970* (New-York: The United Nations, 1970), 16-19.

34 *See*, UNSC Res. 18 (1947), Feb. 13, 1947. (S/268/Rev.1/Corr.1).

35 For the U.S. draft see UNSC S/C.3/SC.3/7. For a partial overview of U.S. considerations in formulating the draft and its inclusion of biological weapons *see*, Nina Tannenwald, *The Nuclear Taboo: The United States and the Non-Use of Nuclear Weapons Since 1945*, (Cambridge: Cambridge University Press, 2007), 102-105. In September 1947, the Commission for Conventional Armament agreed that its jurisdiction covered all weapons other than "atomic explosives weapons, radio-active material weapons, lethal chemicals and biological weapons, and any weapons developed in the future which have characteristics comparable in destructive effect to those of the atomic bomb or other weapons mentioned above." *See*, UNSC S/C.3/SR.3, Sept. 6, 1947.

ent legacy: creation of the category Weapons of Mass Destruction (WMD).[36] The fact that "organizational reasons" provoked the inclusion of chemical weapons, as a leading chemical weapons expert noted, has long since been forgotten. [37] The categorization of all weapons other than WMD (i.e., nuclear, biological, and chemical weapons) as "conventional weapons" was thus established. In January 1952, given the stalemate in both Commissions,[38] they were effectively unified by the UNGA into the UN Disarmament Commission.[39]

The onset of the Cold War snuffed out any meaingful international effort to prohibit or regulate means of warfare, nuclear weapons excluded. An exception to this trend was the attempt made by the ICRC to prohibit the use of "certain specific weapons" in response to the widespread and devastated attacks on civilian populations, carried out during WWII, mainly through indiscriminate aerial bombardments.

After three years of work (1953-56), the ICRC drew up a draft treaty that included an article (Art. 14) entitled "*Prohibited Methods of Warfare*."[40] Other than "delayed-action weapons," the draft explicitly referred to nuclear weapons. [41] In retrospect, the inclusion of nuclear weapons was one of the main reasons for the entire draft's failure to secure adequate support at the 1957 annual XIX ICRC Conference.[42]

36 Its decision was approved by the UNSC in 1948. *See*, UNSC S/C. 3/32/Rev. 1 (1948).

37 Julian Perry Robinson, "Chemical Weapons Proliferation in the Middle East", in: Efraim Karsh *et al.*, (eds.), *Non-Conventional Weapons Proliferation in the Middle East: Tackling the Spread of Nuclear, Chemical, and Biological Capabilities*, (London: Clarendon Press, 1993), 69-98, at 79.

38 The Atomic Energy Commission, for instance, held its last meeting in July 1949 as the Soviet Union withdrew from the discussions in January 1950. *See*, *The United Nations and Disarmament*, *supra* note 33, at 23.

39 *See*, UNGA Res. 502(VI) Jan. 11, 1953. After dissolving the Atomic Energy Commission and recommending that the UNSC dissolve the Commission for Conventional Armament as well, this important resolution created the Disarmament Commission in its place. Three weeks later, on January 30, UNSC dissolved the Commission for Conventional Armament in accordance with the UNGA's recommendation. *See*, UNSC Res. 97 (1952), Jan. 30, 1947, (S/2506).

40 ICRC, "Draft Rules for the Limitation of the Dangers Incurred by the Civilian Population in Time of War" (1956), Chap. IV, Art. 14, *Laws of Armed Conflicts*, 339, at 342.

41 Art. 14 reads: "… the use is prohibited of weapons whose harmful effects – resulting in particular from the dissemination of incendiary, chemical, bacteriological, radioactive or other agents – could spread to an unforeseen degree or escape, either in space or in time, from the control of those who employ them, thus endangering the civilian population". Id. It should be noted that Art. 15 obliged the mapping of minefields when mines were employed.

42 For objections to these rules *see*, R.R. Baxter, "Humanitarian Law or Humanitarian Politics? The 1974 Diplomatic Conference on Humanitarian Law", *Harvard International Law Journal 16* (1975) 1, at 3.

## B. Initial Efforts to Ban "Certain" Conventional Weapons (1966-1970): The "High Speed Barbed Missiles"

Undoubtedly, it was the Vietnam War that returned the legality of certain weapons to the international arena and prompted the international community's renewed interest in this "almost forgotten topic"; conventional weapons were already on the table.[43] Professor Kalshoven has argued that all in all, the means employed in the Vietnam War were not entirely different from previous and concurrent armed conflicts having similar characteristics (i.e., extensive guerrilla fighting). The war did differ, however, from all previous wars in the "ample media coverage it received and the acute political controversies it aroused".[44] The use by the U.S. and its allies of tear gas, herbicides, incendiary weapons, cluster munitions, and *flechette projectiles* caused an international outcry. Whether the responses derived from genuine concern or – as U.S. proponents argued – the pursuit of alternative arguments against the War by its opponents,[45] the conflict provided the first major impetus to negotiate restrictions on conventional weapons since the end of WWII. Nonetheless, much time and effort would still be required before any restrictions – however limited and weak they would prove to be – were negotiated.

In May 1967, the ICRC sent a memorandum, entitled "Protection of Civilian Population from the Dangers of Indiscriminate Warfare", to all states party to the four Geneva Conventions[46] and the 1907 *Hague Convention.*[47] In a survey of existing law, annexed to the memorandum, its authors openly speculated whether, *inter alia*, "new weapons [such] as napalm and high velocity rockets" should be included in the category of weapons inflicting "needless suffering."[48] Both the memo and this specific suggestion drew very few immediate responses.[49]

A year and a half later, in December 1968, the UNGA requested that the UN Secretary-General (UNSG), U Thant, study "(b) [t]he need for additional humani-

43 *See, e.g.*, Frits Kalshoven, "Arms, Armaments and International Law", *Recueil Des Cours* 19 (1985[II]) 181, at 225. *See also*, Parks, "Conventional Weapons..." *supra* note 11, at 69; R.R. Baxter, "Conventional Weapons under Legal Prohibitions", *International Security* 1(3) (Winter 1977), 42, at 45.

44 Kalshoven, *Id.*, at 225.

45 Parks, "Conventional Weapons..." *supra* note 11, at 69.

46 *See*, Convention (I) for the Amelioration of the Condition of the Wounded and Sick in Armed Forces in the Field. Geneva, 12 August 1949, *Laws of Armed Conflicts*, 459; Convention (II) for the Amelioration of the Condition of Wounded, Sick and Shipwrecked Members of Armed Forces at Sea. Geneva, 12 August 1949, *Laws of Armed Conflicts*, 485; Convention (III) relative to the Treatment of Prisoners of War. Geneva, 12 August 1949, *Laws of Armed Conflicts*, 507; Convention (IV) relative to the Protection of Civilian Persons in Time of War. Geneva, 12 August 1949, *Laws of Armed Conflicts*, 575.

47 *Laws of Armed Conflicts*, 55.

48 *See*, Kalshoven, "Arms, Armaments..." *supra* note 43, at 226; *ICRC Commentary*, at 400, para. 1410.

49 For the few immediate government responses to the moves *see*, Roger S. Clark, "Human Rights Strategies of the 1960s Within the United Nations: A Tribute to the Late Kamleshwar Das", *Human Rights Quarterly* 21 (1999) 308, at 340.

tarian international conventions ... to ensure ... the prohibition and limitation on the use of certain methods and means of warfare."[50] The UNSG's report, submitted the following year (November 1969), indeed recommended reviewing the legality of napalm bombs.[51] A month later, the UNGA, although it had yet to properly review the report, requested that the UNSG continue its work.[52] This request yielded an additional report, submitted in September 1970.[53]

At that point, the UN's progress was integrated with ICRC activity to extend the Geneva Conventions so as "to cover certain supplementary matters and to clarify some important points."[54] Having received few reactions to its 1967 Memorandum, the ICRC convened a meeting of experts in February 1969 to discuss "reaffirmation and development of the laws and customs applicable in armed conflicts." In a preliminary document, the ICRC invited the experts' opinions, *inter alia,* on questions of "napalm", "fragmentation bombs" and, most interestingly, "high speed barbed missiles" (flechettes). However, both flechettes and fragmentation bombs (cluster munitions) were left unmentioned in the meeting report, which focused solely on napalm.[55] In September 1969, the twenty-first ICRC Annual International Conference in Istanbul recommended to the ICRC to draft a "prohibition of 'non-directed' weapons or weapons causing unnecessary suffering".[56] The conference adopted a resolution urging the ICRC to actively pursue its efforts with a view toward "proposing, as soon as possible, concrete rules which would supplement the existing humanitarian law" and to invite experts for this purpose. This "Resolution of major importance"[57] shifted the movement's center of gravity to the Geneva Conventions and the ICRC; the latter had no intention of allowing this opportunity to slip away. A new period, that would culminate in the 1977 conclusion of the Additional Protocols to the Geneva Conventions of 12 August 1949, was about to begin.

---

50 *See,* UNGA Resolution 2444 (XXIII), Dec. 16, 1968, UN Doc. A/7218 (1969). The UNGA's request was the result of a resolution adopted by The international Conference on Human Rights held in Teheran on the initiative of the UN. *See,* Res. XXIII (entitled "Human rights in armed conflicts"), Final act of the Int'l Conf on Human Rights, UN Doc. A/conf.32/41, at 18 (1968).

51 *Respect for Human Rights in Armed Conflicts: [First] Rep. of the Secretary-General,* UN Doc. A/7720 (1969), at 59-63, paras. 183-201. For the recommendation to review the legality of using napalm *see, Id.*, at.62-63, para. 200.

52 *See,* Res. 2597 (XXIV) Dec. 16, 1969.

53 For the Report *see, Respect for Human Rights in Armed Conflict:* [Second] Rep. of the Secretary-General, 25 UN GAOR, UN Doc. A/8052.

54 *See, ICRC Commentary,* at XXX.

55 The report referred to a variety of opinions regarding napalm's legality and "the desirability and feasibility" of a prohibition on its use. *See,* Frtiz Kalshoven, *Reflections on the Law of War: Collected Essays* (Leiden: Martinus Nijhoff, 2007), 149 [the chapter incorporating this page was originally published in Netherlands Yearbook of International Law (NYIL) 6 (1975), 77-102].

56 Res. XIII, A/7720, Annex I, Sec. D as reprinted in ICRC, *International Red Cross Handbook* 11th ed. (Geneva: ICRC 1971), 449-50.

57 *See, ICRC Commentary,* at XXX.

## C. Flechettes Weapon during Consolidation of the Additional Protocols, 1971-1977

### *CI. The ICRC Activities and the Swedish Foreign Ministry Report (1971-1973)*

In May 1971, the ICRC, "after due consultation" with the UNSG, convened a Conference of Government Experts on the *Reaffirmation and Development of International Humanitarian Law Applicable in Armed Conflicts*, with participants from 40 states.[58] However, the issue of weapon legality was soon marginalized. Use of WMD was explicitly excluded from the agenda as most experts agreed that this subject was in the purview of other bodies, mainly the UN and the sitting Conference of the Committee on Disarmament (CCD).[59] As for other "conventional" weapons, the ICRC, which confined itself to studying aspects of this problem, refrained from submitting any concrete proposals of its own.[60] Instead, it introduced basic principles, already expressed in the 1868 *St. Petersburg Declaration* and the 1907 *Hague Convention,* into its drafts. Alongside the issue of incendiary weapons (napalm in particular), which attracted some attention, a group of experts submitted a proposal that included, *inter alia*, restrictions on the use of weapons with a delayed effect, such as mines and – more importantly for here – bombs that fragment into a great number of small caliber pellets.[61]

In May-June 1972, the ICRC convened the second session of the Conference of Experts, with participants from 77 states.[62] Much progress was made. In its wake, the ICRC concluded two draft Additional Protocols to the Geneva Conventions, one for cases of International Armed Conflict (IAC), the other for non-IAC. These were to serve as a basis for the discussions to be held by the Diplomatic Conference on the Reaffirmation and Development of International Humanitarian Law Applicable in Armed Conflicts (hereinafter, the Diplomatic Conference) to be convened by the Swiss government as the depository of the Geneva Conventions.

---

58 Conference of Government Experts on the *Reaffirmation and Development of International Humanitarian Law Applicable in Armed Conflicts*, (24 May-12 June 1971), Report on the Work of the Conference (ICRC, 1971) [hereinafter, *1971 ICRC Conference*]. For a detailed overview on this conference *see,* Kalshoven, *Reflections on the Law of War, supra* note 55 at 33-55, for the full quotation *see, Id.*, at 36.

59 *1971 ICRC Conference*, para. 477. *See, also,* Kalshoven, "Arms, Armaments…" *supra* note 43, at 227-28.

60 Kalshoven, *Id.*, at 228.

61 For a summary of the proposals, *see,* the Secretary-General's report (Sept. 2, 1971) *Respect for Human Rights in Armed Conflict: [Third] Report of Secretary-General,* UN Doc. A/8370, at 47-50, paras. 105-07 (1971). For the proposal on fragmentation weapons, *see, id.*, para. 106. We should add that the Brazilian delegation, with the support of Spain, even suggested banning the use of all types of weapons which "by their nature, affected indiscriminately both military objectives and non-military objects, or both armed forces and civilian populations ... as well as of 'blind' weapons," *Id.*

62 For a detailed overview on this Conference *see,* Kalshoven, *Reflections on the Law of War, supra* note 55, at 56-99 (originally published in NYIL 3 (1972), 18-61).

However, all the suggested amendments to a draft article on the means of combat that pertained to prohibitions on the use of certain weapons (e.g. incendiary weapons) were opposed by delegates who thought that this was beyond the Conference's mandate.[63] Humanitarian law was viewed as the Law of Geneva, while restrictions on the use of weapons were considered to be part of the Law of The Hague.[64] As Prof. Baxter noted, the meeting was plagued by anxieties over concluding a convention on this issue in general, and fears about retarding work on the protocols in particular.[65]

Because these anxieties were shared by the ICRC as well, it refrained from suggesting any specific prohibitions. Instead, the ICRC confined itself to proposing a reaffirmation of the principles expressed in Articles 22 and 23(e) of the 1907 Hague Regulation – i.e., that belligerents have no unlimited choice of means of warfare and that it is forbidden to use weapons which inflict unnecessary suffering.[66] In rationalizing its reticence, the ICRC noted the following: (a) other fora deal with arms limitation and disarmament (i.e., the UN and the CCD); (b) no attempts have yet been made to include prohibitions against specific weapons whose use is liable – in accordance with international law – to reprisals or even reciprocity in the "absolute nature" of the Geneva Conventions;[67] and (c) if restrictions on specific weapons were necessary, it would be preferable to incorporate them into a separate instrument.[68]

Among the amendments suggested was a proposal to add a clause "forbidding indiscriminate, delayed-action weapons, incendiary weapons … and fragmentation weapons".[69] It was submitted by a group of experts from 15 states, headed by "their undisputed leader"– as Professor Kalshoven recalled – Sweden's Dr. Hans Blix.[70]

63 These proposals are summarized in *Human Rights in Armed Conflict – Respect for Human Rights in Armed Conflicts, [Fourth] Report of the Secretary-General*, UN Doc. A/8781, Sep. 20, 1973, at 51-54, paras. 145-55.

64 For the reasons for rejecting these proposals, *see*, Ashley J. Roach, "Certain Conventional Weapons Convention: Arms Control or Humanitarian Law?", *Military Law Review* 105 (Summer 1984), 2 at 10; Richard John Erickson, "Protocol I: A Merging of The Hague and Geneva Law of Armed Conflict", *Virginia Journal of International Law* 19 (1979) 557.

65 *See* ,Baxter, "Conventional Weapons …", *supra* note 43, at 50.

66 Kalshoven, *Reflections on the Law of War*, *supra* note 55, at 67. On the ICRC's qualms over whether assuming a position regarding the legality of specific weapons would detract from its long-time humanitarian mission as well as require work in an area where it lacked expertise *see*, Parks, "Conventional Weapons…", *supra* note 11, at 72.

67 Instead, it prefers to limit the use of weapons indirectly (through the Conventions and the Protocol) by imposing greater respect for certain categories of persons or objects. As Kalshoven noted, this "argument was particularly weak", a fact acknowledged by the ICRC itself, because the ICRC was by no means certain that the protocols "would eventually assume the character of 'rules of an absolute nature'"; Kalshoven, *Reflections on the Law of War*, *supra* note 55, at 151.

68 *Id.*, and also *see*, *id.*, at 150-151; Bothe, Michael, Karl J. Partsch and Waldemar A. Solf (eds.) *New Rules for Victims of Armed Conflicts: Commentary on the Two 1977 Protocols Additional to the Geneva Conventions of 1949* (Leiden: Martinus Nijhoff, 1982), 197 [hereinafter, *New Rules*].

69 *See*, UN Doc. A/8781, *supra* note 63, at para. 150.

70 Kalshoven, *Reflections on the Law of War*, *supra* note 55, at 151.

Inspired by public outrage in its own country over the Vietnam War, Sweden had begun an international campaign to prohibit or restrict selected types of conventional weapons.[71] After all the amendments were rejected, only a few governments, mainly Sweden, remained adamantly in favor of such limitations. And so, at Dr. Blix's initiative, experts from 19 countries requested that the ICRC convene a special meeting of legal, military, and medical experts to discuss imposing prohibitions or restrictions on those weapons, including fragmentation weapons.[72]

*Session of Experts, Geneva (1973)*: The ICRC promptly invited experts (physicians, jurists, and a few military personnel), all autonomous and, unlike past practices, not representing governments, together with selected government experts, to two working sessions in Geneva, held during March and June of 1973 (hereinafter, the Geneva Sessions).[73] Since the experts already possessed the October 1972 UNSG's comprehensive report on incendiary weapons, they could avoid replication of the work and hold extensive discussions on other means of warfare, such as antipersonnel mines and, more relevant to this book's subject, fragmentation weapons.[74]

It was in this context that flechette came up for discussion, in the context of those characteristics of fragmentation weapons viewed as responsible for causing "unnecessary suffering".[75] Since the meeting's purpose was to collect and analyze the pertinent details – legal, military, and medical – about the means of warfare that were perceived as causing such unnecessary suffering or having indiscriminate effects, no proposals for prohibiting specific weapons were submitted. Yet, the report concluded as follows:

> It is clear... that several categories of weapon tend to cause excessive suffering and particularly severe injuries or may.., strike civilians and combatants indiscriminately ... Another example is provided by the fragmentation weapons. Weapon of this category cause the greatest number of casualties in modern armed conflicts... It is obvious that the trend towards weapons which fragment into vast numbers of small

---

71 For the Swedish contribution at the beginning of this process, *see*, Baxter, "Conventional Weapons..." *supra* note 43, at 46; Eric Prokosch, *The Technology of Killing: A Military and Political History of Antipersonnel Weapons*, (London: Zed Books Ltd, 1995), 148-9. For allegations that Sweden's objections to U.S. weapons "were not entirely humanitarian", *see*, Parks, "Conventional Weapons...", *supra* note 11, at 70.

72 *See*, UN Doc. A/8781, *supra* note 63, at 54, para. 155.

73 Representatives of international organizations were invited as well. It should be noted that the participation of the main military powers in these sessions was negligible. This left room for the expression of a bias in favor of sweeping prohibitions. *See, e.g.*, Kalshoven, "Arms, Armaments..." *supra* note 43, at 231.

74 For the report submitted in accordance with UNGA Res. 2852 (XXVI), Dec. 20, 1971, *see*, *Napalm and Other Incendiary Weapons and All Aspects of their Possible Use: Report of Secretary-General*, UN Doc. A/8303, Oct. 9, 1972. For the response of 21 states to the Report, disseminated amongst the UNGA members, *see*, A/9207, Oct.11, 1973.

75 ICRC, *Weapons that may Cause Unnecessary Suffering or Have Indiscriminate Effects* (Report of the Work of Expert) Geneva, (1973), 12 [hereinafter, *1973 Geneva Report*], Chap. IV(2). For a discussion of 155mm beehive artillery shells *see*, *Id.*, para. 132.

fragments, and are susceptible of covering large areas, increases the risk of multiple injuries and the possibility that civilians will be affected.[76]

That same year, the Swedish Ministry of Foreign Affairs published a comprehensive and elaborate report recommending the prohibition or restriction of the use of various weapons (e.g., incendiary weapons, land mines, booby traps) while *singling out flechettes* – historically, the first proposed ban specifically targetting flechettes.[77] In addition to multiple injuries, the report cites flechette warheads as possibly having high-velocity effects (i.e., a striking velocity exceeding 900 m/sec.), and that the darts "tumble on impact or create shock waves", damaging tissues along their track, and causing bone splinters to act as secondary projectiles.[78] Hence, the report concluded, "a construction rule and operational rules are necessary."

The first offered rule required that flechette warheads meet three requirements: (a) maximum permissible impact velocity of about 900 m/sec; (b) no use of material that may have serious toxic effects; and (c) production of detectable material (x-rays). The proposed operational rule, however, had two versions. The first banned all anti-personnel flechette warheads on the ground of "needless suffering" due to multiple injuries. The second restricted their use "against well-identified and well-localized military targets".[79] Yet, in order to avoid the likely indiscriminate effects, the second version proposed: "(a) actions against areas larger than 1 square kilometer may only be carried out if it has been established that no non-combatants are present in the area".[80]

A month later, in November 1973, the UNSG submitted another comprehensive report (2 volumes) on the international law applicable to weaponry prohibitions and restrictions.[81] In that report, the issue of fragmentation weapons was only briefly mentioned. One of the opinions mentioned argued that fragmentation weapons, which break into thousands of small-caliber pellets, *could potentially be considered*

76 Emphasis added. *1973 Geneva Report*, at 71, paras. 244, 246.

77 It should be noted that the ICRC Feb. 1969 document (the first legal document to mention flechettes) only *invited* experts' opinions on the question of flechette's legality. For the ICRC Doc. *see*, Kalshoven, *Reflections on the Law of War, supra* note 55, at 149. As mentioned at the time: "By the publication of this valuable, both realistic and idealistic report, the Swedish Government has once again made a major contribution to the work in this area of conventional arms limitations…". *See,* Kalshoven, *id.*, at 118 (the chapter on the first session of the CDDH was originally published in 5 NYLR [1974] 3-44. Indeed, the fact that the U.S. delegation to the 1974 Lucerne Conference made sure to report on the Swedish delegation's approach to this study reflected its salience. See Robblee Paul A., Jr. "The Legitimacy of Modern Conventional Weaponry" *Military Law Review* 71 (Winter 1976) 95, at 96, fn.4.

78 *See,* the Swedish Working Group Study, *supra* note 2, at 167-8.

79 *Id.*, at 168

80 *Id.*

81 *Existing Rules of International Law Concerning the Prohibition or Restriction of Use of Specific Weapons*, UN Doc. A/9215 (Nov. 7, 1973) (Vol. I, II). The report was submitted pursuant to the request of the UNGA, UNGA Res. 2842 (XXVI) of Dec. 20, 1971, para. 4 and 3032 (XXVII) of Dec. 18, 1972. For the pertinent UNGA Resolution on this matter, up to UNGA Res. 3032 of Dec. 18, 1973, *see, id.*, Annex I.

*illegal*, either by analogy to the already prohibited dum-dum bullets or as a weapon causing unnecessary suffering in and of itself.[82] "Fragmentation bombs", it continued, were included in that class of weapons considered indiscriminate. However, the report's conclusion highlights an ominous issue regarding the international community's prospects for prohibiting the use of such weapons: "Because consideration of this type of weapon is a product of the conflict in Viet Nam, pronouncements on the legality of the weapon are still too sparse to permit the identification of a prevailing view".[83] That is, the novelty of flechette projectiles, coupled with the fact that its developer and sole user at the time (the U.S.) had a keen interest in retaining this weapon, left few chances, *a priori*, for achieving a successful ban.

### *CII. The Ad Hoc Committee on Conventional Weapons (First Session, 1974) and its Aftermath*

In December 1973, the UNGA called upon the Diplomatic Conference to consider the use of "other specific conventional weapons which may be deemed to cause unnecessary suffering or to have indiscriminate effects," in addition to incendiary weapons, for the purpose of obtaining a consensus on rules for prohibiting or restricting their use.[84] Due to pressure exerted by Sweden and its supporters (which became known as the "Swedish Group"[85]), the Ad Hoc Committee on Conventional Weapons [hereinafter, the Ad Hoc Committee] was established in March 1974[86] at the first session of the Diplomatic Conference [hereinafter, First Session].[87] This proved to be the first serious attempt by the international community to discuss restrictions on conventional weapons since the 1930s; as Professor Kalshoven, the

82 A/9215 (Vol. I), at 186, para. 150.

83 *Id.*, at 209, para. 204. Nonetheless, the report made sure to cite sources that mentioned the weapon's military importance; as a result, it advised caution in drawing up general prohibitions or restrictions. *Id.*, at 186, para. 149.

84 *See*, UNGA Res. 3076, 6 Dec. 1973, 28 U.N. GAOR, Supp. (No. 30), at 15, UN Doc. A/9030 (1973).

85 Its core members were Egypt, Mexico, the Netherlands, Norway, Switzerland, Austria and Yugoslavia. *See*, Parks, "Conventional Weapon...", *supra* note 11, at 74.

86 Its task was to "examine the question of prohibition or restriction of use of specific categories of conventional weapons which may cause unnecessary suffering or have indiscriminate effects, and to consider all proposals which are submitted to the Conference relating to such weapons". *See*, Kalshoven, *Reflections on the Law of War*, *supra* note 55, at 126.

87 The committee's very title reflects the hesitation felt by a number of states, especially the major military powers, regarding supervision of conventional weapons. In practice, the Committee, which dealt with five categories of weaponry, beginning with incendiary weapons, served as an auxiliary forum to the three main plenary committees already established. *See*, Baxter, "Conventional Weapons..." *supra* note 43, at 51. Furthermore, the resolution on the Committee's establishment failed to achieve consensus; it was thus put to a vote (68 to zero, with 10 abstentions). For a detailed account of the Diplomatic Conference's First Session *see*, Kalshoven, *Reflections on the Law of War*, *supra* note 55, at 101-130. For the Swedish "instigation" in the establishment of the Ad Hoc Committee *see*, *Id.*, at 126.

rapporteur, noted at that time, it did do useful work.[88] It ought to be mentioned that it would have been impossible to convene the Committee if the U.S. delegation and its allies had not withdrawn their categorical opposition to discussing restrictions on conventional weapons at the ICRC Conference in Tehran, held during November 1973.[89]

The Swedish Group's intention was to draft proposals targeted at specific weapons instead of general articles, which in any case were included in the Protocol's drafts (e.g., Art. 35). Therefore, in the Ad Hoc Committee's first session on February 21, 1974, Sweden and six other states submitted a proposal for banning flechettes (labeled as "small metal arrows or needles"). The proposal, drafted by the head of the Swedish delegation, Dr. Blix,[90] reads: "*Munitions which act through the release of a number of projectiles in the form of flechettes, needles and similar, are prohibited for use*".[91] A month later, on March 22, the issue came up again in the Ad Hoc Committee's first session. Blix described the then-current trend in the weapons industry to manufacture anti-personnel weapons that released a vast number of very small pellets or small aero-dynamic fragments in order to increase the velocity and depth of penetration. In order to accentuate his point, he described the flechette as a weapon inflicting unnecessary suffering and having indiscriminate effects.[92]

*The Lucerne Conference (1974)*: As one participant noted, the Ad Hoc Committee was not particularly enthusiastic about moving forward on drafting the respective prohibitions and restrictions,[93] a feeling apparently shared by the ICRC itself.[94] Yet, the 1973 ICRC's International Conference in Tehran also requested (Res. XIV) that a conference of governmental experts be convened to examine the issue in depth and submit a report to the Diplomatic Conference. Following sensitive negotiations between Dr. Blix and George Aldrich, the head of the US delegation to the Diplomatic Conference, which had preceded this resolution, the Lucerne Conference's purpose was narrowed: "To study in depth, from the humanitarian standpoint, of [sic] the

---

88 Kalshoven, *id.*, at 131.

89 For the Resolution of the 23rd ICRC Conference, which led to the UNGA resolution, *see*, Res. XIV as reprinted in UN Doc. A/9123/Add.2, Annex, at 4 (1973).

90 For Blix's identity as the main drafter of the proposal, and his personal views on this issue, *see*, Hans Blix, "Current Efforts to Prohibit the Use of Certain Conventional Weapons" *Instant Research on Peace and Violence* 4 (1) (1974), 27-28.

91 Emphasis added. *Official Record of the Diplomatic Conference on the Reaffirmation and Development of International Humanitarian Law Applicable in Armed Conflict, Geneva (1974-1977)*, [hereinafter, *CDDH O. R.*] Vol. XVI, (1978), at 9. The six other states were: Egypt, Mexico, Norway, Sudan, Switzerland, and Yugoslavia.

92 For Blix's statements, *see*, *CDDH O. R.* Vol. XVI, *id.*, at 47. For the central role he played in restricting the use of certain means of warfare employed in Vietnam, such as cluster bombs and flechettes, *see, e.g.*, Kalshoven, *supra* note 43, at 231.

93 Baxter, "Conventional Weapons…" *supra* note 43, at 24. Professor Baxter was a member of the U.S. delegation.

94 As Kalshoven noted "The ICRC had shouldered this task without any great enthusiasm" as by that time it had become fully aware of "the political implications of the subject and its close connection with considerations of national security". Kalshoven, *Reflections on the Law of War*, *supra* note 55, at 148.

question of the prohibition or limitation of the use of conventional weapons that may cause unnecessary suffering or have indiscriminate effect."[95]

In October 1974, the Conference of Governmental Experts on the Use of Conventional Weapons convened in Lucerne, Switzerland, attended by 150 experts from 49 states and organizations.[96] The agenda included various categories of weapons, each of which were examined from the military and medical (i.e., the medical effects of the various weapons, including the degree of suffering or injury they cause) points of view.[97] The Swedish Group's proposal to ban flechette weapons was submitted as a "working paper" at the conference's outset.[98]

During the discussion on blast and fragmentation weapons – in which flechette projectiles (then known as beehive munitions) were defined as "the nearest to being pure fragmentation weapons"[99] – the experts focused on the latter type of weapon. However, a disagreement broke out pitting neutral European countries (Sweden and Norway) and developing countries (Egypt and Mexico), against the main weapons producers and their supporters (the U.S. and its allies in NATO and the USSR and its allies). While the experts called by the Swedish Group argued that pre-fragmented and flechette weapons of various types both have indiscriminate effects and inflict multi-injuries that cause unnecessary suffering, their opponents claimed that the question of indiscriminateness depended on the use to which the weapons were put rather on their nature. As for unnecessary suffering, they argued that the evidence seemed to show that the degree of suffering inflicted by the modern cluster bombs or flechettes was often less than that resulting from the use of those weapons' predecessors.[100]

---

95 *See*, Rules of Procedure, Rule 8, para. 3, *Conference of Government Experts on the Use of Certain Conventional Weapons (Lucerne, 24.9-18.10.1974) ICRC Report* (1975) [hereinafter, *1974 Lucerne Report*], 99: Annex 2, Rules of Procedure. For the negotiations *see*, Kalshoven, *Id*, at 153.

96 As Kalshoven noted, there was "a marked over-representation of experts from countries which had already expressed their support for the idea of far-reaching prohibitions on the use of conventional weapons…moreover, the contribution from qualified military experts, especially from the major Powers, had been virtually nil". *See*, Kalshoven, *Id*, at 153. For a detailed overview on this Conference *see*, *Id.*, 147-170. For the number of participants *see*, *Id.*, at 158.

97 *Id.*, at 163. Some states, mainly Sweden, sought to use the *1973 ICRC Report* as the conference's authoritative documentary cornerstone, but several delegations (mainly those of Canada and the U.K.) objected; their position was sustained by the Conference chairman. *See*, the U.S. delegation report (U.S. Dep't of Army Message RUEKJCS/9337, 271455Z, Sept. 1974, at 3) as cited in Robblee, "The Legitimacy of Modern…", *supra* note 77, at 95.

98 For the list of documents submitted at the Conference, *see*, *1974 Lucerne Report*, at 3, para. 8. It should be mentioned that the invitations to this Conference were sent to all those invited to the Diplomatic Conference's first session, which meant that representatives of national liberation movements were invited as well.

99 *See*, *id.*, at 49, para. 156.

100 1974 *Lucerne* Report, at 61, para. 215. It is no surprise that one of the experts found it appropriate to comment that "[a]t times he had had the impression that he was at an armaments conference rather than at a conference on humanitarian law." *Id.*, para. 218.

### *CIII. The Ad Hoc Committee on Conventional Weapons (Second Session, 1975) and its Aftermath*

At the Committee's second session, in February-April 1975 (parallel to the Diplomatic Conference's session), Blix repeated his proposal to ban flechette weapons. However, this time he presented empirical data, taken from extensive studies done by the U.S. army Surgeon-General and the Chief of Military History, to substantiate his claims.[101] In light of the discussions in Lucerne and with the support of 11 states, Blix submitted a new working paper that essentially proposed prohibiting the use of the majority of weapons employed by the U.S. in Vietnam, including incendiary weapons, cluster bombs, and flechettes. The proposed prohibition underwent no revisions in its wording until the end of the CCW process.[102]

*The Lugano Conference (1976)*: The Lucerne Conference concluded with a general sense of the need for further study of the subject; hence, the ICRC – with permission from the Ad Hoc Committee in its second session – convened another session of governmental experts in Lugano, Switzerland, during January-February 1976.[103] Although the same categories of weapons were on the agenda, it was decided that, contrary to the former session in Lucerne, the "focus would be on such conventional weapons as had been, or might become, the subject of proposed bans or restrictions of use."[104]

At this stage, with some 40 states represented, several Western Powers submitted working papers containing proposals for a range of less-stringent prohibitions.[105] As for fragmentation weapons in general, those who opposed their use not only repeated the claims already heard in Lucerne, but added a new argument: "The public concern which the use of these weapons has aroused".[106] While acknowledging fragmentation weapons' military effectiveness, proponents of the prohibition did express their wish "to take some of the brutality out of warfare."[107] The proposals therefore do not refer to anti-materiel or even to dual-effect fragmentation weapons but, rather, to anti-personnel fragmentation weapons (e.g., flechettes) exclusively. Similarly, the weapons' proponents repeated arguments made at Lucerne as well as

101 For his comments (Feb. 26, 1975) *see CDDH O. R.* Vol. XVI., at 121. The data were not specifically related to flechette weapons, based on the contention that one 'regular' artillery shell (155mm), which can disperse over 3000 fragments, each weighing up to half a gram, was sufficient for modeling the medical impact of multiple injuries. For these data *see, ICRC 1973 Geneva Report*, at 41, para. 124.

102 For the position paper, *see*, CDDH/IV/201, para. III. For the proposal, *see*, *CDDH O. R.* Vol. XVI, at 614.

103 For the detailed overview on this Conference *see*, Kalshoven, *Reflections on the Law of War, supra* note 55, at 171-179.

104 *See*, *Conference of Government Experts on the Use of Certain Conventional Weapons* (Lugano, 28.1- 26.2.1976) ICRC Report (1976) [hereinafter, *1976 Lugano Report*], at 2, para. 6.

105 For the proposals, *see*, *Id.*, Annex B.

106 *Id.*, at 17, para. 45.

107 *Id.*, at 31, para. 2.

their confidence that in humanitarian terms, weapons based on controlled fragmentation constituted improvements over weapons that fragmented randomly.[108]

As for flechette weapons, the Conference report[109] concluded that "there was little specific reference during the debate in plenary to this particular type of weapon".[110] In fact, following prolonged (one month) discussions, only one proposal achieved broad approval, the prohibition on non-detectable fragmentation weapons.[111] However, even this proposal, submitted by experts from Mexico and Switzerland, was significantly diluted by the requirement that the fragments' production be considered the weapon's "primary effect". In consequence, noted the rapporteur for this Conference, "with one exception [the proposal on non-detectable weapons - E.B.], it would be an exaggeration to say that anything critically important had been contributed".[112] Eventually, with a slight change in wording, this proposal became Protocol I to the CCW.[113] Yet, despite mistaken claims, the Protocol has nothing to do with any prohibition on the use of flechette weapons.[114]

### *CIV. The Ad Hoc Committee (Third Session, 1976) and its Aftermath*

A few months after the end of the Lugano session (May 27, 1976), a Mexican representative to the *Ad Hoc Committee*, Reyes Perez, again raised the possibility of banning flechette weapons. Noting that 14 states had already supported the ban on these weapons, he justified such a ban on the grounds that "victims of [these weapons] ... suffered multiple injuries which caused intense pain". He likewise expressed his delegations' hope that the proposal would be accepted.[115] His speech, however, like

108 *Id.*, at 72, para. 6.

109 As the conference's rapporteur commented, it was "about twice as thick as that of the session at Lucerne". Yet, as he added, it was "in inverse ratio to the importance of the affair". Kalshoven, *Reflections on the Law of War, supra* note 55, at 390.

110 This reference includes technical information on the properties and modes of delivery of these devices, and was transmitted by an expert who had expressed his concern at the superfluous injury that flechettes might cause. *1976 Lugano Report*, at 18, para. 48. For the discussions, *see*, Kalshoven, *id.*, at 69-78. For the proposal that was submitted to the Conference, *see*, Doc. CDDH\IV\201, *1976 Lugano Report*, Annex 21, at 198.

111 The text reads: "The use of weapons producing fragments which in the human body escape detection by the usual medical methods shall be forbidden", *1976 Lugano Report*, Annex 11, at 188 (COLU/212). The poor results upset some participants. *See, id.*, at 95.

112 Kalshoven, *Reflections on the Law of War, supra* note 55, at 390.

113 The final text reads: "It is prohibited to use any weapon the primary effect of which is to injure by fragments which, in the human body, escape detection by X-rays". *See*, the Protocol on Non-Detectable Fragments, Oct. 10, 1980, *Laws of Armed Conflicts*, 190.

114 For a critique on this erroneous perception, as presented in a well-known textbook (Ingrid Detter, *The Law of War*, [Cambridge: Cambridge University Press, 1987], at 182-3), *see*, Howard S. Levie, "The Law of War", *AJIL* 82 (1989) 198 (book review). This mistake has been addressed in the second edition of this textbook. *See*, Ingrid Detter, *The Law of War*, (Cambridge: Cambridge University Press, 2000), 217.

115 *See, CDDH, O. R.*, Vol. XVI, at 328.

countless others during the Committee's sessions, failed to lead to any discussion on the issue.[116]

### *CV. The Ad Hoc Committee and the Conclusion of the Diplomatic Conference (Fourth Session, 1977)*

In this last session (April-May 1977), 16 out of the 98 states, which eventually signed the Final Act of the Diplomatic Conference on June 10, 1977, supported the proposed prohibition on the use of flechettes. However, the conference records show that the issue was in effect sidelined. Even delegates from states that supported such a proposal did not mention the flechette weapon in their speeches. The Egyptian representative (Dr. Sayed Anwar Abu-Ali), for instance, expressed his sorrow that the Diplomatic Conference was ending without "any tangible achievements by the Ad Hoc Committee". Furthermore, he complained that there was not enough time for adequate discussion of "other important weapons". As an example, he chose to mention only "small caliber projectiles" (hereinafter, small caliber weapons systems – SCWS) and fuel-air explosives.[117] The Swiss representative, Mr. Pierre De Graffenried, expressed his sorrow as well about the lack of "true progress" on the discussion on other types of weapons.[118] However, as with the Egyptian representative, when given the opportunity to discuss these categories, even briefly, flechettes – as opposed to incendiary weaponry, fuel-air explosives and SCWS – remained unmentioned.[119]

Therefore, when the Diplomatic Conference adopted the drafts of the Additional Protocols, no consensus over the prohibition or restriction of a single (significant) means of warfare was achieved.[120] AP-I does, however, contain an important section on "Methods and Means of Warfare" (Part II, Section I). The section reaffirms

116 As Prof. Kalshoven, who served as the *Ad Hoc Committee*'s rapporteur (as well as at both the Lugano and Lucerne conferences), wrote, "Its meetings were generally short…it happened quite often that a particular item on the agenda – say, a proposed ban on the use of incendiary weapons – was introduced in an elaborate exposé by the Swedish delegates; they were followed by a series of like-minded speakers, and after that: silence. Their opponents, in other words, simply refused to let themselves be drawn into a debate on the merits or demerits of proposals they were not at that stage prepared seriously to consider anyway." *See*, Kalshoven, "Arms, Armaments…" *supra* note 43, at 248.

117 *See*, *CDDH, O. R.*, Vol. XVI, at 431. Fuel-air explosives (FAE), referred to as "enhanced blast munitions" by some experts, disperse an aerosol cloud of fuel that, once ignited, produces a highly destructive shock wave (*See*, *e.g.*, Erich Prokosch & Ernst Jan Hogendoorn, "Antipersonnel Weapons," in: Ilkka Taipale *et al.*, (eds.) *War or Health? A Reader* (New-York: Zed Books 2002) 65, at 75). Anyone in the vicinity of the explosion faces being crushed by the blast or being killed by having air sucked out of their lungs (due to the vacuum created by the explosion).. For more on the first generation of fuel-air explosives s*ee*, *e.g.*, *SIPRI*, *Anti-Personnel Weapons* (New York: Taylor and Francis, 1978), 171-175.

118 *CDDH, O. R.*, Vol. XVI, at 444.

119 *Id.*, at 445-6.

120 On this lack of consensus *see*, also, Robert J. Mathews, "The 1980 Convention on Certain Conventional Weapons: A useful framework despite earlier disappointments", *International Review of the Red Cross* (*IRRC)* 844 (2001) 994.

the general obligations regarding weapons' legality (i.e., Articles 35(1)&(2), reiterating, *mutatis mutandis*, regulations 22 and 23(e) of the 1907 Hague Convention, respectively)[121] in addition to two new innovative provisions. The first imposed an obligation on each state party to take environmental considerations into account when selecting a weapon (Article 35(3)).[122] The second obligated States Parties to monitor the development of new weapons in light of their obligations under IHL (Article 36).[123] Furthermore, in the chapter on "Civilians and civilian population"

121 Only France demurred regarding its inclusion and declared that this article was the first of a series of articles which went beyond the strict confines of humanitarian law and in fact had "direct implications for the defence and security of states". *See,* Hans Blix "Means and Methods of Combat" in: *International Dimensions of Humanitarian Law,* (Geneva: UNESCO, 1988), 135, at 136.

122 The text reads: "It is prohibited to employ methods or means of warfare which are intended, or may be expected, to cause widespread, long-term and severe damage to the natural environment". Yet, at the time that AP-I was concluded, several states made sure to clarify that this article "applies only to states which have accepted those provisions" (i.e., Art. 35(3)) (*see,* U.S Declaration upon Signature, *Laws of Armed Conflicts,* 221; *see* also France's reservation upon signature, that was not confirmed upon ratification; *id,* at 217). In 1996, however, the ICJ stated that environmental considerations must play a role in the targeting of any armed attack. *See, Nuclear Weapons Advisory Opinion,* at para. 30. The ICRC and many jurists currently hold the opinion that this article, coupled with Article 55(1), in which the general rule is elaborated and adapted to the protection of civilian objects, have established a new customary rule in weapons selection. *See,* ICRC Study Report, Vol. I., at 151-8. However, as Prof. Dinstein clarified, the three nuclear powers (the U.S., the U.K. and France), which have taken the position that Rule 45 of the ICRC Study Report [i.e., the repetition of art. 35(3) adding that: "Destruction of the natural environment may not be used as a weapon"] does not reflect customary international law, act as "States whose interests are specially affected". Given the meaning of this status as constructed by the ICJ in its *North Sea Continental Shelf* cases (see North Sea Continental Shelf Cases [1969], ICJ Report. 3, at 43), "no custom can emerge". *See,* Yoram Dinstein, "The ICRC Customary International Humanitarian Law Study" in: Yoram Dinstein and Fania Domb (eds.) *Israel Yearbook on Human Rights* 36 (2006) 1, at 13. For the complexities created by these two articles *see,* Dinstein, *The Conduct of Hostilities, supra* note 23, at 181-185. The other instrument directly addresses the protection of the environment with respect to weaponry is the 1976 Convention on the Prohibition of Military or Any Other Hostile Use of Environmental Modification Techniques (ENMOD) (1978) [hereinafter, E*NMOD Convention*], *Laws of Armed Conflicts,* 163. For the history of its negotiation *see,* Juda Lawrence, "Negotiating a Treaty On Environmental Modification Warfare: The Convention on Environmental Warfare and its Impact Upon Arms Control Negotiations", *International Organization* 32 (1978), 975-991; for a brief review of its features *see,* Jozef Goldblat, *Arms Control: the New Guide to Negotiations and Agreements* 2nd ed. (Washington D.C.: Sage Publications 2002), at 158-162. For a brief overview of supplementary international law texts within the context of environmental considerations and weapons legality *see,* Dinstein, "The ICRC Customary", *id.,* at 185-88. Yet as one jurist rightly noted, given the interpretation of "long-term" in years, rather than months, coupled with the greater degree of discrimination expected of modern weaponry, the main concern is likely to be that of a weapons' toxic components. *See,* Justin McClelland, "The Review of Weapons in Accordance with Article 36 of Additional Protocol I" *IRRC* 850 (2003) 397, at 409.

123 As the 1982 Commentary noted, the obligation in Article 36, though new to IHL, had already been incorporated into the internal legislation of some states. Furthermore, as

(Part IV, chapter II), AP-I contains provisions having significant implications for the freedom that states exercise when selecting methods and means of warfare in terms of possible indiscriminate effects (Article 51 and 57).[124] Therefore, while none of the two Additional Protocols deals with specific weapon, Blix claimed in reference to the above provisions that:

> [I]n fact Part III, covering methods and means of warfare, and Part IV covering civilian population, contain vitally important rules which may be said to supplement the Hague Conventions more than the Geneva Conventions.[125]

Yet, Cassese noted at the time:

> Many States tended to believe that such principles need to be concretely implemented by specific prohibitions. Furthermore, many States tended to consider that modern conventional weapons such as...anti-personnel fragmentation weapons, etc., do not fall as of now under the prohibitory scope of those principles. ... Consequently, for many States, the only realistic and proper approach to such weapons should consist in considering whether it is feasible to elaborate new rules forbidding or restricting their use.[126]

As no specific weapon was addressed, pressures mounted for a compromise to keep the overall efforts of the Diplomatic Conference on track. [127] In June 1977, the Diplomatic Conference recommended to the UNGA that a Conference of Governments should be convened no later than 1979, to be preceded by consultative meetings during September-October 1977. The Conference's primary purpose was clear:

> ...reaching:(a) agreements on prohibitions or restrictions on the use of specific conventional weapons including those which may be deemed to be excessively injurious or have indiscriminate effects, taking into account humanitarian and military considerations;[128]

---

its three authors stated, "This procedure appears to be an obvious and indispensable corollary of the somewhat abstract principle stated in Art. 23 (e) of the Hague Regulations of 1907, and in Art. 35, para. 2, of the Protocol". *See, New Rules*, at 199. For an extend analysis of the article *see, e.g.*, Parks, "Conventional Weapons...", *supra* note 11, at 113-121. For its implementation *see,* also McClelland, *Id.*

124 For a thorough discussion of its provisions see *infra* chapter 6, and mainly Chapter 8.

125 *See,* Blix, "Means and Methods..." *supra* note 121, at 135.

126 Cassese Antonio, "Means of Warfare: The Traditional and the New Law", in: Antonio Cassese (ed.) *The New Humanitarian Law of Armed Conflict*, (Napoli: Editoriale Scientifica, 1979), 161, at 181.

127 As Prof. Kalshoven noted, such a consensus was made possible after some measure of rapprochement between the parties had come about, mainly as a result of the "lively discussions" at the Lucerene and Lugano Conferences where the same people participated. Kalshoven, "Arms, Armaments..." *supra* note 43, at 248-9.

128 The second purpose was "(b) agreement on a mechanism for the review of any such agreements and for the consideration of proposals for further such agreements". *See,*

Given this Resolution – Prof. Kalshoven noted – not to convene such a conference had become virtually impossible.[129]

## D. Flechettes during Formulation of the Convention on Certain Conventional Weapons (CCW Convention), 1978-1980

### *DI. The Preparatory Conferences (August-September 1978; March 1979)*

In December 1977, as expected, the UNGA accepted the recommendation of the Diplomatic Conference and requested that the Preparatory Conference be convened in 1978.[130] The first preparatory session was held in August-September 1978 with representatives of 74 states. The participants were presented with the report of the Ad Hoc Committee, which included the proposal to prohibit the use of flechettes weapons within all the proposals submitted.[131] In addition, Mexico presented a draft of a comprehensive treaty to deal with conventional weapons (hereinafter, the Draft Treaty), containing an optional protocol on "blast and fragmentation weapons." More importantly, it introduced a "draft clause relating to the prohibition of the use of flechette", using the same wording that had been reiterated since 1974.[132]

However, no consensus was reached on the substantive issues. Furthermore, a debate erupted over whether the UNGA's rules of procedure should apply (i.e., majority vote decision making) as opposed to a consensus on substantive matters. Finally, it was decided to refer this question to further examination at a second session, to be held in March-April 1979, subject to UNGA approval.[133] In September 1978, the UNGA endorsed both the request for another preparatory session and the Preparatory Conference's recommendation that the UN Conference be held in Geneva in September 1979.[134]

*Second Preparatory Session (1979)*: With sparse participation (68 states), the second session opened in March 1979. Attendees were given the same documents as in the

---

Resolution 22(IV) – Follow-up Regarding Prohibition or Restriction of Use of Certain Conventional Weapons, 9 June 1977. *Laws of Armed Conflicts*, 827.

129 Kalshoven, "Arms, Armaments..." *supra* note 43, at 250.

130 UNGA Res. 32/152 dated 19 Dec. 1977, 32 U.N. GAOR, Supp. (No. 45), at 57.

131 Originally, in May 1978, the Secretary-General invited only the participants in, and observers of, the Diplomatic Conference (i.e., members of the 1948 Geneva Convention and "national liberation movement" representatives). However, immediately afterwards, all UN members were invited to attend. *See, Reports of the Preparatory Conferences for the United Nations Conference on Prohibitions or Restrictions on Use of Certain Conventional Weapons which may be deemed to be excessively injurious or to have indiscriminate effects*, U.N. Doc. A/33/44 (1978) Nov. 3, 1978 [hereinafter, *1978 Rep. of the Preparatory Conference*] at 2. For a list of proposals, *see, id.*, at 5.

132 *See* A/conf.95/Prep.Conf./L. For the proposal, *see, also id.*, at 12.

133 *See 1978 Rep. of the Preparatory Conferences*, at 3. Eventually, this decision, together with the decision regarding the official languages, were the only decisions taken at this preparatory conference. *See*, Rep. of the Preparatory Conferences, UN Doc. A/Conf.95/3, May 25, 1979, [hereinafter, *1979 Rep. of the Preparatory Conference*], at 9.

134 *See*, UNGA Res, 33/70, Sept. 28, 1978.

previous session, specifically, Mexico's Draft Treaty with the Optional Protocol on fragmentation weapons and the *specific proposal for the prohibition on the use of the flechette.*[135] First, it was decided that a working group would be established to consider the draft proposal on non-detectable fragmentations weapons and on the regulation of the use of landmines and other devices. An informal working group to examine SCWS was established as well.[136] Later on, after a free plenary discussion regarding incendiary weapons, another working group was established to draft a prohibition or restriction on these weapons.[137]

With respect to flechette, it seems that this might have been the moment when these weapons (along with fuel-air explosives and cluster bombs) stood a chance of becoming prohibited. Yet, the essence of the session's final report resembled that of the previous session:

> Concerning the question of fuel-air explosive, anti-personnel fragmentation weapons, *and flechettes*, brief discussions took place at plenary meetings. *Time did not allow for their discussion in more detail, therefore agreement thereon could not be reached.* These questions are recommended for further study by countries, with a view to their discussion during the United Nations Conference.[138]

In fact, no agreement was reached even on the rules of procedure to be applied in the proposed UN Conference.[139] And so, the new draft proposals submitted related only to the use of incendiary weapons and to SCWS in addition to those weapons for which the working groups had already been established.[140] Finally, only a draft treaty on non-detectable fragments was unanimously agreed upon as few of the proposed restrictions on the use of landmines and other devices had yet to reach a broad consensus. [141]

### *DII. The UN Conference on the Use of Certain Weapons (September 1979; October 1980)*

*UN Conference, First Session (1979):* On September 10, 1979, the UN Conference on the Use of Certain Weapons finally convened for three weeks of discussions; 82 states attended, in addition to a number of observers. Four "basic proposals" for the prohi-

135 Emphasis added. *See, 1979 Rep. of the Preparatory Conference*, Annex I, G, 6. As in the first session, representatives of national liberation movements, international organizations, and NGOs were also present.

136 *Id.*, at 12, paras. 34, 36.

137 *Id.*, at 12, para. 36.

138 Emphasis added. *Id.*, at 13, para. 37.

139 *1978 Rep. of the Preparatory Conference*, at 13, para. 38.

140 *1979 Rep. of the Preparatory Conference*, at 13-14, para. 42.

141 *Id.*, at 12, para. 34. Toward the session's end, Mexico raised the proposal of a draft treaty for discussion (*see*, A/conf.95/Prep.Conf./L.8 and Corr.1). Several delegations considered such a general and universally applicable treaty, containing optional protocols or clauses (an "umbrella" arrangement), as quite useful (*see, 1979 Rep. of the Preparatory Conference*, at 13, para. 39).

bition or restriction of certain means of warfare were submitted to the Conference by the *Preparatory Conference,* all of which had previously been discussed extensively.[142] However, as in the case of the Ad Hoc Committee, once it was decided to adopt decisions by consensus during the first plenary meeting, it was quite clear that very few of the proposals would actually be accepted.[143] Not only did every delegation have veto rights, in many delegations the representatives were high-ranking arms control and disarmament officials rather than ordinary representatives, as in the Diplomatic Conference.[144]

In light of the existing "solid basis" for the proposals' agreed text, the first meeting of the plenary committee decided to establish two working groups: one on Landmines and Booby Traps and the other on Incendiary Weapons.[145] However, given the fact that few delegations still wished to discuss "other categories of weapons," a general discussion was held in the fifth plenary meeting (Sept. 21) on fuel-air explosives, SCWS and fragmentation weapons, including flechette weapons.[146] While SCWS merited special attention,[147] according to the Conference's summary report (Oct. 8, 1980), "no conclusions were reached on proposals on anti personnel fragmentation weapons, on flechettes and on fuel-air explosive weapons, which were not examined extensively."[148]

Finally, since much work needed to be done and there were considerable differences within the working groups (incendiary weapons and mines and booby-traps), the conference recommended that the UNGA convene an additional session in September 1980 – which the UNGA approved in December 1979.[149] Yet, it was

---

142 UN Conference on Prohibitions or Restrictions of Use of Certain Weapons which May be Deemed to be Excessively Injurious or to Have Indiscriminate Effects, *Geneva, 10-28 Sept. 1979,* UN Doc .A/Conf.95/8 (Oct. 8, 1979), *[hereinafter Rep. of the UN Conference, First Session*], at 6, para. 17.

143 *See, id.*, at 6, para. 15.

144 *See,* Roach, "Certain Conventional Weapons…" *supra* note 64, at 14.

145 *Rep. of the UN Conference, First Session*, at 9, para. 3. In addition, it was decided by the conclusion of the first plenary session to establish a working group on a General Treaty (i.e., the comprehensive treaty proposed by Mexico) *id.*, at 5, para.10.

146 *See, Rep. of the UN Conference, First Session*, at 11, para. 14.

147 Finally, it was decided that additional research in the area of wound ballistics, including the SCWS field test, was needed in order to develop a standardized assessment methodology, relevant to the ballistic parameters and medical effects of such systems. In other words, to establish a scientific basis concerning the harm caused by these weapon systems. See Resolution on Small-Calibre Weapon Systems. Geneva, 28 September 1979. *Id.*, Annex III. For the wording of the final summary report, see *UN Conference on Prohibitions or Restrictions of Use of Certain Weapons which May be Deemed to be Excessively Injurious or to Have Indiscriminate Effects, Geneva, 10-28 Sept. 1979; 15 Sept.-10 Oct. 1980, Final Rep. of the Conference to the General Assembly*, UN Doc. A/Conf. 95/15, Oct. 27, 1980, [hereinafter, *final Rep. of the UN Conference*], Annex I, at 18, Appendix E.

148 *Id.*, at 7, para. 20.

149 *See,* UNGA Res. 34/82 (Dec. 11, 1979), A/34/PV. 97.

agreed that no general debate at the opening of that session would be held and issues on which agreement had already been achieved should not be reopened.[150]

*Second Session (1980), and the Signing of the CCW Treaty*: In this second, month-long session, Mexico again submitted its draft clause regarding the prohibition on the use of flechettes.[151] However, most efforts were nevertheless devoted to the land-mines and booby-trap issues, with special attention devoted to incendiary weapons[152] as well as to SCWS.[153] Finally, a framework for dealing with weapons considered problematic by the international community, in the form of a convention, was drafted. Its Amendment (Art. 8) provided a further mechanism for monitoring the legality of weapons, albeit for conventional ones only, and considered an addition to the one already provided in 1977 by AP- I (Art. 36).

However, from the above description, we can conclude that the prohibitions outlined in the final drafts of the three protocols were few and mostly marginal.[154] In fact, only the first protocol on non-detectable fragments[155] contained any sweeping prohibition. The two additional protocols adopted (Prohibitions or Restrictions on the Use of Mines, Booby-traps and Other Devices,[156] and another on Incendiary Weapons[157]) actually included use restrictions exclusively. Thus, while some means of warfare (i.e., non-detectable fragments, antipersonnel mines and incendiary weapons) were prohibited or, more accurately, regulated (the issue of SCWS was *apparently* addressed in the September 1979 Resolution), claims against flechette weapons remained unaddressed. As the Conference's final report reads:

---

150 *Rep. of the UN Conference, First Session*, at 7, para. 22.

151 *See*, UN Conference on Prohibitions or Restrictions of Use of Certain Weapons which May be Deemed to be Excessively Injurious or to Have Indiscriminate Effects, *Geneva, 15 Sept.-10 Oct. 1980*, Rep. of the Committee of the Whole, A/CONF. 95/11, Oct. 9, 1980, Annex II.

152 *Id.*, at 1, para. 3. As the ban of non-detectable fragmentation weapons was agreed upon in the previous session, the text of the draft protocol was subsequently transmitted to the Drafting Committee. *Id.*, para. 4.

153 On the basis of the Swedish working paper on the subject, informal consultations were held among interested delegations, with their respective conclusions introduced to the Committee of the Whole (i.e., the Plenary Committee). *Id.*, at 2, para. 9.

154 For a comprehensive discussion of the Convention and the First Protocol, *see*, Ashley Roach (a member of the U.S. delegation to the CCW deliberations), "Certain Conventional Weapons..." *supra* note 64, at 18-72. For a similar discussion on the Second Protocol *see*, Burrus Carnahan, "The Law of Land Mine Warfare: Protocol II to the United Nations Convention on Certain Conventional Weapons" *Military Law Review* 105 (Summer 1984) 73-95. For a general discussion about the three protocols, *see, e.g.*, Frits Kalshoven, "The Conventional Weapons Convention: Underlying Legal Principles", *IRRC* 279 (1990), 510-520.

155 CCW Protocol on Non-Detectable Fragments (Protocol I), Oct. 10, 1980, *Laws of Armed Conflicts*, 190.

156 CCW Protocol on Prohibitions or Restrictions on the Use of Mines, Booby-Traps and Other Devices (Protocol II), Oct. 10, 1980, *Id.*, at 191.

157 CCW Protocol on Prohibitions or Restrictions on the Use of Incendiary Weapons (Protocol III), Oct. 10, 1980, *Id.*, at 210.

> The Conference also noted that time had not allowed for the consideration of questions concerning fuel-air explosives, anti-personnel fragmentation weapons, and flechettes, and that, consequently, no agreement could be reached thereon.[158]

Given the weighty and constant efforts invested by the Swedish Group since the early 1970s, this disregard should not be lightly dismissed. Ultimately, while noting the consensus on the prohibition on weapons with non-detectable fragments and the broad agreement regarding landmines and booby-traps, the 1977 Resolution 22 of the Diplomatic Conference *explicitly* refers to "the effects of the use of other conventional weapons, such as small calibre projectiles and *certain* blast *and fragmentation weapons*," in addition to mentioning that it had "begun the consideration of the possibility of prohibiting or restricting the use of such weapons."[159] Yet, as Professor Kalshoven stated, "(t)he results obtained at Geneva in 1980 ..., represent the very limit of what could be reached at the time".[160]

### E. The Flechette Weapon and the History of CCW Review

While the final report of the UN Conference (Ocober 1980) stated that insufficient time prevented an adequate consideration of these categories, it did add the following important statement:

> The Conference further noted that many delegations felt that those questions could be taken up in due time in the context of the follow-up mechanism provided for in article 8 [i.e., Review and Amendments] of the Convention on Prohibitions or Restrictions on the Use of Certain Conventional Weapons...[161]

Given the current absence of any specific rule addressing the use of flechettes weapon, this statement was in fact an invitation to review the history of the CCW Review Conferences since the Convention entered into force (Dec. 1983) with respect to this issue. After all, as Professor Kalshoven commented regarding the UN Conference's failure to progress in these categories of weapons:

> Some governments had been *strongly convinced* that prohibitions or restrictions on the use of these categories of weapon *were both urgently required and feasible*, and they could not but feel disappointed that their endeavors had come to naught.[162]

158 *See, Final Rep. of the UN Conference,* at 7, para. 23.

159 Emphasis added. Res. 22 – Follow-Up Regarding Prohibition or Restriction of Use of Certain Conventional Weapons, 9 June 1977, *Laws of Armed Conflicts*, at 827.

160 Kalshoven, *Reflections on the Law of War, supra* note 55, at 392.

161 *See, Final Rep. of the UN Conference,* at 7, para. 23.

162 Emphasis added. Kalshoven, "Arms, Armaments..." *supra* note 43, at 252.

### *E1. A Convention in Neglect (1980–1995)*

A prolonged, difficult and bitter debate emerged over the matter of a review mechanism at the UN Conference.[163] A complex review system, together with amendments, *including the proposal of additional protocols*, were eventually devised and elaborated in the CCW Article 8, had in fact been recommended by the Diplomatic Conference in its 1977 Resolutions (no. 22).[164] In reality, the question of convening such a review conference was simply removed from the agenda for more than a decade. The problem seems to have been rooted in the expected weakness of its protocols, the lack of any compliance mechanism and the fact that it had originally applied solely to IAC. In consequence, the CCW did not generate much attention in the 1980s; it was in fact neglected.[165] Developing nations did not deem it necessary to join the CCW and prominent states, such as the U.S. (which ratified the Convention only in

---

163 On the one side stood the major military powers and their supporters, who had opposed a UN Conference in the first place, and on the other side most other states. The former saw no need for a new mechanism because they thought the CCD, which had just been replaced by the Conference on Disarmament (CD), was fully competent to deal with such matters. The latter correctly understood that if the issue were to be transferred to the CCD, it would mean the end of any further development in this area and, as might be expected, were opposed to the move. As Parks noted regarding the argument put forward by some states at the outset of the Diplomatic Conference, to the effect that weapons issues should be addressed in the UN CCD, "Individuals familiar with this process were aware that referral to the UN Committee on Disarmament would have been tantamount to a death sentence for the proposal". Parks, "Conventional Weapons...", *supra* note 11, at 72, fn. 61. Echoes of this debate are noticeable in the Preamble to the Convention, which mentions the UN Disarmament Commission with respect to the "question of a possible broadening of the scope of the prohibitions and restrictions contained in this Convention and its annexed Protocols" and "the Committee on Disarmament" on "the question of adopting further measures to prohibit or restrict the use of certain conventional weapons". *See, Laws of Armed Conflicts*, 181, at 184. For the struggle *see*, Kalshoven, "Arms, Armaments..." *supra* note 43, at 261-3.

164 As para. 3 reads: The Diplomatic Conference... "Recommends that a Conference of Governments should be convened not later than 1979 with a view to reaching: ...(b) agreement on a mechanism for the review of any such agreements and for the consideration of proposals for further such agreements". *See*, Res. 22 – Follow-Up Regarding Prohibition or Restriction of Use of Certain Conventional Weapons, 9 June 1977, *Laws of Armed Conflicts*, 827, at 828.

165 In fact, this was anticipated by prominent experts. As Prof. Kalshoven noted in 1985: "(t)he results obtained at Geneva in 1980...represent the very limit of what could be reached at the time, and in all probability they will remain so for a considerable amount of time to come". Kalshoven, *Reflections on the Law of War, supra* note 55, at 392.

March 1995),[166] accorded it little importance.[167] And so, new means of warfare, as well as those which had previously been ignored (cluster bombs, fuel-air explosives and flechettes), were left untouched until the mid 1990s. Even Sweden, which had led the campaign against flechette weapon use in the 1970s, shifted its focus in the subsequent decade to prohibiting fuel-air explosives and other means of warfare, such as blinding laser weapons and naval mines, which had received very little attention in the past.[168]

The turning point in this state of neglect came on February 11, 1993, when the legality of antipersonnel mines entered the international arena during an official visit to Cambodia by the French President, Francois Mitterrand. Under pressure from human rights NGOs to ban this weapon, Mitterrand called for a Review Conference of the CCW, proclaimed an indefinite French moratorium on the export of landmines, and called on other states to join this initiative.[169]

### *E.II. First Review Conference (1995-1996)*

On December 16, 1993, the UNGA approved the request made by the UN Secretary-General, the depository of the CCW, to convene a Review Conference (in 1994 if possible); concurrently, the UNGA called upon the Secretary-General to establish a Group of Government Experts (GGE) in preparation for the event.[170] This GGE

166 On March 24, 1995 the U.S. Senate gave its advice and consent to the ratification of the Convention and the first two Protocols, subject to four conditions. President Clinton deferred action on Protocol III, pending further examination of its acceptability from a military point of view. For an analysis of whether ratification was in the best interests of the U.S. *see,* Howard S. Levie, "Prohibitions or Restrictions on the Use of Conventional Weapons", *St. John's Law Review* 68 (1994) 643. For the conditions, *see, id.*, at 665-66.

167 For a survey of the distressing condition of the Convention up until the first Review Conference, *see, e.g.*, Mathews, "The 1980 Convention..." *supra* note 120, at 996-99.

168 *See*, Ove Bring, "Regulating Conventional Weapons in the Future- Humanitarian Law or Arms Control?" *Journal of Peace Research* 24 (1987) 275-286.

169 For a personal account of the effect of the French civil society on Mitterrand's request, *see* Philippe Chabasse, "The French Campaign", in: Maxwell A. Cameron *et al.* (eds.), *To Walk Without Fear: The Global Movement to Ban Landmines*, (Oxford: Oxford University Press, 1999), 60-67. For an analysis of the influence of international civil society to the campaign against anti-personnel *Landmines*, until the drafting of the 1997 Anti-Personnel Mine Ban Convention (Ottawa Convention) *see, e.g.*, Richard Price, "Reversing the Gun Sights: Transnational Civil Society Targets Land Mines", *International Organization* 52 (1998), 613-644.

170 Res. 48/79, Dec. 16, 1993. The resolution was passed by a majority of 162 countries, no opposing votes, and three abstentions. In practice, since over a decade had passed since the CCW has entered into force (Dec. 2, 1983), the Secretary-General, as the depository of the Convention, and in accordance with Article 8(3)(a) of the Convention, was required to grant the French request. This was accomplished by ignoring the regular procedure of seeking State Parties' views on the issue, and without attaining a majority vote.

convened for three sessions in 1994, and a fourth in January 1995.[171] As expected, three of the four French objections, submitted within a proposed draft for an Amended Protocol II together with its request to convene a Review Conference,[172] revolved around land-mine related issues.[173] Nonetheless, during the GGE's last two sessions, three additional types of weapons were included within the review: (a) blinding laser weapons,[174] (b) naval mines,[175] and (c) SCWS.[176] However, no proposal was submitted regarding fragmentation weapons, particularly flechette weapons. In any case, once the previous UN Conference practice of making decisions by consensus was continued, any successful agreement on heavy restrictions was precluded, *a priori*.[177]

One day after the Review Conference opened in Vienna on September 25, 1995 (44 party states attended with another 40 non-party states present as observers), Antonio de Icaza, the delegate of Mexico–which had been one of the states active in attempting to prohibit flechettes at the end of the 1970s–declared that:

> It was regrettable that, except in the case of laser weapons, no proposal had been submitted concerning prohibitions on the use of small-calibre weapons, cluster bombs, *fléchettes* or fuel-air explosives. With other delegations, *his delegation would*

171 For the reports of the sessions *see*, *The Group of Government Experts to Prepare the Review Conference of the States Parties to the Convention on Prohibitions or Restrictions on the Use of Certain Conventional Weapons Which May Be Deemed to Be Excessively Injurious or to Have Indiscriminate Effects*, First Sess. (28 Feb.-9 Mar. 1994): CCW/CONF.I/GE/4 and Corr.1; Second Sess. (16 – 27 May 1994): CCW/CONF.I/GE/8; Third Sess. (8 – 19 Aug. 1994): CCW/CONF.I/GE/21; Fourth Sess. (9-20 Jan. 1995): CCW/CONF.I/GE/23.

172 For the French draft *see* Draft amended protocol on prohibitions or restrictions on the use of mines, booby traps and other devices (Protocol II), CCW/CONF.I/GE/3.

173 Strengthening the existing restrictions on anti-personnel mines; expanding the applicability of Protocol II to internal conflicts; establishing new rules relating to the transfer of anti-personnel mines. The fourth objective was the creation of an effective verification mechanism for the Convention. For the French objectives *see e.g.*, Karin Calvo-Goller, "Legal Background and Analysis of the 1997 Convention Banning Anti-Personnel Mines", *Israel Law Review* 32 (1998), 645, at 654-55.

174 For the ICRC's intensive activity regarding these weapons in the preceding years *see*, e.g., Louise Doswald-Beck, (ed.), *Blinding Weapons: Reports of the Meetings of Experts Convened by the International Committee of the Red Cross on Battlefield Laser Weapons, 1989-1991* (Geneva : ICRC 1993). Hence, there was a substantive discussion about the drafting of a new Protocol, based on a Swedish proposal that gained broad consensus. For the proposal *see* CCW/CONF.I/GE/11.

175 Declarations were made regarding the draft protocol proposed by Sweden and France. For their proposals *see*, CCW/CONF I/GE/12 and CCW/CONF.I/GE/CRP.30 respectively.

176 A Swiss proposal for the prohibition of SCWS submitted in the third session; *see*, CCW/CONF.I/GE/16 (Aug. 11, 1994). For the discussion *see also*, Eric Prokosch, "The Swiss Draft Protocol on Small-Calibre Weapon Systems", *IRRC 307* (1995), 411-425. In reality, only the topic of the laser weapons came up for discussion in the Review Conference. For a list of documents submitted to the Conference *see*, CCW/CONF.I/1, July 10, 1995.

177 For the rules of procedure suggested to the Review Conference *see*, CCW/CONF.I/GE/CRP.53/Rev.1 as amended (Annex III).

> *consider the desirability of submitting proposals in that regard, in the belief that public opinion must be alerted to the effects of those weapons*".[178]

During the discussions, in which the issue of landmines dominated in the form of a rolling draft for an amended Protocol II, a consensus was reached over a new protocol regarding "blinding laser weapons," which was adopted on October 13, 1995.[179] In its final draft, the protocol prohibited the use of or trade in:

> laser weapons specifically designed, as their sole combat function or as one of their combat functions, to cause permanent blindness to unenhanced vision, that is to the naked eye or to the eye with corrective eyesight devices.[180]

Without having reached a consensus on the Amended Mines Protocol, a decision was made on October 12 that the Review Conference would henceforth be suspended until it could be reconvened in Geneva (January and April 1996); the amended Second Protocol was then adopted.[181] Other issues were sidetracked due to a lack of support, the time devoted to drafting the Laser Protocol and the effort to reach a consensus on the CCW Amended Second Protocol.[182] Although the reasons for doing so may be debated, delegates suggested that the two additional pro-

178 Emphasis added. Summary Rec. of the 2nd Meeting (26 Sep.1995) CCW/CONF.I/SR.2, Sep. 29, 1995, at 14, para. 69. *Final Report of the First Review Conference of the States Parties to the Convention on Prohibitions or Restrictions on the Use of Certain Conventional Weapons Which May Be Deemed to Be Excessively Injurious or to Have Indiscriminate Effects*, CCW/Conf.i/16 (part II), *Documents and Summary Records of the Conference, Geneva 1996* [hereinafter, *Final Rep. of the First Rev. Conf.*].

179 For the consensus reached on October 6 in the Main Committee III *see id.*, CCW/Conf.I/4, para.6. For the importance of the mines question *see, e.g.*, the statements made by the Russian representative (Mr. Krylov) to the Conference on September 27, 1995, Summary Rec. of the 4th Meeting CCW/CONF.I/SR.4.

180 *See*, CCW Protocol IV on Blinding Laser Weapons (entered into force July 30, 1998), Oct. 13, 1995, *Laws of Armed Conflicts*, 212. For the Draft Protocol *see*, *Final Rep. of the First Rev. Conf*, CCW/CONF.I/7. For a legal analysis of the draft *see*, Burrus M. Carnahan and Marjorie Robertson, "The Protocol on 'Blinding Laser Weapons': A New Direction for International Humanitarian Law", *AJIL* 90 (1996), 484-490. For the reasons behind considering the weapon as causing unnecessary suffering or superfluous injury, *see*, Bengt Anderberg *et al.*, "Blinding Laser Weapons and International Humanitarian Law", *Journal of Peace Research* 29 (1992), 287, at 294-5. It should be noted, however, that legal scholars disputed the ICRC study (Rule 86) according to which the above mentioned prohibition reflects "a norm of customary international law (*see*, *ICRC Study Report*, Vol. I, at 292). *See*, Parks, "Conventional Weapons...", *supra* note 11, at 85, fn.118; Turner, "Weapons in the ICRC Study ..." , *supra* note 53, at 233.

181 For the Assembly's confirmation *see*, UN Res. 50/74 12 Dec. 1995. For a comprehensive analysis of the amended Protocol, *see*, Nash, Marian (Leich), "Contemporary Practice of the United States Relating to International Law" *AJIL* 91 (1997) 325, at 329-344.

182 The Swiss proposal on SCWS, for instance, failed to garner the required interest to promote the proposal further. As one writer noted, the proposal "was withdrawn before conclusion of the conference after receiving virtually no support from States Parties". Donna Marie Verchio, "Just Say No! The SIrUS Project: Well-Intentioned, But Unnecessary and Superfluous", *Air Force Law Review* 51 (2001) 183, at 197.

posals (on naval mines and SCWS) discussed in the Preparatory Conference be deferred to the next Review Conference.[183] The types of weapons that the Mexican representative had mentioned – including flechette weapons – did not appear at the time to merit such consideration. References to flechette weapons were also absent from the speeches delivered by NGO representatives, who were allowed (as customary) to attend the public sessions. When one of the observers for *The Friends World Committee for Consultation*, Diane Hendrick, declared that there still remained "a large unfinished agenda concerning specific classes of weapons", she was referring to SCWS, incendiary weapons,[184] fuel-air explosives, and cluster bombs.[185] In other words, her subject was all the categories for which a ban had been sought in the 1970s, save for one: flechette weapons.

### *EIII. Second Review Conference (2001)*

In applying the lesson learned from the CCW's poor record on the issue, the first Review Conference decided "to convene a further Conference five years following the entry into force of the amendments it had adopted, but in any case not later than 2001, with preparatory expert meetings starting as early as 2000, if necessary".[186] Following the November 2000 UNGA recommendation to convene a Second Review Conference in December 2001,[187] three preparatory expert meetings were held. In the first meeting (December 2000), true to form, it was again decided that decisions would be made by consensus.[188]

On November 11, 2001, the Review Conference was opened with the attendance of 65 party states, 4 signatory states, and 18 non-party states as observers. Five main issues were on the agenda: extending the scope of CCW applicability to non-IAC; Anti-Vehicle Mines (hereinafter, AVMs);[189] establishment of a compliance mechanism; Explosive Remnants of War (hereinafter, ERW); and SCWS (a Swiss

---

183 *See Final Rep. of the First Rev. Conf.*, Annex C, Final Declaration, Review of the Articles, Art. 8.

184 A complete ban as opposed to the current restriction on aerial use against a "military objective located within a concentration of civilians". *See,* CCW Protocol on Prohibitions or Restrictions on the Use of Incendiary Weapons (Protocol III), Oct. 10, 1980. *Laws of Armed Conflict*, at 210.

185 For the speech, dated Sept. 28, 1995, *see Final Rep. of the First Rev. Conf,* Summary Rec. of the 6th Meeting, Oct. 5, 1995, CCW/CONF.I/SR.6, at 15, para. 70.

186 *See id.*, Annex C, Final Declaration, Rev. of the Articles, Art. 8. CCW/CONF.I/WP.1/Rev.1, (1 May 1996), at 6.

187 *See*, UNGA Res. 55/37, 20 Nov. 2000, para. 4.

188 The other meetings were held in April 2001 and September 2001. For the Conference agenda as concluded in the third session *see*, *Rep. of the Second Review Conference of the States Parties to the Convention on Prohibitions or Restrictions on the Use of Certain Conventional Weapons which May Be Deemed to Be Excessively Injurious or to Have Indiscriminate Effects,* [hereinafter, *Final Rep. of the Second Rev. Conf.*). CCW/CONF.II/PC.3/1 – Annex IV.

189 We should note that in the Conference's documents, AVMs were sometimes referred to as Mines Other than Anti-Personnel Mines (MOTAPM).

proposal).[190] The Conference adjourned after ten days, having achieved a successful outcome regarding the first issue (Amendment of Article 1) and establishing two working GGEs on the subjects of AVM and ERW.[191] The latter remained the primary focus of the CCW until the Second Working Group produced the Fifth Protocol on ERW, adopted on November 28, 2003.[192]

As in the First Review Conference, other issues were derailed. Even Switzerland's intense efforts to promote the issue of SCWS (e.g., hosting three International Workshops on Wound Ballistics in October 1997, March 1999 and March 2001) after the First Review Conference were to no avail; despite the presence of its proposal on the agenda, the issue received marginal attention.[193]

Flechette weapons were therefore again left outside the Conference's discussions. In fact, unlike the First Review Conference, Mexico's current delegate, Gustavo Albin, failed to include flechettes on the list of weapons that Mexico believed that should be banned:

> Mindful that there were other weapons whose effects were such that they should be controlled under international humanitarian law and considered by future Review Conferences, Mexico would support initiatives on the prohibition of such weapons as cluster bombs, munitions containing depleted uranium, fuel-air explosives and naval mines.[194]

In other words, flechettes were deleted from the three categories of weapons (flechettes, cluster bombs, and fuel-air explosives) that Mexico had once promoted as sub-

190 For the list of proposals, including the final text as it appears in Annex III to the Report of the Preparatory Conference in their III Meeting, *see, id.* CCW/CONF.II/PC.3/1. For a forecast, that was indeed realized, regarding the proposals that would be adopted *see,* Mathews, "The 1980 Convention…" *supra* note 120, at 1002-1104.

191 For a survey article on this Conference and its results *see,* David Kaye and Steven A. Solomon, "The Second Review Conference of the 1980 Convention on Certain Conventional Weapons", *AJIL* 96 (2002), 922-936. At the GGE's Meeting in December 2002, it was agreed to negotiate an "instrument" regarding aspects of ERW and to explore issues regarding MOTAPM. See CCW/MSP/2002/CRP.1, paras. 21 & 22.

192 Protocol on Explosive Remnants of War (ERW) (*entered into force* Nov. 12, 2006); *see,* U.N. Doc. CCW/MSP/2003/2 (Nov. 27, 2003). On the negotiations over this Protocol, *see* Peter Herby and Anna Nuiten, "Explosive Remnants of War: Protecting Civilians Through an Additional Protocol to the 1980 Convention on Certain Conventional Weapons", *IRRC* 83, (2001), 195-205. For an analysis of this Protocol *see, e.g.,* Louis Maresca, "A New Protocol on Explosive Remnants of War: The History and Negotiation of Protocol V to the 1980 Convention on Certain Conventional Weapons", *IRRC* 86 (2004), 815–835.

193 *See,* Parks, "Conventional Weapons…", *supra* note 11, at 88-89. Interestingly, in contrast to strong U.S. opposition, the European Union was – as the Belgian representative, Mr. Jean Lint, stated on its behalf – "… prepared to study the question of small-arms ballistics". S*ee, Final Rep. of the Second Rev. Conf.*, at 64, para. 39. For the U.S. position *see, e.g.*, statement by the U.S. representative, Edward R. Cummings, *id.*, at 67, para. 61. According to Parks, however, in mid 2002 the Swiss government withdrew its support for the proposal, "drawing the initiative to a close". *Id.*, at 89.

194 For his speech made during the third session, December 12, 2001, *see, id.*, at 82, para. 7.

jects of prohibition or restriction, and replaced by a more contemporary and burning issue in the disarmament field: depleted uranium.

### *EIV. Third Review Conference (2006)*

After three preparatory sessions held during 2006, the CCW Third Review Conference opened on November 7, 2006. In attendance were 77 states parties, 2 signatory states and 19 observers.[195] Other than AVMs, for which few states, including the U.S., were hopeful of completing a draft protocol, the spotlight was directed to cluster munitions due to Israel's extensive use of the weapon in the 2006 Lebanon War, which had ended only a few months earlier.[196] However, the U.S., together with like-minded states, blocked a call by 25 states, led by Norway, to prohibit the use of cluster munitions "that pose serious humanitarian hazards because they are, for example, unreliable and/or inaccurate".[197] Ultimately, it was decided to convene a Meeting of Experts in June 2007 to further consider the application and implementation of existing IHL "to specific munitions that may cause explosive remnants of war, with particular focus on cluster munitions".[198] In fact, this Review Conference produced nothing substantial. Even the AVMs draft protocol failed to garner full support, irrespective of U.S. pressure.[199] Flechettes were not mentioned even once throughout the discussions.

---

195 For the Conference agenda *see Third Review Conference of The States Parties to the Convention on Prohibitions or Restrictions on the Use of Certain Conventional Weapons which May Be Deemed to Be Excessively Injurious or to Have Indiscriminate Effects, Final Document*, [hereinafter, *Final Doc. of the Third Rev. Conf.* ], Doc. CCW/CONF.III/1. The Review Conference's documents are available at <http://www.unog.ch>. The proposals and documentation related to the preparatory work for this Review Conference, are available at <http://www.disarmament.un.org/ccw/index.html>.

196 For more on these developments see Eitan Barak, "None to be Trusted: Israel's Use of Cluster Munitions in the Second Lebanon War and the Case for the Convention on Cluster Munitions", *American University International Law Review* 25, (2010, 423, 430-434).

197 *See*, Doc. CCW/CONF.III/11 (Part III), at 41.

198 *See*, Doc. CCW/CONF.III/11, Part II, at 6, Decision 1.

199 It was decided to devote up to two days to the issue in the next annual States Parties meeting (Nov. 2007). *See, Third Review Conference of The States Parties to the Convention on Prohibitions or Restrictions on the Use of Certain Conventional Weapons which May Be Deemed to Be Excessively Injurious or to Have Indiscriminate Effects. Final Declaration*, at 4, Decision 2.

# Chapter 3

## Five Reflections

More than anything else, the first part of the book indicates how vague the relationship between the scant literature on flechette weapons, legal and non-legal alike,[1] and their actual use (mainly in the early stages of WWI and the Vietnam War) has been. Nonetheless this military and diplomatic history's detailed overview, important questions are yet to be further explored. Five of them are addressed in this chapter as follows: the disappearance of free-fall air-dropped flechettes; the legality of their use in WWI giving the then treaty law; the effectiveness of firearm flechettes in the Vietnam War; the involvement of U.S. political authorities in their introduction into that War; the reasons behind the failed attempts to ban flechette weapons.

### A. The Disappearance of Free-fall Air-dropped Flechettes

As previously detailed, flechette weapons continued to be employed after 1914-1915 although, like all other means of warfare, these "primitive aerial arms"[2] soon gave way to more sophisticated and far more deadly instruments. Gravity bombs having either blast and/or incendiary effects were increasingly introduced into warfare, with various types of air-dropped cluster munitions appearing as well.[3] Contrary to flechettes, the other aerial weapons were also effective against "hard targets".[4]

---

1 The mainstream historical literature (even that dealing with WWI) overlooked the development of air-dropped flechettes' in the pre-WWI era, together with their employment during the war. For the total absence of flechette weapons from general histories of WWI *see*, e.g., Cyril B. Falls, *The First World War* (London: Longmans, 1960); Jay Winter and Blaine Bagget, *The Great War and the Shaping of the 20th Century* (New-York: Penguin Studio, 1996); Roger Chickering and Stig Forster (eds.) *Great War, Total War: Combat and Moblilzation on the Western Front, 1914-1918* (Cambridge: Cambridge University Press, 2000). Even among books focusing solely on the French army, the flechette was only mentioned briefly – if at all – and only in the context of 1914, and then subsequently treated as an obsolete weapon. *See*, e.g., Anthony Clayton, *Paths of Glory: The French Army, 1914-1918* (London: Cassell, 2003), 119.

2 See Ezra Bowen, *Knights of the Air*, (New-York: Time-Life Books, 1980), at 53.

3 For instance, the Splitterbombe SD-2, employed by the German Air Force, was an anti-personnel device jettisoned from air-dropped Italian-produced containers containing either 32 or 42 bomblets. For its use and specifications *see*: Wolfgang Fleischer, *German Air-Dropped Weapons To 1945*, [translator: Ted Oliver], (London: Midland Publishing, 2004) 15, 141. For its classification as a cluster munition see the *2008 Convention on Cluster Munitions (CCM)*, art 2, available at <http://treaties.un.org/doc/ Publication/ CTC/26-6.pdf>. Surprisingly, the use of cluster munitions as early as WWI has been overlooked by many writers. *See, e.g.*, Thomas J. Herthel, "On the Chopping Block: Cluster Munitions and the Law of War – Unexploded Submunitions from Cluster Bombs" *Air Force Law Review* 51 (2001) 229, at 234-39.

4 As one writer commented in reference to the rapid technological advances made in air warfare during WWI: "The aerial bombs used in 1914 were so small they could be carried in the pocket; toward the end of the conflict,

an R-plane [*Riesenflugzeug* [giant aircraft]) named for the giant WWI German aircraft used to bomb London during the War-E.B.] managed to release over London a mon-

Flechette weapon use thus became increasingly sporadic. As one expert concluded, despite their deadly image, "they soon became regarded as next to useless".[5] In fact, it seems that the severe flaws inherent in flechettes hurried their replacement. "[T]heir accuracy– concluded Martel– was not good; they were susceptible to atmospheric conditions that were impossible to calculate, especially the wind, as well as differences in pressure between layers of air during the summer".[6]

At first glance, the disappearance of free-fall air-dropped flechettes from the WWI battlefields after 1915 seems quite natural given the introduction of new aerial bombs, so much so that their continued, albeit sporadic, use throughout the war may even appear surprising. Yet, France's use of free-fall air-dropped flechettes at the Battle of Dien Bien Phu in the first Indochina War, as well as the U.S.'s similar use in Vietnam, indicate that the gradual abandonment of flechettes during WWI was far from self-evident. After all, the U.S. development of free-fall darts ("HAIL missiles", later renamed *Lazy Dog* missiles) followed an analysis showing the inefficiency of other types of aerial bombardments against infantry troops.[7] In other words, the lack of systematic analysis of the effectiveness of all available aerial weapons might have led to the wrong conclusion. As mentioned, the U.S. anti-personnel free-fall air-dropped darts project was abandoned only in 1969.[8] When the weapon reemerged in the 1990s, the darts had been redesigned as anti-material – rather than anti-personnel – weapons.[9]

## B. Legal Restraints on the Use of Free-fall Air-dropped Flechettes in WWI

The text of the 1970s proposal to ban flechettes ("munitions which act through the release of a number of projectiles in the form of flechettes")[10] can be applied to free-fall air-dropped flechettes as well. Yet, the proposal's "Explanatory Memorandum" made it clear that despite its use of the same term ("flechettes") for both methods

ster bomb thirteen feet long, weighing over a ton". Lee Kennett, *A History of Strategic Bombing* (New-York: C. Scribner's Sons 1982), at 35.

5 *See,* H. F. King, "Armament", *Flight* (June 28, 1934), 638–640.

6 Martel, *French Strategic and Tactical Bombardment Forces of World War I,* (Lanham: Scarecrow Press, 2006) [first published, *1918*], 14. The accuracy problem referred to the high risk involved with sending a pilot and his aircraft on a mission that could obtain such meager results.

7 *See,* Jean-Pierre Leroy, *Les Fléchettes D'avion De La Grande Guerre À Nos Jo*urs [The Aerial Flechettes of the Great War of Our Time], (Paris: Emotion Primitive, 2007), *at 141.*

8 *See, id.*, at 148. Reliable information on the reasons for the project's discontinuation is still missing is still missing.

9 As mentioned (*supra* Chapter I, B), some of the weapons (e.g., CBU-107) were developed with the clear intention to minimize collateral damage.

10 *See, e.g., Official Record of the Diplomatic Conference on the Reaffirmation and Development of International Humanitarian Law Applicable in Armed Conflict, Geneva* (1974-1977), [hereinafter, *CDDH O. R.*] Vol. XVI, (1978), 9. The entire working paper (CDDH/Dt/2, dated Feb. 2, 1974), which relates to "High Velocity Effects and Dumdum Effects", was revised following the 1974 Lucerne Conference (CDDH/IV/201), although no revision was entered in the section on flechettes. For the revised proposal see *CDDH O. R.* Vol. XVI, at 614.

and the high risk for collateral damage associated with each, Dr. Blix, who drafted the proposal, *had no intention* of making such an application.[11] Yet, as flechettes seem to fall within the broad definition of "projectiles", a question arose as to the applicability of the 1907 Hague Declaration *(XIV) Prohibiting the Discharge of Projectiles and Explosives from Balloons* to the use of this weapon during WWI.[12]

Concerns that launching projectiles and explosives from the air would be imprecise and "pose for the time being an incalculable risk for civilian population"[13] led participants at the 1899 Hague Peace Conference to prohibit the "the launching of projectiles and explosives from balloons, or by *other new methods of a similar nature*"[14] for a five-year term.[15] At the 1907 Second Conference, the prohibition was renewed, almost verbatim, "for a period extending to the close of the Third Peace Conference". As the conference was never convened, legal experts have argued that the "prohibition is theoretically still applicable".[16]

---

11 For instance, the 1974 working paper on this proposal reads: "They can be used in ammunition *for rifles and guns* or instead of fragments in warheads. *At detonation* these latter weapons eject a great number of flechettes" [emphasis added]. *See,* CDDH/DT/2, Feb. 2, 1974, at 9.

12 Declaration (XIV) Prohibiting the Discharge of Projectiles and Explosives from Balloons. The Hague, 18 October 1907, Dietrich Schindler and Jiri Toman, (eds.) *The Laws of Armed Conflicts,* (Leiden: Martinus Nijhoff, 2004), 309 [hereinafter, *Laws of Armed Conflicts*].

13 *See,* Frtiz Kalshoven, "The Soldier and His Golf Clubs", in Christophe Swinarski (ed.), *Studies and Essays in International Humanitarian Law and Red Cross Principles in Honour of Jean Pictat* (Leiden: Martinus Nijhoff, 1984), 369, at 373.

14 Emphasis added. *See,* Declaration (IV, 1), to Prohibit, for the Term of Five Years, the Launching of Projectiles and Explosives from Balloons, and Other Methods of Similar Nature. The Hague, 29 July 1899, *Laws of Armed Conflicts,* at 309. Given the Russians' motives, the proposal was probably meant to cover dirigibles that could be maneuvered and controlled in all but unfavorable weather conditions.

15 A permanent prohibition was voted on following a Russian proposal, made after Russia's own dirigible-related efforts had failed. However, at the next meeting, the U.S. delegate, Gen. William Crozier, raised an objection after consulting with his naval counterpart (Adm. Alfred Thayer Mahan), stating that while they were proposing to forever ban a weapon with which they had no experience, "new developments might soon make airships dirigible, enabling them to be steered by motor power over the area of battle and to take part at a crucial moment with possible decisive effect, thus in the long run sparing lives and shortening the conflict". Crozier therefore proposed a five-year ban, at the end of which all concerned would have a better idea of the capabilities of airships. *See,* Barbara W. Tuchman, *The Proud Tower: A Portrait of the World Before the War, 1890-1914.* (New-York: Bantam Books, 1967), 306-07. *See,* also Kalshoven, "The Soldier…" *supra* note 13, at 371-2. For the Russian motives *see,* W. T. Mallison, "The Laws Of War And The Juridical Control Of Weapons Of Mass Destruction In General And Limited Wars", *George Washington Law Review* 36 (1967), 308, at 320.

16 *See, e.g.,* Detlev F. Vagts, "The Hague Conventions and Arms Control", AJIL 94 (2000) 31, at 35. In the words of Schindler and Toman, the Declaration "is still formally in force today", *Laws of Armed Conflicts,* 309. However, as Parks pointed out, legal writers during WWI or shortly thereafter perceived this agreement as "a dead letter". *See,* W. Hays Parks, "Conventional Weapons and Weapons Reviews", *Yearbook of International Humanitarian Law* 8 (2005) at 61, fn.17. Parks stressed that with the first employment

In practice, balloons were not used for flechettes' dispersion, with various types of dirigibles also making little use of them.[17] Furthermore, among the major powers, only Great Britain and the U.S. ratified the 1907 Declaration.[18] Because the latter had entered the War in April 1917 without employing any dirigible or airship,[19] whereas the use of flechettes had already been discarded by the major belligerents, Great Britain was left as the only WWI belligerent whose use of flechettes might have been construed, *a priori*, as a violation of the Declaration. Yet, in reality, the *Royal Flying Corps* (RFC) had flown only airplanes – which were not equipped with flechette weapons – to France.[20] Furthermore, from a legal perspective, even if the RFC had used either its balloons (it did operate an observation balloon squadron) or dirigibles for dropping flechettes, the Declaration included the so-called "General Participation Clause (GPC)".[21] In other words, under this Declaration, Great Britain

---

of airplanes for bombing attacks during the Italo-Turkish War in Libya (Nov. 1911), the Convention had lost any practical meaning. For this use *see,* Kennett, *History of Strategic Bombing, supra* note 4, at 12-14. In any case, as Prof. Baxter noted in 1975, the Convention "is now only a curious antiquity". *See,* Baxter, R. R. "Criteria of the Prohibition of Weapons in International Law", in: Horst Ehmke *et al.*, (eds.), *Festschrift fur Ulrich Scheuner – Zum 70. Geburtstag,* (Berlin: Duncker & Humblot, 1973), 41, at 42.

17 *See, supra* Chapter 1, BII.

18 By 1907, substantial technical improvements of dirigibles had been made. A few European military powers, notably France and Germany (the Zeppelins), had already introduced dirigibles into their fleets. While dirigibles had no actual wartime experience, their military potential was significant. However, as Mallison noted, Great Britain (an island), which foresaw the diminution of the dirigible's military advantages, urged for the Declaration's renewal, with smaller states recognizing their inherent inferiority in this field, joined her in these calls. The continental powers, such as France, Germany and Italy, had no similar interest to do so and "were eager to retain and improve the airship". See Mallison, "The laws of..." *supra* note 15, at 320. For a list of signatures, ratifications and accessions see *Laws of Armed Conflicts*, 313. It should be noted that Austro-Hungary and Turkey signed but did not ratify the Declaration.

19 With the U.S. declaration of war (April 6, 1917), American pilots were sent to France. Starting in September, they began flying French seaplanes while escorting convoys from Great Britain to France. U.S. airplanes (not dirigibles) arrived only in 1918 and mainly took part in the fighting from June 1918 and onwards. For U.S. participation in the aviation forces during WWI *see,* Treadwell C. Terry, *America's First Air War: The United States Army, Naval and Marine Air Services in the First World War,* (New-York: Airlife, 2000); John H. Morrow, *The Great War in the Air: Military Aviation from 1909 to 1921* (Washington, D.C.: Smithsonian Institution Press, 1993), at 265-73 & 336-43.

20 *See,* Morrow, *id.*, at 74-81.

21 Under this clause, treaties are valid only when all belligerent States to a conflict have ratified them. The GPC's inclusion in the various 1907 Hague Conventions was the source of much uncertainty as to its application during WWI. For a succinct discussion of the GPC and its effect on treaties *see*, for example, Nagendra Singh and Edward McWhinney, *Nuclear Weapons and Contemporary International* Law 2d rev. ed., (Leiden: Martinus Nijhoff, 1989), 42-3. Interestingly, as the 1907 Hague Convention included such a clause (see Art. 2), it was argued in the 1989 *Finta* Case regarding a man charged in Canada with crimes against humanity and war crimes committed during WWII, that "since some of the belligerents...had not been party to that Convention, the convention had not been applicable" (*See* [Regina V. Finta, Canada, High Court of Justice. July 10, 1989], 82 ILR (1990), 425-464, at 425). However, the High Court of Ontario

was released from its commitment made at the war's onset because Germany was a "non-Contracting Power". In any case, it is worth noting that almost all belligerents were members of the 1907 Hague Convention, with its Article 25 deliberately expanded to cover aerial attacks on "open cities"; the article thus remained a dead letter.[22]

### C. Defeating "A Swarming Foe": On the Effectiveness of Flechettes in the Vietnam War

As to firearm flechettes, the Vietnamese theatre was, interestingly, the first and only case of massive firearm flechette use. In fact, when it came to anti-personnel weapons, the Vietnam War was the field laboratory for all anti-personnel weapons developed in the post-Korean War period. As Professor Kalshoven noted, several novel versions of conventional weapons were introduced during the Vietnam War.[23] A few applied the newly formulated principle of controlled fragmentation.[24] Yet, few – if any – of these weapons aroused the disproportionate worldwide interest that flechette did given *the actual scope of their use* in the war. Whether this was due to the weapon's "abnormal" way of killing (i.e., darts instead of "regular" bullets or blast effects) or the fact that its appearance coincided with the new ploy of "dramatic reports about the use and effects of means of warfare"[25] has yet to be explored. As to the facts, while the highest number of flechette projectiles used in a single battle during the prolonged Vietnam involvement stood at 32, the U.S. and its allies expended, in total, approximately *15 million tons of ammunition*. Even within the category of 105mm artillery shells or armor rounds, only 93,135 flechette rounds were actually procured by the U.S. between 1965 and 1973. For the sake of comparison, we should note that estimates of the 105mm "regular" projectiles used by the U.S. and its allies reached some 250 million![26]

No doubt, it was the *comparatively* rare use of flechette weapons during the war that contributed to its impressive effectiveness given the endemic battlefield condi-

---

(Associate Chief Justice Callaghan) rejected this claim and accepted the reasoning of the Nuremberg trial, according to which the Convention's rules "were recognized by all civilized nations, and were regarded as being declaratory of the laws and customs of war" by 1939. *Id.*, at 439.

22 The expansion ("It is forbidden to attack or bombard *by any means whatsoever*, towns, villages, dwellings or buildings that are not defended") was the idea of the French delegation. France's delegates opposed a total ban, arguing that bombing should be restricted to legitimate military targets rather than prohibited altogether. *See,* James Brown Scott (ed.) *The Hague Conventions and Declarations of 1899 and 1907* (Oxford: Oxford University Press, 1915), XVII.

23 Frits Kalshoven, "Arms, Armaments and International Law", *Recueil Des Cours* 19 (1985[II]) 225.

24 *See, e.g.*, the Claymore mine. This type of mine, according to Parks, is "a shotgun-like multiple-fragmentation anti-personnel munition". Parks, "Conventional Weapons..." *supra* note 16, at 63.

25 Kalshoven, "Arms, Armaments..." *supra* note 23, at 225.

26 For the data *see,* SIPRI, *Anti-Personnel Weapons*, (New York: Taylor & Francis, 1978), 26-27, 37.

tions, notably – as U.S. writers put it – combat against "a swarming foe".[27] The exact proportion of flechettes casualties among the 266 North Vietnamese bodies found after the battle in Landing Zone BIRD (December 26, 1966) is unknown. However, "the most important benefit derived from this battle" – as summed up in the formal history of the US Artillery Corps – "'was recognition that the beehive round was a tremendously valuable asset to the overall fire-base defense program".[28] Within a month, the formal history continues, "it [beehive rounds-EB] had earned the confidence and respect of artillerymen and infantrymen".[29] Given such a successful introduction, the amount of beehive rounds shipped to Vietnam the following year increased dramatically, up to six times higher.[30] According to an official U.S. report on the flechettes' accomplishments in Vietnam, "Human wave ground attacks" on semi-isolated artillery bases were "soundly defeated" with beehive ammunition on "a number of occasions" as it "literally decimated the onrushing hordes of attackers".[31]

### D. The Introduction of Flechettes into the Vietnam War and the Question of Political Sanction

The above account raises a salient question regarding whether the introduction of this new means of warfare[32] into the Vietnam War was approved or, at least, considered by any U.S. political authority. After all, the latter might have elaborated – as one scholar noted in reference to the introduction of cluster munitions (CMs) into the war – a "proper account of the political and humanitarian considerations that surround this type of weapon".[33] Relying on information that has become available with respect to CMs, it appears that the introduction of flechette weapons only

27 Edgar C Doleman *et al.*, *Tools of War (Vietnam Experience)*, (Boston: Boston Publishing, 1985), 48.

28 David Ewing Ott, *Vietnam Studies: Field Artillery, 1954-73*, (Washington DC: Department of the Army, 1975), at 110.

29 *Id.*

30 63,000 rounds of M546 type, which was used in the famous battle in Landing Zone BIRD, were procured in 1967, in comparison to 10,000 rounds in 1966. For US Department of Defense procurement data, *see*, SIPRI, *Anti-Personnel Weapons*, *supra* note 26, at 36-7.

31 *See,* Picantery Directorate Office, Summary Sheet, *MUCOM ENSURE* Program, Mar. 17, 1972, at 3.

32 In this context it should be noted that the author disagrees with scholars who claim that the flechette should not be considered a novel weapon or a U.S. innovation given its appearance in WWI. *See*, for instance, W. Hays Parks, "Means and Methods of Warfare", *The George Washington International Law Review (GWILR) 38* (2006), 511, at 512-13. As chapter II detailed, the substantial differences between WWI *free-fall* air-drooped flechettes to *firearm* "beehive munitions", beyond the different platforms (airplanes as opposed to armor and artillery projectiles) render the cases non-comparable. Unlike the 8,000 darts contained in the common flechette projectile employed in Vietnam (M546), for instance, a box of the Renault-made French flechettes contained only 500 darts. Release of a second box was not assured and, in any case, the attacked forces had already been warned.

33 Michael Krepon, "Weapons Potentially Inhumane: The Case of Cluster Bombs", *Foreign Affairs* 52 (1974) 596.

in mid 1966 (its first combat use was recorded as November 1966) prevented such speculation. CMs – by far more indiscriminate in their effects than flechettes, to the point where they earned the epithet "the most indiscriminate and lethal area weapons developed for the Vietnam War"[34] long before their post-conflict dimensions (i.e., unexploded bomblets) became acute – were introduced in the same year.[35] The U.S. Joint Chiefs of Staff (JCS), according to the sources, *refrained* from making any formal request regarding their use to any political authority. It turned out that only a year before (on March 9, 1965), the JCS had finally won a prolonged battle with civilian officials at both the Defense and the State Departments to lift restrictions on the use of napalm.[36] The object of these officials' concern, however, was not the humanitarian question associated with CMs but the possible adverse "world opinion".[37] Therefore, as Michael Krepon concluded following an interview with the JCS's Director of Operations, Vice Admiral Lloyd M. Mustin, "…when CBUs [Cluster Bomb Units-E.B.] were ready for operational use a year later, the Chiefs were in no mood to engage in another debate over ordnance selection".[38] If CMs were categorized and explained as standard, off-the-shelf, weapons, as "conventional ironmongery"[39] in Mustin's views, it seems probable that no political figure was involved in the decision to introduce flechettes. In fact, as Krepon found, civilian assistant secretaries for research and development (R&D) in both the navy and the air force had more pressing problems to consider, while "(c)ontrolled fragmentation, from an R-and-D standpoint, was purely a technical matter".[40]

### E. The Failure to Ban Flechettes: The Political Dimension of the Diplomatic Conference and the CCW's Frameworks

The extensive use of napalm following the lifting of restrictions on its use, together with the concurrent introduction of CMs, pushed aside the issue of flechettes and prevented any significant uproar during the initial stages of its use.[41] This state of

34 This phrase was used by a retired U.S. Army colonel, as cited in Michael Krepon, "Blanket Coverage: Two Case Studies of Area Weapons in Indochina", in: (eds.) Mary Kaldor and Asbjorn Eide, *The World Military Order: The Impact of Military Technology on the Third World* (London: Macmillan Press, 1979), 49, at 56.

35 For an overview of CMs use in Vietnam *see,* Eric Prokosch, *The Technology of Killing: A Military and Political History of Antipersonnel Weapons* (New-York: Zed Books, 1995) 81-105.

36 *See,* Krepon, "Blanket Coverage…" *supra* note 34, at 49-56.

37 *Id.*, at 52. Indeed, an example of the concern aroused by CM use is the status it received in the 1967 discussion before the "International War Crimes Tribunal". *See,* John Duffett (ed.), *Against the Crime of Silence: Proceedings of The International War Crimes Tribunal* (New-York: Simon & Schuster, 1970) 249-68.

38 Michael Krepon, "Weapons Potentially Inhumane…" *supra* note 33, at 600.

39 *Id.*

40 *Id.*, at 601.

41 This course of events is well-reflected in the fact that the May 1969 ICRC Report on the Feb. 1969 experts' meeting made no mention of flechette weapons (labeled as "high speed barbed missiles"), although *the question of their legality was raised* before the experts in a preliminary document (put before the meeting by way of a provisional

affairs changed dramatically in the early 1970s, when the international community began to concentrate its efforts on prohibiting or restricting the use of certain conventional weapons.

Although every development, protocol or review merits much discussion, the scope of this book confines us to a few general comments to be made for the purpose of shedding light on why this specific weapon escaped any explicit restrictions, let alone prohibitions, despite the Swedish Group's persistent efforts to introduce them. The flechette was, after all, a perfect candidate. As Professor Kalshoven recalled a few years after the CCW's conclusion:

> Two charges were levelled against these munitions [controlled-fragmentation anti-personnel munitions-EB]: they acted *indiscriminately*, and they caused multiple wounds and, hence, *unnecessary suffering* [i.e., as one such wound is sufficiently to put a soldier *hors de combat*-E.B.]. It was of course an undisputed fact that the fragments of an exploding "fragmentation weapon", depending on its characteristics, spread over a more or less wide area of terrain. As a logical consequence, these fragments were apt to hit whomever or whatever was in that area and the chances were, moreover, that more than one fragment would hit a single "target"... was all of this sufficient ground for a prohibition of the new munitions ... .[42]

Yet, we cannot avoid the gloomy conclusion that *non-inclusion* of a particular weapon in the CCW protocols bears no relation to the question of whether – as noted in the CCW's title – it was *"deemed to be excessively injurious or to have indiscriminate effects"*. First, as the U.S. negotiator of three important CCW protocols, Hays Parks, has recalled in reference to these negotiations, "(t)he legality of weapons was a field not easily entered. Governments historically viewed determinations of legality of weapons as within their prerogative ...".[43] Indeed, as Blix commented with respect to the twentieth century: "International examinations of different categories of weapons have been few and far between".[44]

The 1970s discussions therefore contributed little to the issue at hand. Prof. Cassese stated regarding his search for the failure of the Diplomatic Conference to agree on any explicit prohibition:

> It is apparent... that the primary reason for the failure... lies in the non-cooperative attitude and the delaying tactics of two groups of States: those of NATO (minus Norway) and those of the Warsaw Pact (minus Romania). This strong coalition succeeded in nullifying all the generous efforts of a group of States, made up of some

agenda). Instead, as mentioned, the report focused solely on napalm. See *supra* Chapter 2, B.

42 Emphasis added. Kalshoven, "Soldier and his Golf Clubs...", *supra* note 13, at 382.

43 Parks, "Means and Methods...", *supra* note 32, at 514.

44 As Blix noted, other than the 1925 Geneva Protocol's negotiation and the 1932-34 League of Nations Disarmament Conference, discussions of specific weapons took place only in the mid 1970s (i.e., the 1974-1977 Geneva Conference). Hans Blix, "Means and Methods of Combat" in: *International Dimensions of Humanitarian Law* (Geneva: UNESCO, 1988), 135, at 141.

> Afro-Asian countries, a few Latin American countries and some Western States such as Sweden and Norway.[45]

In that sense the 1970s discussions expressed the same political competition that had underlined most – if not all – international negotiations over weapons limitations or restrictions. Many military powers effectively shared a stance described by one member of the U.S. delegation, Prof. Baxter, as follows:

> Quite aside from the humanitarian aspects of the matter, the negotiations involve a certain squaring-off in advance of war, in which an attempt is made to get states with certain weapons to renounce their use, thereby enhancing the relative military power of the have-nots.[46]

Baxter's reflections make clear why the military powers and their allies argued, from the very outset, that the Diplomatic Conference – where, as "generally accepted in codification conferences",[47] a two-thirds majority could overrule them on matters of substance raised by the plenary Assembly – was not the appropriate forum for dealing with weapons in general and smart weapons (e.g., cluster bombs and flechettes) in particular.[48] Once discussions did take place, they tended to fend off the "humanitarian attack" by stressing the need of further study before ascertaining whether any type of weapon is really cruel or indiscriminate. However, as Professor Cassese noted:

> Those very States which had advanced it [the need for further research] had not put forward any constructive suggestion nor given any major technical contribution on the matter over the past four years. The argument, in the opinion of Third World countries, served only to hide the substantial unwillingness of the States bound by military alliances to achieve any real humanitarian progress on the matter.[49]

45 Cassese Antonio, "Means of Warfare: The Traditional and the New Law", in: Antonio Cassese (ed.) *The New Humanitarian Law of Armed Conflict*, (Napoli: Editoriale Scientifica, 1979), 161, at 184.

46 *See,* R. Baxter, "Humanitarian Law or Humanitarian Politics? The 1974 Diplomatic Conference on Humanitarian Law", *Harvard International Law Journal 16* (1975) 1, at 24.

47 Claude Pilloud *et al.*, (eds.) *Commentary on the Additional Protocols of 8 June 1977 to the Geneva Conventions of 12 August 1949*, (Leiden: Martinus Nijhoff, 1987), at xxxiii. In fact, in the three main plenary Committees only a simple majority was required.

48 In the powers' opinion, the more appropriate forum was the CCD, where they had quasi-veto rights and where questions of production, stockpiling, etc., were also taken into account. However, many states believed that the Diplomatic Conference constituted the appropriate forum as the discussions' starting point was the humanitarian dimension, This was reflected in the fact that they dealt solely with restrictions or prohibitions on use, which they argued should be separated from disarmament issues. *See Conference of Government Experts on the Use of Certain Conventional Weapons* (Lugano, 28.1- 26.2.1976) ICRC Report (1976) [hereinafter, *1976 Lugano Report*], at 5, para. 2.

49 Cassese, "Means of Warfare ..." *supra* note 45, at 186.

The impact of this political context on the CCW discussions substantially increased once the consensual format for decision-making was reached. In principle, as the case of flechette weapon and the U.S. (its producer and main user at the time) indicates, states having means of warfare that they perceived as effective would forestall any attempt to ban or restrict their use.[50] While apparently self-explanatory, the practice of reaching a consensus transformed the interests of a lone State into a "universal" interest if the respective state (e.g., UNSC permanent members) was able to withstand external political pressures. In other words, a single state's veto was sufficient to block any proposal to ban or restrict a weapon regardless of the international community's position against it.

In consequence, the prohibitions outlined in the final drafts of the three protocols were few and often marginal in their effect. As mentioned, only Protocol I (undetectable fragments[51]) contained a sweeping prohibition; the others (Protocol II on Mines, Booby-traps and Other Devices,[52] and Protocol III on Incendiary Weapons[53]) included use restrictions that, by their very mention, implicitly endorsed the weapons' legality.

In light of this tepid outcome, the first protocol's "instantaneous and unanimous" acceptance requires explanation.[54] It appears that the protocol's proponents believed that the U.S. already possessed a weapon having undetectable fragments (e.g., shells filled with glass) and had used them in the Vietnam arena. However, because no state actually possessed or completed development of such a weapon (there was and still is no military need for this weapon), it was easy for the U.S. and other major military powers to acquiesce to its prohibition. In fact, as Professor Cassese noted, its acquiescence allowed them "to show their readiness to make concessions".[55]

50 This regrettable state of affairs indirectly points to the importance of AP-I, article 36 (review of new weapons) that, at the time it was firstly proposed by the ICRC during the 2nd session of the Conference of Government Experts in 1973, was criticized by Prof Kalshoven as follows: "The effect of this clause would be to shift the discussion about the legality or illegality of certain means of warfare from the courtrooms to the drawing table, that is, from the judging of weapons actually used to the decision whether to produce them at all". Fritz Kalshoven, *Reflections on the Law of War: Collected Essays* (Leiden: Martinus Nijhoff 2007), at 68. Unfortunately, once weapons are used, there is often no one ready to discuss the possibility of their ban or restriction.

51 CCW Protocol on Non-Detectable Fragments, *Laws of Armed Conflicts*, 190.

52 CCW Protocol on Prohibitions or Restrictions on the Use of Mines, Booby-Traps and Other Devices, *Laws of Armed Conflicts*, *Id.*, 191.

53 CCW Protocol on Prohibitions or Restrictions on the Use of Incendiary Weapons, *Laws of Armed Conflicts*, *Id.*, 210.

54 For the complete text, *see*, David Turns "Weapons in the ICRC Study on Customary International Humanitarian Law" *Journal of Conflict and Security Law* 11 (2006) 201, at 227.

55 *See* Antonio Cassesse, *International Law* (New-York: Oxford University Press, 2005) 412. For the non-existence of such a weapon, *see also*, Ove Bring, "Regulating Conventional Weapons in the Future- Humanitarian Law or Arms Control?" *Journal of Peace Research* 24(1987), 275, at 277. The situation of a theoretical ban (i.e., a ban on non-existing weapons) remained unchanged in the 1990s, as the U.S. Administration reported to the Senate upon requesting its advice and consent to the CCW ratification. *See*, US Senate Treaty Doc. 103-25 (103d Cong., 2d Sess.), Message from the President

Even the circumstances behind the successful conclusion of Protocol III on Incendiary Weapons leaves little doubt as to the potential for success of a proposed ban on flechette weapons. As the U.S. negotiator for this protocol, Hays Parks, noted, prohibition or regulation of incendiary weapons "was for many the raison d'être for the CCW".[56] Yet, Protocol III, originally considered the Conference's main achievement, only prohibits use of air-delivered incendiary weapons whenever the military objective is located "within a concentration of civilians". As to attacks on such targets with "other than air-delivered incendiary weapons", it only imposes restrictions.[57] Hence, the Protocol far from categorically prohibits the use of incendiary weapons.[58] Few states, but mainly the US strenuously objected to even this partial ban.[59] Only during the Conference's closing session, when the prolonged diplomatic efforts to prohibit or restrict certain conventional weapons seemed to be on their way to oblivion given the few delegations (e.g., Mexico) that had insisted on the ban, did the U.S. retract its objections.[60] Parks later revealed that despite the Soviet Union's repeatedly declared consent to a total ban on incendiary weapons, the Soviet ambassador had acknowledged to him, after the protocol was concluded, that "the Soviet Union had no intention of accepting a prohibition on incendiary weapons" given its huge stockpiles. As both the USSR and the Warsaw Pact states never thought the U.S. would compromise, they simply preferred to hide behind the U.S. position.[61] Although Parks stated that the ban "is an acknowledgement of the legality of their [Incendiary weapon] use against military objectives and enemy combatants",[62] only in January 2009, nearly 30 years later and despite its ratification of the CCW in March 1995, did the U.S. finally accede to the Protocol. Yet, it continued to reserve the right

---

of the US, May 12, 1994, VI. In any case, this provision's current significance is debatable: not only have x-ray detection capacities changed immeasurably, alternate methods for detecting fragments in the human body that are not based on ionizing radiation, have also been developed (e.g., ultrasound or Magnetic Resonance Imaging [MRI]).

56 Parks, "Conventional Weapons…", *supra* note 16, at 77.

57 *See*, "Protocol on Incendiary weapons", Art 2(2) and (3), *Laws of Armed Conflicts*, 210.

58 The prohibition "in all circumstances to make the civilian population as such, individual civilians or civilian objects the object of attack by incendiary weapons" (Art. 2(1)) is in effect by virtue of the laws of war.

59 Mediation efforts, mainly by Australia and Holland, to introduce a new definition that would exclude some types of weapons commonly perceived as incendiary weapons (i.e., "heat munitions") failed. *See*, Kalshoven, "Arms, Armaments…" *supra* note 23, at 259. For their proposal *see*, UN Doc.A/CONF.95/CW/L.3, Sept. 27, 1979. It should be noted that munitions having incidental incendiary effects, such as illuminants, tracers, or smoke bombs, were excluded. *See*, Art. 1(b)(i), (ii). For a personal account of these crucial events *see*, W. H. Parks, "The ICRC Customary Law Study: A Preliminary Assessment", *American Society of International Law Proceedings* 99 (2005) 208, at 209.

60 For the Addendum, submitted three days before the Conference's adjournment, see UN Doc.A/CONF.95/CW/6/Add. 1, para 3; for the U.S.' opposition to this protocol prior to its 2009 acceptance *see*, Marian Nash (Leich), "Contemporary Practice of the United States Relating to International Law", AJIL 91 (1997), 325, at 326-7.

61 *See*, Parks, "ICRC Customary Law Study…", *supra* note 58 at 209. For the history of the Protocol's negotiation, *see*, W. Hays Parks, "The Protocol on Incendiary Weapons'" 30 *International Review of the Red Cross* 279 (1990) 535-550.

62 *Id.*

to "use incendiary weapons against military objectives located in concentrations of civilians where it is judged that such use would cause fewer casualties and/or less collateral damage than alternative weapons".[63]

Two main factors appear to be responsible for the failure of the proposed ban on flechettes to sustain the attention it had enjoyed during the 1973-1976 Government Experts Conferences and the ICRC-sponsored meeting during the 1978-1980 CCW process: (a) insufficient public awareness; and (b) the U.S.'s successful attempts to undermine arguments for the weapon's illegality (e.g., the severity and the extent of damage caused). Indeed, as Parks once stated: "the success the United States has enjoyed within the conventional weapons arena" is the result of both "a substantial amount of preparation" and "law of war expertise within the Departments of State and Defense".[64]

The issue of flechette weapons – despite the proposed ban – was consequently brushed aside while other weapons, particularly incendiary weapons, caught the public's eye.[65] As various reports indeed indicate, reaching agreements on other categories of weapons (non-detectable fragmentation excluded) meant that no sufficient time would be left for an adequate discussion of flechette.

Ultimately, and irrespective of how they were addressed in the original three CCW protocols, only four additional categories of weapons were either repeatedly proposed for banning or restriction, or reached the negotiating table during the 1970s and early 1980s: fuel-air explosives,[66] anti-personnel cluster munitions (CMs), small caliber weapons systems (SCWS) *and flechettes*. This fact should not be overlooked. It represents one major lesson to be learned from this historical survey: Flechette weapons are anything but "regular" weapons. Whether or not the respective claims are based on solid scientific grounds in practice, flechettes were and still are considered a *dubious weapon*. One category (CMs) among these four has already been addressed in a specific ban Convention (the CCM); another (SCWS) merited a special September 1979 UN Resolution.[67] Fuel-air explosives and flechettes

63 The reservation and Declaration of the U.S. are available at <http://www.icrc.org/ihl.nsf/NORM/3AB9E36D37F951 ECC1257558003E6A3F?OpenDocument>.

64 *See*, Parks, "Means and Methods…", *supra* note 32, at 535.

65 Anti-personnel mines, for instance, aroused much concern, largely based on the perception that they violated the criterion of indiscrimination and involved elements of perfidy (certain usages were likely to attack civilians, particularly children). *See, e.g.*, R. R. Baxter, "Conventional Weapons under Legal Prohibitions", *International Security* 1(3) (Winter 1977) 42, at 58.

66 Given the potentially severe collateral damage caused by fuel-air explosives (FAE), and their devastating impact on combatants, two proposals were submitted. One sought to ban "the anti-personnel use of weapons which for their effects rely exclusively on shock waves in the air," (see COLU/202, submitted by Sweden). The other proposal prohibited their detonation "for military purposes gas-air and dust-air mixtures which release gas pressure." (see COLU/209, submitted by Switzerland). For the proposals see *1976 Lugano Report, supra* note 47, Annex A.1, at 167. For their explanation see Id., at 121-2. For more on the need to ban this weapon *see, e.g.*, Bring, "Regulating Conventional Weapons…" *supra* note 5, at 278-280.

67 Resolution on Small-Calibre Weapon Systems. Geneva, 28 September 1979, *Laws of Armed Conflicts*, 237. However, Russia and other Warsaw Pact delegations blocked its

have been ignored. Importantly, despite the absence of any explicit prohibitions on fuel-air explosives – similar to the current case of flechette weapons – these are still considered sensitive weapons by some military authorities.[68] Yet, as these concluding remarks strongly suggest, the ban on flechette weapons had no chance from the outset against the still-fresh experience of their high effectiveness against "a swarming foe" in Vietnam.

As for the Review Conferences, one could have expected this issue to be raised again following the report of the Second Session of the CCW Conference (1980); yet, no proposals on flechette weapons were raised during the First Review Conference, nor were the weapons mentioned in the following two conferences, held in 2001 and 2006. However, as the cases of the Amended Protocol II on Mines, Protocol V on Explosive Remnants of War (ERW),[69] and the recent discussions over Protocol VI on CMs clearly indicate,[70] the Review Conference's agenda is *significantly* affected by international public opinion, expressed mainly through activities initiated in the sphere of civil society.[71] As one writer correctly noted in 2005:

> Over the course of the life of the CCW, many types of weapons have been suggested for its attention (fuel-air explosives, direct energy weapons, napalm, flechettes). Whilst these would more neatly fit the framework and intention behind CCW, they have not, however, captured the NGO and therefore the public imagi-

renewal the subsequent year and as Parks noted "It had a life span of the one-year period between the 1979 and 1980". *See,* Parks "Conventional weapons..", *supra* note 16, at 79, fn. 93.

68 For a succinct survey of fuel-air explosives and their use, by both U.S. and Russia, *see, e.g.*, Douglas Huber, "Fuel-Air Explosives Mature", *Armor* 110, (Nov.-Dec. 2001), 14-17. For the continuing concern regarding this weapon *see, e.g.*, HRW News Release, *Don't Use Fuel-air Weapons in Chechnya* (an open letter sent to the Acting Russian President, Vladimir Putin, on Feb. 14, 2000), available at <http://www.hrw.org/en/ news /2000/02/17/ dont-use-fuel-air-weapons-chechnyaa>. However, the U.S. employed its six-ton bomb, the BLU-82 (Known as the "Daisy Cutter") as an anti-personnel weapon in Afghanistan during 2001. Although technically not a fuel-air explosive (it is filled with a 12,600-pound explosive "slurry" mixture), such use of a bomb originally designed to clear vegetation while creating small craters for helicopter landing zones raised much concern among legal experts. *See,* Robert Cryer, "The Fine Art of Friendship: Jus In Bello in Afghanistan", *Journal of Conflict and Security Law* 7 (2002) 37, 58-60. Yet, as "(r) epulsion is unfortunately not, in and of itself, enough to render these weapons illegal", Cryer concluded that at best, "...the legality of these weapons may be open to question". *Id.*, at 60. Technical details on this bomb are available at <http://www.nationalmuseum.af.mil/factsheets/ factsheet.asp?id=1013>.

69 Protocol on Explosive Remnants of War (ERW) (entered into force Nov. 12, 2006); *see,* U.N. Doc. CCW/MSP/2003/2 (Nov. 27, 2003).

70 For a succinct overview on these negotiations effective as of late 2009, *see,* Eitan Barak, "Doomed to be Violated ?The U.S.-Israeli Clandestine End-User Agreement and the Second Lebanon War: Lessons for the Convention on Cluster Munitions", *Denver Journal of International Law and Policy* 38 (2009) 1, at 26-34.

71 The only non-original protocol to be accepted in isolation from significant public pressure was Protocol IV on Blinding Laser Weapons. For its negotiation *see, e.g.*, Burrus M. Carnahan & Marjorie Robertson, "The Protocol on 'Blinding Laser Weapons': A New Direction for International Humanitarian Law", *AJIL* 90 (1996), 484-490.

> nation sufficiently to provide the international impetus needed to register on the CCW scale.[72]

Flechette weapon non-use in "traditional" conventional war between regular armies and its rare use in Non-IAC in the post Vietnam era have kept the issue far from public view. Once large-scale use of flechette weapons had desisted, its non-appearance on Review Conference agendas was almost assured. As detailed in the following chapter (Chapter 4), Israeli use of flechette in South Lebanon and the Gaza Strip was intentionally low profile. This stance collapsed only in late 2002, once the Petition against its use was submitted to Israel's HCJ by two Israeli and Palestinian NGOs protesting the civilian casualties incurred.[73] Hence, the only Review Conference in which it could possibly be raised was the third one. After it disengaged from the Gaza Strip in 2005, Israel has minimized its use of flechette, and the weapon was again driven to the sidelines long before the Third Review Conference convened in November 2006. The uproar over Israel's extensive use of CMs in the July-August 2006 Second Lebanon War subsequently overshadowed the entire conference. Until the December 2008 Signing Ceremony of *the Convention on Cluster Munitions* (CCM) in Oslo and Israel's "*Cast Lead*" operation in the Gaza Strip, carried out that same month, CMs were the only armaments positioned high on the international civil agenda.

As the survey of the three CCW review conferences demonstrates, the very same twists and turns endured by the flechette's legal status during the 1970s were replicated at an accelerated pace in the last decade. Until the September 2009 appearance of the Goldstone Report and its call for an urgent discussion on the weapon's legality, the flechette was a forgotten issue, with no international interest bent on lifting it from obscurity, or drafting a protocol banning its use.

---

72 Justin McClelland, "Conventional Weapons: A Cluster of Developments" *International Law and Comparative Law Quarterly* 54 (2005) 755 at 767.

73 Physicians for Human Rights & The Palestinian Centre for Human Rights v. O.C. Southern Command & The State of Israel – Ministry of Defense, Petition for an Interim Decree, available at <http://www.btselem.org/hebrew/Legal_Documents/8990_02_Flechette_Appeal.doc>.

# Part II.

## Flechette Weapons: The Israeli Case

# Chapter 4

## Israel's Policy

This chapter, which focuses on the legal debate surrounding flechette use in Israel (2002-2003), provides the requisite factual background for framing the extensive legal analysis presented in Chapter 5 and in the last part of the book. The chapter begins with a brief survey of Israel's flechette weapons procurements along with a description of subsequent improvements to the weapon's efficiency, followed by an extensive survey of flechette use in Lebanon (1985-2000). It then reviews Israel's employment of the weapon in the Israeli-Palestinian conflict in the Gaza Strip during the period preceding the legal debate (2001-2003). Special attention is paid to the incidents that involved the killing or injury of civilians, which acted as a catalyst for the Petition to the HCJ by two human rights NGOs in October 2002.

### A. Procurement and Indigenous Improvements

In the absence of relevant archival materials,[1] the exact date of the arrival of the first U.S.-made flechette rounds to Israel is unknown. Some media reported their arrival during the U.S. massive emergency airlift (12,000 tons) (Oct. 13-Nov. 12) during the 1973 War.[2] However, while this claim may find some support in unofficial Egyptian complaints about its soldiers falling victim to flechette weapons,[3] accord-

1 Relevant documentation in this regard (e.g., IDF Ordnance Corps' internal correspondence or minutes of IDF General Staff meetings) can be found in the Israel's Ministry of Defense and IDF Archives, Tel-Hashomer. However, such documentation is formally classified for a period of 50 years.

2 *See*, for instance, Itzik Saban, "The Lethal APAM – [Israel] Military Industries is Developing a Round to Replace the Flechette", *Yedioth Ahronoth*, (Oct. 25, 2002), at A8 (in Heb.). For Israel's main military requests (flechette weapons are missing from this list) *see,* Zach Levey, "Anatomy of an Airlift: United States Military Assistance to Israel During the 1973 War", *Cold War History* 8, (2008), 481, at 489.

3 For such a complaint by an Egyptian expert during the discussion of the legality of the use of flechettes at the Sept.–Oct. 1974 Lucerne Conference *see,* Eric Prokosch, *The Technology of Killing: A Military and Political History of Antipersonnel Weapons*, (London: Zed Books Ltd., 1995) 155, at 178. Prokosch attended the conference as the *Friends World Committee for Consultation*'s representative – almost the sole NGO to attend. The complaint, which lacks any direct mention of either the state employing this weapon or the specific war in question, given the Conference rules, is not mentioned in the Official report. *See, Lucerne Conference of Governmental Experts on the Use of Certain Conventional Weapons* (Lucerne, 24.9-18.10.1974) (Geneva: ICRC, 1975), 59-60 [hereinafter, *Lucerne Conference*]. Yet, this affair was later recalled by Julian Perry Robinson,

ing to Brigadier-General Elazar Barak, the IDF's Chief Ordnance Officer during the war, flechette rounds were not part of the IDF's arsenal during that war nor were they included in the U.S. airlift.[4] Instead, tens of thousands of 105mm flechette tank rounds (Model-M494) were sent by the U.S. from the late 1970s for Israeli tanks equipped with 105mm main guns:[5] the US-Made M-60 and M-48 (both known in the IDF as the *Magach* model) as well as the *Merkava* ("Chariot" in Hebrew) Mark I (in service since 1979) and Mark II (in service since 1983).[6]

Nevertheless, with the introduction of Israel's improved *Merkava* model Mark III in 1990, followed by the Mark IV (operational since 2002),[7] the IDF faced a problem regarding the use of flechette rounds. Both models were equipped with 120mm smoothbore main guns and, as such, lacked the ability to fire the flechette rounds designed for the rifled 105mm gun. Hence, various NGOs and even some defense news reporters were led to believe that Israel had initiated a production line of 120mm flechette rounds that contained a larger number of darts than did the older 105mm.[8] In practice, only 120mm fin-folded *casings* for the rounds were produced by the state-owned Israeli Military Industries (IMI), while the rounds themselves remained intact.

Another indigenous Israeli improvement concerned the flechettes round's fuse. In mid 1990, the original U.S.-made mechanical time fuse (the M571) was replaced by an Electronic Time (ET) fuse (Omega M-127) in the 105mm model.[9] The new ET fuse enables automatic, in addition to the traditional manual settings.[10] The manual setting is performed with "range to target" data: range indication is internally trans-

---

then the World Health Organization's (WHO) representative. *See,* email correspondence, Feb. 2, 2008, Feb.21, 2010, Feb. 23, 2010 (on file with the author).

4 Telephone interviews, Oct. 26, 2006; Nov. 9, 2006.

5 There is no reliable data as to the exact amount of ammunition provided. However, the IDF has purchased some 50,000 fuses, as of mid 1990s, that are designated exclusively for these rounds. Telephone interview with Ehud Lahav, member of senior management, Reshef Technologies LTD, Oct. 27, 2009.

6 Telephone interviews with Brig.-Gen. Elazar Barak, Oct. 26, 2006; Nov. 9, 2006. Regarding their absence in the 1973 War, *see,* also Steve Rodan, "Israel's Military Debates Use of Flechette Round", *Jane's Defence Weekly,* (May 23, 2001), at 4. It should be mentioned that Barak's argument was made during a second interview, after he had held consultations with other senior IDF Ordnance Corps officers.

7 For a concise history of the Israeli Armored Corps, *see,* David Eshel, *Chariots of the Desert: Story of the Israeli Armoured Corps,* (London: Brasseys, 2004). For more details on the Merkava Mark I & II see Sam Katz, *Merkava Main Battle Tank Mks I, II & III,* (Oxford: Osprey Publishing, 1997).

8 *See,* for instance, HRW, "Civilian Pawns: Laws of War Violations and the Use of Weapons on the Israel-Lebanon Border", HWR, May 1996, at 74, fn. 96, [hereinafter, *Civilian Pawns*] also available online at <http://www.hrw.org/reports/1996/ Israel. htm.>; Rodan, "Israel's Military...", *supra* note 6, at 4.

9 For the fuse's development *see,* Barbara Starr, "US test Programme to assess ASRAAM," 22(17) *Jane's Defense Weekly* 22(17) (Oct. 29, 1994), at 18; Rodan, id. The fuse's technical specifications are detailed at the manufacturer's (Reshef Technologies, Ltd.) website; *see,* "Omega M127- technical specifications", available at <http://www.resheffuzes.com/RESHEF/products/m127_3.html>.

10 Range data is derived from the tank's fire control system, then translated into a time-setting that is transmitted to the fuse by the fuse setter. *Id.*

lated into a time-setting ranging from 210 meters to 4990 meters in 10-meter increments. As with the original fuse, the range in which the fuse was set to operate, normally 6 meters high, could be set at 60 meters before the requested distance to ensure optimal flechette dispersion. In addition, the new ET fuse enabled muzzle action for emergency use from zero to 200 meters. This fact is relevant to the discussion in part III in that the fuse's increased accuracy, according to the manufacturer's claim, meant a maximum range deviation of ten meters at a fixed muzzle velocity.[11]

Naturally, the 120mm smoothbore gun is more powerful than the 105mm rifled gun. As such, a distinctive fuse (M-120) was developed for Israeli production by the same manufacturer of the Omega M-127 fuse.[12] While the new fuse was much more reliable, the "killing zone," was increased as the darts disperse at a 7° angle 150 meters ahead of the target instead of a 17° angle at 60 meters, as with the 105mm model. Yet, since Israel has not had a heavy armor battle on either the Egyptian or the Syrian front since November 1973 and May 1974 (The Israeli-Syrian Disengagement Agreement), respectively, Israel's first opportunity to employ these weapons was in Lebanon.

## B. Use in Southern Lebanon (1982-2000)

### *BI. The Period of Silence (1982-1985)*

It appears that at the time that Israel launched its first large-scale invasion into Lebanon, in the March 1978 *Litani Operation*, flechette rounds were not yet deployed. However, by June 1982, the time of Israel's second invasion (*Operation Peace for the Galilee* or its more popular name: *the First Lebanon War*), the situation had changed.[13] Nonetheless, flechette weapons were employed mainly against the Syrian infantry during Israel's attempt to capture the town of Jezzine and during the fierce battle to take over the Syrian-held town of Sultan Yacoub.[14] This being the case, no large-scale collateral damage (if any) was inflicted by use of such an area weapon. The fact is that flechette rounds are neither designed for nor capable of harming civilians hiding in bomb shelters.[15] Furthermore, unlike cluster munitions, once they have hit a target they are no longer capable of harming the surroundings.

In effect, the international outcry against Israel's use of cluster munitions during the 1982 War (albeit minor in comparison to the criticism following the 2006 Second Lebanon War), in contrast to the silence regarding its concurrent use of flechettes is an indication of IDF caution given flechette rounds' great potential to inflict collat-

---

11 The standard deviation of the delay error, however, is smaller than 0.01 seconds (in the automatic setting mode). *Id.*

12 Telephone interview with Ehud Lahav, member of senior management, Reshef Technologies, Ltd, Oct. 27, 2009.

13 Interview with S. H. (full name with author), tank crew commander during the First Lebanon War, Oct. 30, 2009.

14 *Id.*

15 This is a result of the dart's low average penetration capability given a standard shelter's thick cement wells.

eral damage in certain settings. Since the cluster munitions then used – like flechette rounds – were American made, their use by Israel led to a diplomatic crisis with the U.S.[16], during which the Reagan administration issued a temporary suspension of scheduled CMs shipments to Israel[17], later extending the suspension "indefinitely."[18] Only in December 1988, following prolonged Israeli lobbying, was an understanding reached and the ban revoked.[19]

### *BII. Keeping a Low Profile (1985-1993)*

In contrast to the deployment of CMs, flechette rounds arrived at the front only in April 1985, after Israel's withdrawal to its self-proclaimed "Security Zone" along the Israeli-Lebanese border.[20] Once the defense line was established, the need arose to block hostile armed groups, mainly Hezbollah-affiliated,[21] from attacking Israel. Attacks along the border were to become increasingly effective despite the presence of Israel's ally, the South Lebanon Army (SLA). The absence of a proper electronic

16 The crisis began subsequent to a July 2, 1982 news report from Beirut on Israel's use of U.S.-made CMs (confirmed by the technical data and serial numbers on duds that were found). The reporter was the well-known London Times Middle East correspondent, Robert Fisk. For Fisk's reflections on his efforts to collect as much evidence as possible on Israel's use of US-made CMs, *see,* Robert Fisk, *Pity the Nation.* 4th edition, (London: Nation Books, 2002) 277-78. Already in June 30, 1982, however, President Reagan had voiced concern over reports of Israeli use of CMs, saying that the U.S. "is reviewing the question" of whether its laws were violated. *See,* Jack Nelson, "Reagan Denies Giving Israelis a 'Green Light'", *Los Angeles Times*, (July 1, 1982), at B1.

17 *See,* Michael Getler, "President Halts Shipment of 'Cluster' Shells to Israel", *Washington Post,* (July 20, 1982), at A1; George Skelton, "Cluster Bombs for Israel Held Back", *Los Angeles Times*, (July 20, 1982), at A1.

18 *See,* Judith Miller, "U.S. Bars Cluster Shells for Israel Indefinitely", *New York Times*, (July 28, 1982), at A16; Richard F. Grimmett, "U.S. Defense Articles and Services Supplied to Foreign Recipients: Restrictions on Their Use", *CRS Report*, (2005) at 5.

19 *See, e.g.*, Nancy Dunne, "US Lifts Ban on Cluster Bombs for Israel", *Financial Times* (Dec. 8, 1988), at A4; Yoav Karni, "The U.S. Lifts the 1982 Embargo on Selling Israel Cluster Bombs", *Haaretz,* (Dec. 6, 1988), at 1 (in Heb.).

20 On the Israeli considerations leading to the emergence of the Security Zone see Beate Hamizrachi, *The Emergence of The South Lebanon Security Belt* (New York: Praeger 1988); The Israeli government's decision (No. 291, Jan. 14, 1985) to withdraw from the Lebanese hinterland and redeploy in the Security Zone is available at <http://www.mfa.gov.il/MFA/Foreign%20Relations/Israels%20Foreign%20Relations %20since%20 1947/1984-1988/37%20Remarks%20by%20Prime%20Minister%20Peres-%20Vice%20 Premier%20a>.

21 Hezbollah means "party of God" in Arabic. Although in most of the Arab and Muslim states it is referred to as a resistance movement, some states such as the US, the UK and Canada consider it to be a terrorist organization. However, for the sake of objectivity, the term "armed group" would seem to be the least partisan way to describe the organization. For more on the movement founded in Southern Lebanon in 1982 mainly as a response to Israel's invasion, *see, e.g.*, Hala Jaber, *Hezbolla: Born with a Vengeance.* (New York: Columbia University Press, 1997); Augustus Richard Norton, *Hezbolla: A Short History.* (Princeton: Princeton University press, 2007); for an inside account *see,* Naim Qassem, *Hizbullah: The Story From Within.* [Dalia Khalil trans.] (London: SAQI, 2005).

fence along the Security Zone demarcation line meant that a chain of military posts and infantry patrols would have to serve as substitutes.[22] The IDF, however, quickly learned to take advantage of its superior conventional military power to organize an additional defense: tank ambushes.[23] Various anti-tank munitions were useless against hostile armed groups, which meant that there was a need for effective anti-personnel weapons; flechette rounds offered a perfect solution and seemed a natural choice.

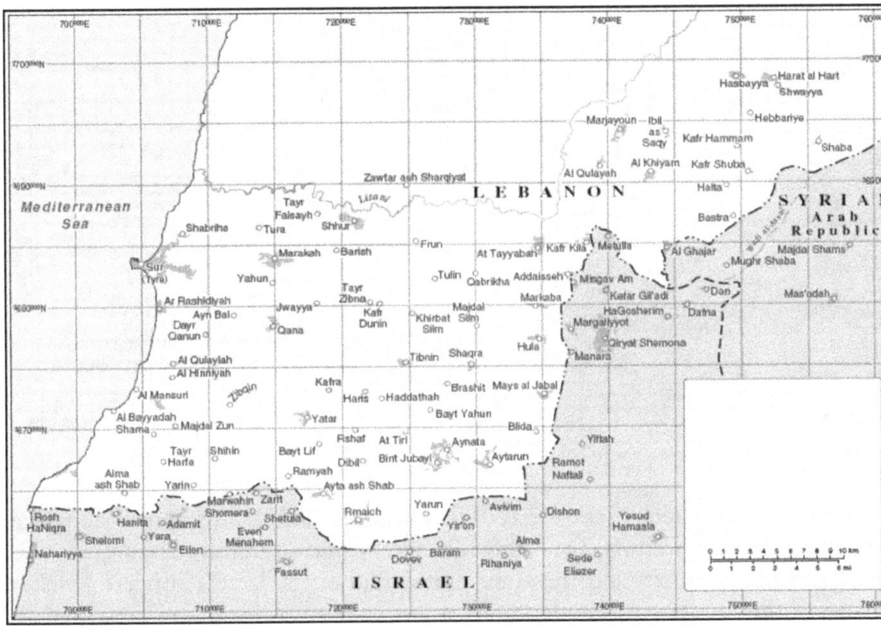


*Illustration 12: South Lebanon*[24]

22 On Israeli security arrangements in the Security Zone *see, e.g.*, Avraham Sela, "Civil Society, the Military, and National Security: The Case of Israel's Security Zone in Lebanon", *Israel Studies* 12(1) (Spring 2007): 53-78; Gal Luft, "Israel's Security Zone in Lebanon – A 'Tragedy'?", *Middle East Quarterly* 7(3) (Sept. 2000) 13-20. For a rare insider's perspective, *see*, Yuval Halamish, *Israel's Security Concept in Southern Lebanon 1985-1998* [Hebrew] (Ramat-HaSharon: National Security College, 1998). Halamish, the IDF's former Chief Intelligence Officer, served as the Northern Command's Chief Intelligence Officer during 1995-1998. On Israeli-Hezbolla relations with regard to the Security Zone see: Lior Avni, *Israel-Hezbolla Relations from "Operation Accountability" to the "Second Lebanon War" (July 1993-June 2006): Limited Security Regime or Violent Dialogue?* [Heb., unpublished manuscript, on file with author] (2007).

23 *See, e.g.*, David Eshel, "Armored Anti-Guerrilla Combat in South Lebanon", *Armor* (July-Aug. 1997), 26, at 27.

24 The map is available at <http://upload.wikimedia.org/wikipedia/commons/8/80/South_lebanon_map .gif>.

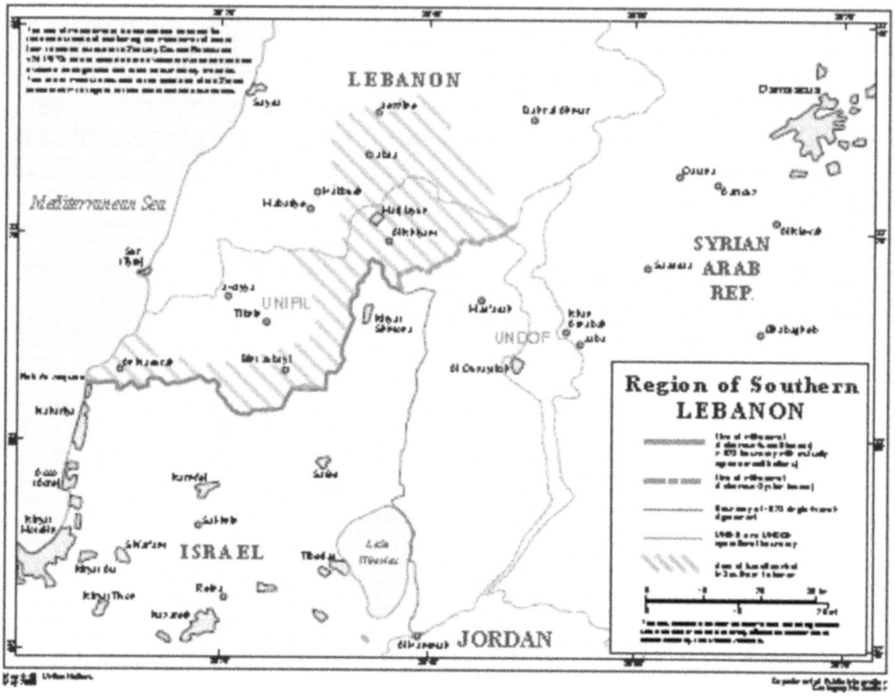


*Illustration 13: Israel's Security Zone in South Lebanon*[25]

As reported in the Israeli press, the IDF defined so-called "killing zones" along the perimeter of Israel's Security Zone, in which any suspicious figures discovered were presumed to be terrorists and therefore legitimate targets. Israel's armored combat doctrine in this regard was simple: "Once a movement was identified", reported an Israeli journalist, "the tanks fired flechette rounds. Three flechette rounds were to be followed by a single High Explosive Squash Head (HESH, an explosive warhead that creates a powerful explosion equivalent to 40 grenade-EB) round".[26]

However, as long as no human collateral damage was inflicted, the employment of flechettes failed to attract public attention. Therefore, leaving aside a single incident in December 1991 in which a flechette round accidently hit a (temporarily unmanned) UNIFIL position near Taire Village in the vicinity of Bint Jbeil – an incident that led the UNIFIL command to lodge a strong protest – the weapons

25 UN Dept. of Public Information, Cartographic Section (May 2000), available at <www.un.org/Depts/ Cartographic/map/profile/lebanon.pdf>.

26 *See,* Meron Rapoport, "They Walked in the Fields" *Yedioth Ahronoth*, (Mar. 8, 2002), (Weekend Magazine) at 2 (in Heb.). Following an ambush in which 'terrorists' were killed, the soldiers participating used to wear a flechette on their uniform. It should be noted, however, that no HESH rounds were ever available for the 120mm Merkava tanks, implying that High-Explosive Anti-Tank (HEAT) rounds were probably used instead.

remained in the shadows until the mid 1990s, when a series of incidents involving civilians transpired.[27]

### *BIII. The Media Wakes Up (1994-2000)*

The first major incident occurred on October 19, 1994, when a flechette round fired by 120mm *Merkava* tank located on the mountain ridge overlooking Nabatiya al-Fakwa, a suburb of the Shia market town of Nabatiya, targeted a crowd of civilians, killing five and injuring six. This incident, followed by another incident that same day in which two women were killed as a result of SLA shelling, sparked harsh reproaches by Lebanon's President Elias Hrawi and Prime Minister Rafik Hariri.[28] However, while they avoided referring directly to the weapons used, it soon became a source of interest for the international media, excluding that of Israel.[29] News agencies hurried to report on the use of "banned shells,"[30] while the British journalist Robert Fisk interviewed eyewitnesses regarding the "shell with the nails inside" stating that "[t]he flechette canister round is a ruthless anti-personnel weapon…"[31]

The next incident, on July 9, 1995, caught considerable attention, at least in the Israeli media. This incident resulted from what the IDF Chief of Staff, Major General Amnon Lipkin-Shahak defined as "human error."[32] Three flechette rounds – as required by the IDF's combat doctrine – were fired at the same suburb from the same outpost as in the 1994 incident, killing two Lebanese girls and injuring four other children. Sources from the Lebanese army quickly issued a statement declaring that "[s]uch shells…are banned under [the] terms of the Geneva Convention".[33] They were supported by UN officials who revealed that "The UN has previously protested to Israel against their use in Lebanon".[34] In Israel, the weapons used were

---

27 David Rudge, "UNIFIL Protests Shelling in Lebanon", *The Jerusalem Post*, (Dec. 24, 1991), at A2.

28 David Rudge and News Agencies, "Five Lebanese Killed by IDF/SLA Shelling", *The Jerusalem Post*, (Oct. 21, 1991), at A4.

29 While the incident itself was mentioned in the Israeli press, no reference to the type of weapons employed appeared in any of the related articles. *See, e.g.*, Aluf Benn & Eitan Rabin, "Rabin: Lebanese Civilians Were Killed Due to Misfire by IDF Tank in the direction of Nabatiya", *Haaretz*, (Oct. 24, 1994), at A11 (in Heb.); *see, e.g.*, Goga Kogan, "Tense Quiet in the Northen Border", *Yedioth Ahronoth*, (Oct. 23 1996), at A6 (in Heb.).

30 *See, e.g.*, "Lebanon: Israel Uses Banned Shells in South Lebanon", *Reuters*, (Oct. 20, 1994).

31 Robert Fisk, "Seven Die in Israeli 'Revenge' Raid", *The Independent* (London), (Oct. 22, 1994), at 11.

32 Aluf Benn *et al.*, "Rabin to Discuss Ways to Improve understandings Following Operation Accountability", *Haaretz*, (July 10, 1995), at A1 (in Heb.). According to officers in the Northern Command, who justified the firing, the tank crew saw suspicious persons and believed them to be terrorists. Unfortunately, the firing coordinates were incorrectly calculated by the tank crew. *Id.*

33 David Rudge, "2 Katyusha Attacks Hit Galilee", *The Jerusalem Post*, (July 10, 1995), at A1.

34 Kamal Jaber, "Hizbollah Hits Back after Israeli Shells Kill Girls", *The Independent* (London), (July 10, 1995) at 10.

described by an Israeli journalist as "cluster bombs";[35] Hezbollah's harsh response (two Katyusha salvos were fired into northern Israel), led to a discussion of the event during the Israeli cabinet's weekly meeting. Israel's Prime Minister, Yitzhak Rabin, severely criticized the Israeli tank crew's firing of the rounds.[36]

A few months later, on December 11, 1995, three Norwegians UNIFIL soldiers were wounded by flechette rounds near the village of Blatt after their patrol was mistakenly identified as a Hezbollah unit. UNIFIL lodged an official protest against the IDF who opened an investigation and hurriedly issued an official apology, expressing deep regret over the incident.[37] However, this as well as a similar incident three weeks later (Dec. 29) in the village of Qabrikha, during which "fléchette anti-personnel munitions" – as the UNSG termed these rounds[38]– killed one civilian and injured four others, awakened the Israeli media to the seemingly new weapons being employed in Lebanon.

In a radio interview a few days later, Ephraim Sneh, the sitting Minister of Health and former (1981-1982) IDF Commander of the Southern Lebanon Area (later known as IDF Lebanon-Liaison Unit), strongly justified the use of these weapons. By doing so he provided the first official confirmation of Israel's use of flechettes, an event that quickly attracted international media.[39] Stressing that the rounds were designed for use against infantry, he defined the flechette as a "very good weapon and most effective for hitting infantry targets." Upon being asked about the attendant legal and moral questions, he maintained that flechette use was completely legitimate, rationalizing his position as follows: "We are, in effect, fighting infantry – terrorists – and they are targets for this [type of] ammunition. I don't see any ethical constraints here".[40]

Five months later, in May 1996, Human Rights Watch (HRW) published a lengthy report on alleged international humanitarian law (IHL) violations by both

35 *See,* Eitan Glickman & Shraga Eshel, "Heavy Fire Exchange", *Yedioth Ahronoth,* (July 10, 1995), at A7; Eitan Rabin & News Agencies, "The UN: The Two Girls from Nabatiya were Killed by Israeli Cluster Bombs", *Haaretz,* (July 10, 1995), at A2 (in Heb.).

36 Aluf Benn *et al.*, "Rabin to Discuss Ways to Improve…", *supra* note 32; David Rudge, "2 Katyusha Attacks Hit the Galilee...", *supra* note 33.

37 David Rudge, "IDF Probes Wounding of Three UNIFIL Men by Friendly Tank Fire", *The Jerusalem Post,* (Dec. 11, 1995), at A12; according to the HRW Report, no less than six flechette rounds were fired. See *Civilian Pawns,* at 50 (based on Telephone Interview conducted by HRW with a UNFIL Spokesman, Jan. 31, 1996).

38 *See,* UN Security Council, *Report of the Secretary-General on the United Nations Interim Force in Lebanon,* S/1996/45 (Jan. 22, 1996), para. 6(b).

39 Israeli officials had previously refused to acknowledge that use. *See, Civilian Pawns,* at 5. For the attention attracted by the incident *see, e.g.* "Israel Confirms it Uses Banned Shells in Lebanon", *Reuters,* (Jan. 1, 1996) as cited in *HRW Report, Id.*

40 David Rudge & News Agencies, "Fighting Continues in Security Zone. Sneh Confirms IDF Uses Dart-Filled Shells", *The Jerusalem Post,* (Jan. 2, 1996), at A2. Interestingly, there was no follow-up to his admission in the Israeli press. *See,* Modi Kreitman *et al.*, "Barak: If Violations of the Understandings Are to be Continued, We Would Need a Reassessment in Lebanon", *Yedioth Ahronoth,* (Jan. 2, 1996), at A2.

(belligerents in the South Lebanon arena.[41] The report contained a chapter entitled "Questions about Some of the Weapons Used in the Conflict", in which only two types of weapons were discussed: phosphorus and flechette shells.[42] While acknowledging that the IDF's long-standing practice in the Lebanese arena had involved employing flechette rounds, "mainly for their ability to penetrate dense foliage" in comparison to "conventional antipersonnel shells," the HRW claimed that reports on civilian flechette casualties and injuries had increased in 1994-1996.[43] The HRW detailed several cases from the period mentioned in the report (e.g., Oct. 19, 1994 and July 8, 1995) while trying to demonstrate the "lethality of flechettes"; it also cited Lebanese doctors on the "devastating effect" the weapons inflicted.[44]

> The crux of the HRW campaign has been the use of weapons with a large "kill radius" in populated areas. Based on Article 51(4) of Additional Protocol I of 1977 (AP-I) – indiscriminate attacks are: ... (b) those which employ a method or means of combat which cannot be directed at a specific military objective; or (c) those which employ a method or means of combat the effects of which cannot be limited as required by this Protocol.[45]

The HRW cast doubts on whether flechette use was lawful. "The flechette shell," it stated, "because of its large 'kill radius,' can be particularly destructive when used in civilian areas, as it appears to have been."[46] It then called Israel to investigate allegations of flechette use and, more importantly, to "make a public commitment to refrain from using such weapons in civilian areas".[47]

Although it seems probable that Israel took further precautions, which lead to a decrease in civilian casualties, the use of flechette rounds continued.[48] As an Israeli journalist commented: "Human Rights NGOs have protested ... but the doctrine

41 *See, Civilian Pawns. See,* also, "Human Rights Watch, Report on Events on the Israel-Lebanon Border", *Journal of Palestine Studies* 25( 4) (Summer 1996), 152-155.

42 *Civilian Pawns*, Chap. VI, (written by Ernst Jan Hogendoorn).

43 *Id.*, at 48.

44 *Id.*

45 *See,* Protocol Additional to the Geneva Conventions of 12 August 1949, and relating to the Protection of Victims of International Armed Conflicts (Protocol I), 8 June 1977, [hereinafter AP-I], Art. 51(4). Dietrich Schindler and Jiri Toman, (eds.) *The Laws of Armed Conflicts*, (Leiden: Martinus Nijhoff, 2004), [hereinafter, *Laws of Armed Conflicts*] 711, at 736.

46 *Civilian Pawns*, at 44.

47 *Id.*, at 5; For the call, *see, Id.*, at 10.

48 *See, e.g.*, "Israeli Soldier Killed, Six Civilians Wounded", *AP* (Dec.7, 1996); David Rudge, "IDF Fears Reprisals after Lebanese Civilian Killed", *The Jerusalem Post*, (July 13, 1997), at A2. Both incidents were reported in the Hebrew language Israeli press as well, but with no reference to the specific weapons employed. *See, e.g.*, Amir Rappaport & Eitan Glickman, "Hezbollah Threaten: We Will Target Israeli Civilians", *Yedioth Ahronoth*, (Dec. 8, 1996), at A2-3; Eitan Rabin, "Lebanese Senior was Mistakably Identified as Terrorist and killed by IDF Tank", *Haaretz*, (July 13, 1997), at A2.

[firing three flechette rounds and then a single HESH round-E.B.] has been proven to be efficient and the IDF has continued to abide by it".[49]

As such, when *B'Tselem* – The Israeli Information Center for Human Rights in the Occupied Territories, decided in early 1999 to deviate from its original objective in order to "combat the denial and disregard of human rights violations committed during Israel's occupation of South Lebanon", an open debate on the issue became unavoidable.[50] However, *B'Tselem's* detailed inquiry over the use of "phosphorous munitions" (its main target), dated February 28, 1999, motivated the IDF's universal and pithy response to weaponry issues – "The IDF uses lawful weapons against lawful targets"[51] – in the following month. It thus became clear that there was no point in addressing the flechette issue separately.

Within six months (September 10, 1999), however, the weapons were back in the headlines as UNIFIL again filed a complaint with the Israel-UNIFIL Liaison Committee against the use of the flechette rounds, a complaint that resulted from an incident the week before in which a number of civilians were wounded.[52] A UNIFIL officer (probably the spokesperson) admitted that flechettes "are not expressly named in any weapons convention"; yet, he stated, in UNIFIL's view, the fact that "they are not aimed at a specific target but scatter over a wide area, inflicting serious damage, including damage to civilian facilities, puts them indirectly into the category of banned weapons".[53] However, once the news reached the Israeli press, IDF officers declined to discuss this subject and even cast doubts over the existence of such a complaint.[54]

In January 2000, *B'Tselem* finally published its final report, which contained a sub-chapter entitled "Prohibited Weapons," designed to address "Israel's use of two particularly deadly weapons employed in populated areas: phosphorous and flechette shells".[55] While acknowledging that "[t]he laws of war do not expressly prohibit or limit the use of flechettes", the organization challenged the IDF's position.[56] Relying heavily on the 1996 HRW report, it held that the flechette round's "large killing range… as proven in Lebanon, is liable to kill many civilians". *B'Tselem* thus

49 Rappaport & Glickman, *id.*, at A2.

50 *B'Tselem* – The Israeli Information Center for Human Rights in the Occupied Territories, *Israeli Violations of Human Rights of Lebanese Civilians,* (Jerusalem: *B'Tselem*, 2000) at 4.

51 Letter from *B'Tselem* to Capt. Avital Margalit, Head, Assistance Department, IDF Spokesperson's Office, Feb. 28, 1999; a reminder was sent on Mar. 28, 1999. For the IDF response *see,* "Letter to *B'Tselem* from Capt. Avital Margalit, Head of Assistance Department, IDF Spokesperson's Office, Mar. 30, 1999 (*B'Tselem* Archive, on file with author).

52 Jihad Saqlawi, "Flechette Shell Wreaks Havoc on Lebanese Civilians", *Agence France-Presse* (*AFP*), (Sept. 10, 1999).

53 *Id.*

54 *See, e.g.,* Arieh O'Sullivan, "IDF Accused of Firing 'Dart' Shells", *The Jerusalem Post,* (Sept. 13, 1999), at A1.

55 *Israeli Violations of Human Rights, supra* note 50, at 55-59. In addition, the report addressed "both sides' use of roadside explosives".

56 *Id.*, at 58. For the call *see, Id.*, at 62.

concluded that flechette use "particularly in populated areas, is equivalent to indiscriminate firing, which is expressly prohibited by the laws of war". The report called for the Israeli Government to prohibit IDF and SLA forces from using flechette rounds.[57] Interestingly, in that same year (2000), a legal scholar who also relied heavily on the HRW report, became concerned that the weapons allegedly violated a different "basic rule" of IHL, specifically, the prohibition on causing superfluous injury or unnecessary suffering to combatants.[58] Flechettes, he stated, "clearly cause unnecessary suffering by virtue of the fact that the head and tail of the dart both spin and continue to do so after entering the body and while traversing it".[59]

Four months later, on May 24, 2000, the last Israeli soldier left Southern Lebanon. This event made all the aforementioned criticisms effectively immaterial, at least until the next confrontation.

## C. Use in The Gaza Strip (2001-2005): High Profile Incidents and Their Aftermath

On March 3, 2001, six months after the Israeli-Palestinian conflict erupted anew (September 2000), a 24-year old resident of the Al-Boreg refugee camp, Mustafa Rimlawi, was killed on the Karni-Netzarim road. Whether he intended to plant a roadside bomb – as Israel claims – or whether he was indeed a mentally disabled youth wandering in the area – as the Palestinian version maintains – Rimlawi became the first known victim of flechette weapons in the Gaza Strip. However, another flechette victim and another ten days were needed before news of the new weapons' use would reach the public.[60]

---

57 *Id.*, at 58.

58 David Turns, "Some Reflections on the Conflict in Southern Lebanon: The 'Qana Incident' and International Humanitarian Law", *Journal of Conflict and Security Law* 5 (2000), 177, at 205.

59 *Id.*

60 This was despite the fact that a Reuters journalist had seen Rimlawi's corpse with dozens of darts all over the body. *See*, Reuters, "Nail-like Ammunition Denounced", *The Toronto Star*, (Mar. 13, 2001), at 1.

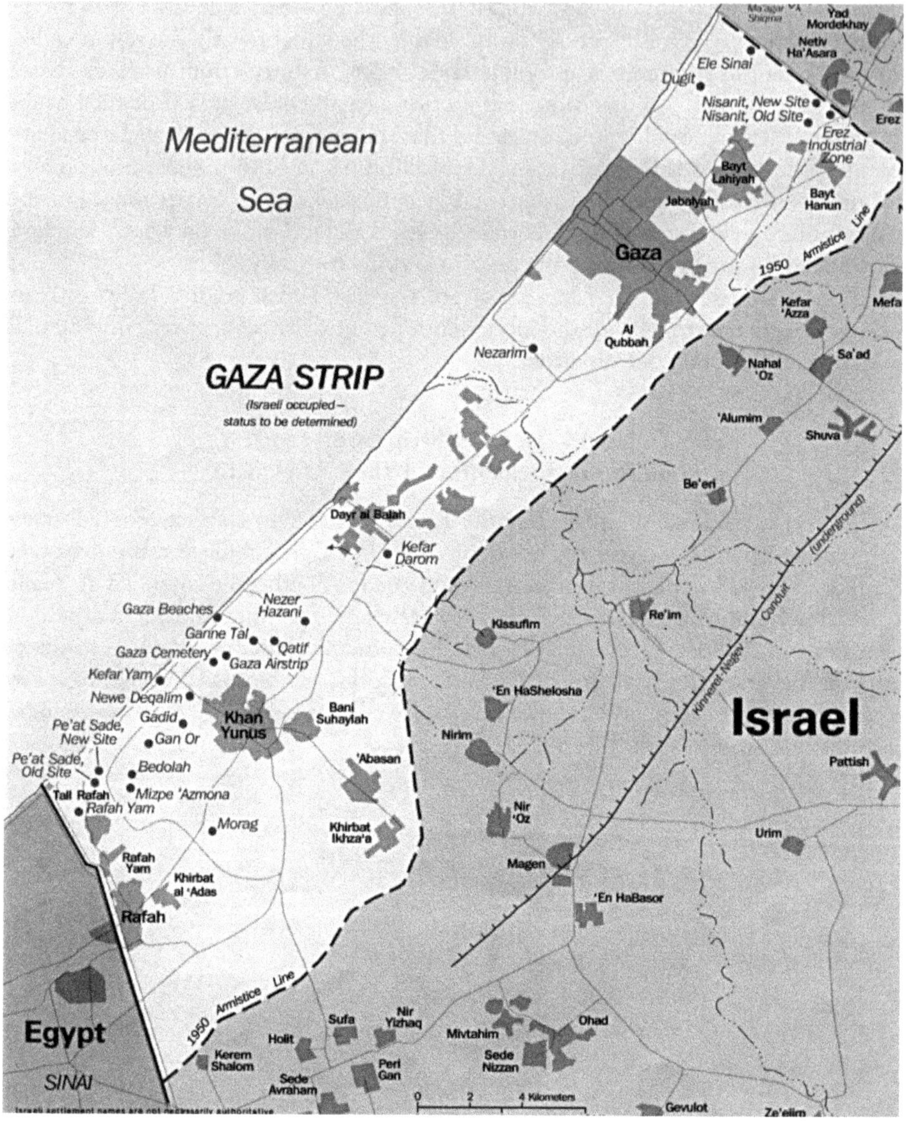


*Illustration 14: The Gaza Strip prior to Israel's Disengagement (2005)*[61]

As in the case of Lebanon, the exact time at which the decision to employ flechette rounds in the dense Gaza Strip was made is still unknown. However, we do know that the weapons' introduction into the Gaza Strip was accompanied by an internal IDF debate. As Steve Rodan, the well-connected *Jane's Defence Weekly*'s correspondent in Israel reported, while some generals pointed to the tactical advantages

61 The map is available at http://lib.utexas.edu/maps/middle_east_and_asia/gaza_1992.jpg.

offered by the use of these weapons, others "warn[ed] of an international backlash".[62] Ultimately, commanders were ordered to use the round "sparingly" and against targets such as mortar crews, targets that cannot be engaged effectively by "automatic fire" (i.e., coaxial machine gun fire). Military sources thereafter – as of May 2001 – insisted that the weapons have been employed on only a "handful" of occasions in the Gaza Strip.[63] As part of this debate, Israel's Military Advocate General (MAG) was requested to sanction employment of such a weapon in the Occupied Territories, which he did while instructing the operational command to employ flechette rounds "only after confirming the target's legality and verifying that no injury to civilian populations will take place."[64]

Concerned over the likelihood of the said "international backlash," IDF policy – which remained unchanged for the eight years following the Rimlawi incident – was to maintain as low a profile as possible regarding use of these weapons. The introduction of flechette rounds as a legitimate means of warfare in the Gaza Strip was therefore concealed from the Palestinian and the Israeli press alike. As the latter had generally relied heavily on information provided by the IDF Spokesperson's Unit in its reports on fighting in the Gaza Strip, the Spokesperson's announcement according to which Rimlawi was shot by soldiers using light weapons was quoted verbatim.[65]

In fact, only a week later, after Ziad Ayad, a Palestinian farmer, was killed on the very same road did the Palestinians themselves become aware of the newly employed weapon. As in Rimlawi's case, while the IDF maintained that "he was seen in an area where terrorists had planted roadside bombs in the past", his family maintained that he was making his way to the family farm. Irrespective of the uncertainty surrounding Ayad's presence on that road, the type of weaponry used was indisputable: flechette.[66]

The Israeli press was surprised as well. As Israel's leading newspaper, *Ha'aretz*, reported, apparently under heavy military censor's restrictions: "[a] Palestinian farmer... was killed by flechettes, dart-like projectiles and anti-personnel ammunition not yet seen in the Intifada…".[67] In its response, the IDF Spokesperson merely

62 Steve Rodan, "Israel's Military Debates...", *supra* note 6, at 4.

63 *Id.*

64 Rappaport, "They Walked in the Field", *supra* note 26.

65 *See,* for instance, Aluf Benn, "Sharon Questions IDF Plans to Quell Intifada", *Haaretz*, (Mar. 4, 2001) at A1 (in Heb.); Roni Shaked, *et al.*, "Hamas: 10 Suicide-Bombers are Just Waiting for the Order", *Yedioth Ahronoth*, (Mar. 4, 2001), at A11.

66 As reported by world media, X-rays from *Al-Shifa* hospital, Gaza's main hospital, showed "a nail-like projectile in his left eye, two in his pelvis and one in his left lung". *See, e.g.* "Sharon Points a Finger at Arafat", *CBS News*, (Mar. 11, 2001), *available* at <http://www.cbsnews.com/stories/2001/03/08/world/ main277261.shtml>.

67 *See*, "Palestinians Claim IDF Using Flechettes", *Haaretz*, (Mar. 12, 2001) (Internet ed., no longer available, on file with author). The heavy military censorship applied can be deduced from the absence of any mention of flechette weapons in the newspaper's main Hebrew edition, which simply reported on a Palestinian death from 'IDF fire'. See Amos Harel, "27-year old Palestinian Killed on the Karni-Netzarim Road", *Haaretz*, (Mar. 11, 2001), at A2. Reports of the use of flechette rounds in Lebanon were ascribed here to Palestinians sources. *See*, "Palestinians Claim…", *Id.*

stated that "the IDF uses means best suited to the overall security conditions and specific threats in the area" – Israel's standard response from then on regarding its use of flechettes.[68]

From that point, flechette weapons – possessed solely by Israel in the Mediterranean arena – have become effective weapons for the Palestinians in a no less important battlefield: the media. A day after Ayad's death (March 12), the Palestinian Authority hurriedly organized a press conference announcing that "new and dangerous nail-like ammunition" had been used by the IDF against Palestinians in the Gaza Strip.[69] This action was the beginning of continuous Palestinian attempts to attribute flechette use to Israel as often as possible. However, while a number of attempts throughout April-May failed to arouse the expected uproar in world opinion, an incident in June 2001 altered this state of affairs.[70]

### *CI. The Bedouin Camp – (June 2001)*

On the night of June 9, 2001, an IDF post at the northeast corner of the isolated Israeli settlement Netzarim, located in the outskirts of the city of Gaza, came under Palestinian fire. Due to the settlement's proximity to large Palestinian populations, large IDF forces (approx. one and a half brigades) guarded its defense perimeter with the assistance of tanks outfitted with flechette rounds. Once IDF lookouts had identified two figures, who seemed to be fleeing the area, three tank rounds were fired, two containing flechettes. Unfortunately, while the figures appeared to be some 600 meters away from the settlement, the rounds split open at 1400 meters, near a Bedouin camp in the Sheikh Ajlin neighborhood of Gaza. Three women from the same family were killed and two more critically wounded.[71]

Unlike the Rimlawi and Ayad cases, there was no question about the victims' status as non-combatants; a prompt investigation within the Southern Command was therefore opened. However, even before its conclusion, senior military sources

68 "Palestinians Claim...", *Id.*

69 The Palestinian Health Minister, Riyad al-Zanoun, described flechettes as "another addition" in the war that Israel is engaging against his people. *See,* "Nail-Like Ammunition Denounced..." *supra* note 60.

70 For such attempts *see*, for instance, "Palestinians Deny Knowledge of Weapons Boat Intercepted by Israelis", *BBC Monitoring Middle East,* (May 8, 2001) (citing the Palestinian news agency website, Wafa, Gaza, in Arabic May 7, 2001). According to the BBC report, which quoted the Palestinian news agency's announcement, the IDF shelled residential areas and a school in Khan Yunis with flechette rounds, killing a baby girl and wounding many others. A photo of a single flechette – supposedly taken out of a Palestinian victim after an incident on May 14, 2001 – was shown in Rodan's article even though there was no published record of such an incident. *See,* Rodan, "Israel's Military Debates..." *supra* note 6.

71 The victims were three members of the al-Malalha family: Selmiya, 37, Hekmat, 17, and Nasra, 61. For eyewitness testimonies as collected by HWR *see,* "Israel: Dart Shells Pose Civilian Threat", HWR Press Release, (June 15, 2001), available at <http://www.hrw.org/ar/news/2001/06/15/israel-dart-shells-pose-civilian-threat>.

announced that "lapses in communication" were to blame and admitted that "this was a serious, professional mistake which should not have happened."[72]

At the investigation's conclusion, the tank company's deputy commander was suspended as he had not been authorized to order the use of flechette rounds – an authorization then reserved for the battalion commander – and had ignored orders prohibiting flechette use beyond a range of 400 meters.[73] Other lapses in military procedures discovered included the failure of the senior commander to confirm the soldiers' findings before they fired and the absence of the Bedouin camp from any of the maps available to the soldiers.[74]

The General Officer Commanding (GOC) Southern Command at the time, Major General Doron Almog, was forced to submit an initial report to the Chief of Staff, Shaul Mofaz, and the Minister of Defense, Benjamin Ben-Eliezer. The latter openly apologized before the Palestinians; he also asked difficult questions of the IDF's senior command.[75]

In anticipation of the international criticism that would be directed at Israel and the IDF given the victims' identity (innocent civilians) in combination with the public uproar over this type of weapon, the IDF initially attempted to conceal its use of flechette rounds. The cover-up, however, became an embarrassment. Two days after the incident, a military official testified in the Knesset's Foreign Affairs and Defense Committee (FADC) that a "regular" (i.e., HESH round) was used. Yet, the following day (June 12) *B'tselem* announced that it had evidence (photos of the wounds indicating darts hits in addition to darts found at the scene) for flechette use and requested the IDF to immediately desist employing the weapons.[76] The IDF Spokesperson was forced to announce, without referring to the specific incident, that "the use of various types of ammunition is dependent on the decisions of the commanders in the field and according to the threats facing them... *In certain instances, flechette shells have been used*".[77] As reporters stressed, it was the first time the IDF admitted using flechette rounds in the Gaza Strip.[78]

---

72 Amos Harel, "Probe Slams IDF Tank Killing of 3 Palestinian Women in Gaza", *Haaretz*, (June 13, 2001), at A2. In effect, while those suspected of carrying out the shootings were located a few hundred meters from the tank, the tank crew mistakenly understood that the targets were figures located at a distance of 1400 meters. The findings of the investigation were transmitted by the Defense Minster, Benjamin Ben-Eliezer to MK Ran Cohen, a member of the Knesset's Foreign Affairs and Defense Committee and reproduced in Akiva Eldar, "The Investigation Was a Success, the Bedouin Women Died", *Haaretz*, (Oct. 25, 2001), at B3.

73 *See*, Aryeh Bendar *et al.*, "IDF Admits: We Lied to the MK's", *Ma'ariv*, (June 13, 2001), at A6; Roni Shaked *et al.*, "Mistaken Identity: Terrorists Shot to Death a Greek Monk", *Yedioth Ahronoth*, (June 13, 2001), at A9.

74 Akiva Eldar, "The Investigation Was a Success...", *supra* note 72.

75 Amos Harel, "Probe Slams IDF Tank Killing...", *supra* note 72. See also Amir Rappaport *et al.*, "IDF Embarrassed by the Killing of Three Bedouin Women in the Gaza Strip, *Yedioth Ahronoth*, (June 13, 2001), at A3.

76 Arieh O'Sullivan, "IDF Admits Using Flechette Shells Against Palestinians", *The Jerusalem Post*, (June 13, 2001), at A3.

77 Emphasis added. *Id.*

78 *Id.*

However, although the affair ended in Israel when the Chief IDF spokesman, Brigadier-General Ron Kitrey apologized to the FADC's members about the initial erroneous report while stressing that the error was unintentional, outside Israel it had only begun.[79] As expected, the Palestinians took advantage of the incident with considerable alacrity, accusing Israel of using various weapons against Palestinian civilians, particularly flechette rounds, forbidden by "international conventions."[80] In these circumstances, however, they were indeed able to stir up international interest.

Objective news accounts of the incident vied for international headlines focusing on the unusual type of weapon used.[81] As the BBC reported "[t]he women were killed by a type of shell which sprays nail-like darts called flechettes, currently banned in international conflicts."[82] Israeli journalists similarly highlighted the problematic nature of flechette weapon use in the Gaza Strip and reported on the harsh internal criticism of the IDF, which focused on the fact that this weapon was banned in Judea and Samaria. [83] Other senior IDF officers stated unanimously that tanks should not be employed in Gaza at all for "when you fire shells into such a densely populated area, it is no surprise that innocent civilians are harmed."[84] Some reporters even adopted the Palestinian interpretation (similar to that of UNIFIL in Lebanon) regarding the weapon's illegality[85] or simply noted that "the use of the flechette, which scatters the arrows around the target rather than hitting it precisely, is much more problematic in Gaza, which is a densely-populated area."[86]

In response, Prime Minister Ariel Sharon expressed his sorrow at the deaths of the women. However, the Military Advocate General (MAG),[87] Major-General

79 Bendar, *et al.*, "IDF Admits…", *supra* note 73.

80 *See, e.g.*, accusations blasted by the Palestinian Security Forces' Head of the northern Gaza Strip, Saeb al-Ajaz as cited in Harel, "Probe Slams.." *supra* note 72. *See,* also Al Ayyam staff, "Injured By Flechettes", *Palestinian Report Online* 8(2), (Jun. 22, 2001), in which it stated "[t]he Aqsa Intifada has been the first occasion in which Israel has used this internationally banned ammunition" (on file with the author). The *Palestinian Report* is published by the Jerusalem Media and Communications Center (JMCC).

81 For news reports lacking any discussion of the type of weaponry employed, *see, e.g.*, Howard Schneider & Lee Hockstader, "Mideast Truce Begins on Tentative Footing", *The Washington Post*, (June 14, 2001), at A1.

82 "Israeli Press Skeptical on Truce", *BBC*, (June 13, 2001), available at <*http://news.bbc.co.uk/2/hi/middle_ east/1386345.stm*>. The official Chinese news agency, for instance, *XINHUA*, began its report by flatly declaring: "The Israeli army has used flechette shells, a highly controversial weapon." *See*, "Israeli Army Admits Using Flechette Shells", *XINHUA*, (June 12, 2001), EMEDIA-ACC-NO: 200106125092.m90. The claim by an IDF official that "the shells were only used on rare occasions…the soldiers chose the flechette shells, rather than a regular tank fire [*sic.*], due to 'its accuracy'" are found in the Chinese report encased in quotation marks.

83 S*ee*, Roni Shaked *et al.*, "Mistaken Identity…", *supra* note 73 .

84 *See*, Amir Rappaport *et al.*," IDF Embarrassed…", *supra* note 75.

85 *The Jerusalem Post*, for instance, began its report with these words: "The IDF admitted yesterday that in fighting the Palestinians its tanks have used flechette shells, which are implicitly banned by international conventions". O'Sullivan, "IDF Admits…", *supra* note 76.

86 Harel, "Probe Slams…", *supra* note 72.

87 *See, e.g.*, HRW, "Israel: Dart Shells Pose Civilian Threat...", *supra* note 71.

Menachem Finkelstein, succumbed to public pressure and press reports citing the call made by senior military officials for a military police investigation "in light of the severe findings and failures involved in the rounds' firing"[88] by in fact initiating one. This was the very first such investigation since fighting with the Palestinians had renewed in September 2000.[89]

The investigation was welcomed by human rights groups, such as HRW, which called for Israel's commitment not to use such weapons in or around populated areas. While it did acknowledge that "flechette ammunition is not banned under international law," the HRW pointed out that "Anti-personnel weapons that have a large 'kill radius' must never be used in areas where civilians live." Employing flechette rounds in a place like the Gaza Strip, the argument notes, is highly questionable since it raises the risk of civilian casualties to a threshold that is intolerable under international law.[90]

On August 27, 2001, two months after the incident, *B'tselem*, having adopted a position similar to that of the HRW (i.e., the flechette itself is legal but "when used in populated areas it becomes an indiscriminate weapon"),[91] requested that the Minister of Defense prohibit flechette use in the Occupied Territories.[92] On November 21, 2001, the Assistant Minister of Defense, Haim Israeli, replied as follows: "the very use of flechette rounds in the field is not forbidden. Regarding the circumstances in which these rounds were used – the IDF is aware of the need to balance between military needs on one hand with humanitarian concerns and minimization of damage to the civilian population on the other. The policy regarding the use of various types of weapons is determined by the balance achieved between these considerations."[93]

### *CII. The Three Youths – (December 2001)*

Some months after the Ministry of Defense's response, the flechette weapon again made headlines, albeit with less impact since it was unclear whether the victims were non-combatants. On December 30, 2001, three Palestinian youths (Muhammad Ahmad Lubad, Muhammad 'Abd a-Rahman al-Madhun; and Ahmad Muhammad Banat) were making their way from the Palestinian town of Beit Lahiya toward the Israeli settlements Elei Sinai and Dugit at the northern tip of the Gaza Strip. Their reasons for being in the area at night and their destination remained unclear; the IDF and the Palestinians, as well as human rights NGOs were thus able to offer con-

88 *See,* Eitan Rabin, "Military Police Investigation will be Opened for Incident of Tank Fire that Killed Three Women in Khan Yunis", *Ma'ariv* (June 22, 2001), at A6.

89 Amos Harel, "An Investigative Officer will be Appointed for the Death of Three Bedouin Women from Shooting in the Gaza Strip", *Haaretz* (Aug. 19, 2001), at A1.

90 HRW, "Israel: Dart Shells Pose Civilian Threat", *supra* note 71.

91 According to the NGO spokesperson, flechette use is forbidden according to Article 51 of AP-I. *See,* O'Sullivan, "IDF Admits…", *supra* note 76.

92 *See,* letter written by Yehezkel Lein, *B'Tselem*, to Benjamin Ben Eliezer, Minister of Defense (Aug. 27, 2001), *B'Tselem* Archive.

93 *See,* letter written by Haim Israeli, Office of the Minister of Defense, to Yehezkel Lein, *B'Tselem* (Nov. 21, 2001), 12400/*Koph*, *B'Tselem* Archive.

flicting explanations. The IDF considered them a "terrorist squad on its way to carry out an operation" against either settlements' residents or the IDF forces protecting them. The IDF buttressed its conclusion by noting, among other things, the "suspicious way" the youths were moving, the two knives found by their bodies (though there were never presented), and their reactions to the IDF's presence once they were spotted. The alternative argument made by the Palestinians was that as residents of the city of Gaza, the youths did not realize that at a certain point they had crossed a quasi-territorial boundary separating Beit Lahiya and the Israeli settlements from Gaza; where IDF forces were free to target anyone crossing it. As an Israeli reporter wrote, "it is not a defined line… but whoever cares about his life in Beit Lahiya does not approach it."[94]

As documented by an IDF night surveillance camera 135 meters south of the Elei Sinai settlement, the three youths were moving toward the settlement's periphery close to an IDF tank stationed to protect the settlement when they broke into a run once the tank's engine was ignited – required to employ its main gun – before being hit by the flechettes.[95] The first youth suffered injury from a large number of darts, which penetrated his body from the back. An autopsy by Israel's National Center of Forensic Medicine (NCFM) revealed no less than 71 dart wounds. As the Chief Medical Examiner and the Center's Director, Prof. Yehuda Hiss, concluded, his death was caused "from serious injuries to the brain from the path made by a metal dart through the head".[96] The second youth suffered eight dart hits, most of which entered through the back into the lungs, leading to a massive and fatal loss of blood.[97] The third Palestinian (Ahmad Banat) suffered from four darts in his right lung and back, two darts entered his back and two entered his left forearm.[98]

The darts' locations and their paths in the last victim's body acquired particular importance once it was revealed that after he was hit, he was run over by the IDF tank, "unintentionally" as the IDF later maintained.[99] His multiple wounds barred the pathologist from determining the cause of death "at the necessary level of certainty."[100] Furthermore, the bodies' condition had aroused suspicions that they

94 Meron Rappaport, "They Walked in the Fields", *supra* note 26. For the IDF version of the incident see also HCJ 8990/02, *A Response by the State Attorney's Office*, para. 46, available at <www.btselem.org/ hebrew/Legal_Documents/8990_ 02_Flechette_ Appeal_State_Response.pdf >.

95 The IDF movie was presented to the Knesset Committee on Foreign Affairs and Defense on Jan. 22, 2002. The meeting's minutes were handwritten by a representative of Physicians for Human Rights-Israel (PHR-I), PHR-I Archive.

96 *See,* National Center of Forensic Medicine, Rep. 3/2295/2001, (Jan. 16, 2002), at 9.

97 National Center of Forensic Medicine, Rep. 3/2294/2001, (Jan. 20, 2002), at 9.

98 National Center of Forensic Medicine, Rep. 3/2293/2001, (Jan. 16, 2002), at 9.

99 According to the IDF, "during the charge against the terrorists, one of them was unintentionally run over by a tank whose operators could not see him." For this version – provided following multiple NGO's requests to clarify the tread marks on the body, *see,* letter from Southern Command Advocate, Lieut.-Col. Baruch Mani, to PHR-I, Aug. 21, 2002, PHR-I Archive.

100 Rep. 3/2293/2001, Jan. 16, 2002, *supra* note 98, at 13. For Prof. Hiss's different opinion *see,* the meeting's minutes, PHR-I Archives.

had been abused. This as well as the suspicion that the third youth was crushed alive by the tank led an Israeli left-wing MK (Tamar Gozansky, Hadash) to summon a special meeting of the Knesset's FADC to discuss the suspected mistreatment on December 30, 2001. After all, the IDF's removal of Palestinian corpses from a site (instead of the common procedure of calling a Palestinian ambulance for their evacuation), especially their transfer to Israel – supposedly for purposes of identification – was unprecedented.[101] Furthermore, a series of puzzling queries over this incident attracted the attention of Israeli human rights NGOs, mainly Physicians for Human Rights–Israel (PHR-I).[102] The most problematic queries concerning Israel's conduct in this affair were directed at the reasons for the autopsies, the attempt to conceal the transfer from the youths' families and the IDF's reluctance to accept assistance from either the ICRC or the Palestinian Authority during identification of the bodies – the formal reason for the transfer.[103]

On January 22, 2002, an extraordinary meeting took place in the presence of two representatives from Israeli human rights NGOs, PHR-I and the Public Committee against Torture in Israel (PCATI).[104] *The issue of flechette weapons – it should be stressed again – was not on the agenda.* Yet, Israel's Health Minister Nissim Dahan, who attended the meeting, while acknowledging indications of "serious crushing," reiterated Prof. Hiss's opinion that death itself was caused by the "penetration of nails and foreign objects" or "winged nails", as Hiss himself described them.

These "winged nails" puzzled several of the participants, who were unaware of the existence of such weapons. Colonel Rannam Hamada, the Northern Gaza Strip Brigade Commander, therefore felt compelled to clarify the matter. His explanation motivated the following comment in the press: "Perhaps Hamada has not noticed that the *flechette is not among the issues dealt with in public.*"[105]

The press coverage and commentary was effective: Flechette weapons returned to center stage, accompanied by acerbic criticism. The December 2001 incident became the subject for an in-depth Israeli newspaper report in which tank crews testified to the fact that "flechette is quite broadly used in the Gaza Strip."[106] One journalist noted the IDF's past denial about the use of flechette weapons and criti-

101 *See,* Nina Gilberg, "IDF: 3 Palestinians Killed were Hit by Shell", *The Jerusalem Post,* (Jan. 25, 2002), at A7.

102 For more on this NGO see its website, available at <http://www.phr.org.il/default.asp?PageID=4>.

103 A list of eight "serious concerns about the conduct of Israeli authorities in treating the issue", is included in a letter from Dr. Ruchama Marton, the President of PHR-I, to MK David Magen, Chairman of the Knesset FADC, entitled "Questions Left Unanswered Regarding the Death of Three Youths From Sheikh Radwan in Gaza Strip", Feb. 14, 2002, PHR-I Archive.

104 For more on this NGO see its website, available at <http://www.stoptorture.org.il/en>.

105 Emphasis added. Rappaport, "They Walked in the Field" *supra* note 26.

106 A tank crewman testified that during his three-month tour of duty in the Southern Gaza Strip, he fired only three flechette rounds, as opposed to 30 "regular" rounds. However, he stated that "tanks stationed closer to the Strip demarcation fence fired many more." *Id.*

cized what he perceived as deterioration in the IDF's moral code, reflected, *inter alia*, in its readiness to use such weapons in the Strip.[107]

At this point, PHR-I decided to take the lead in the campaign against use of these weapons. On February 14, 2002, it asked the Chairman of the Knesset FADC, MK David Magen, to summon another meeting to further discuss the puzzles mentioned. But this time it explicitly raised the weapons issue while specifically mentioning an international ban on flechette missiles.[108] On that very same day, the weapons returned to the headlines as three Palestinian policemen in Deir el-Balah were killed by these rounds.[109]

MK Magen was not the sole official to receive PHR-I correspondence that day (Feb. 14, 2002); the Minister of Health also received a letter,[110] as did the MAG, three months later (May 12).[111] Yet, however incisive their content, the letters did not address the weapons themselves but only matters strongly connected to their medical implications, issues that physicians might naturally address.[112]

For the next six months, correspondence continued over those questions between PHR-I and the Ministry of Health as well as the MAG.[113] Yet, while both provided explanations for the issues within their particular domains, no response was received from MK Magen over the flechette use.[114] The PHR-I then turned once more to the MAG (on July 22, 2002) with direct queries over the employment of "flechette missiles for killing the youths – use banned by International Law."[115] A month later, on August 21, 2002, the Southern Command Judge Advocate, Lt.-Col. Baruch Mani,

107 "Using them [flechette rounds] against people is a violation of the Geneva Convention. But bits of metal don't disintegrate in a body". Gideon Samet, "You Won't be Able to Say We Didn't Know", *Haaretz*, (Jan. 28, 2002), at 5.

108 *See*, letter from Dr. Ruchama Marton to MK David Magen, *supra* note 103.

109 *See, e.g.*, Amos Harel & Uri Nir, "Palestinians: IDF Use Forbidden Shells", *Haaretz*, (Feb. 14, 2002), at A3. For the international echo *see*, e.g., Ben Lynfield, "Straw Backs US Stance as Israeli Tanks Pound Gaza", *The Scotsman*, (Feb. 14, 2002), at 10.

110 Letter from Ms. Hadas Ziv, PHR-I Project Director, to Dr. Isaac Berlovich, Office of the Associate Director-General, Ministry of Health, entitled "Questions Left Unanswered Regarding the Death of Three Youths From Sheikh Radwan in Gaza Strip", (Feb. 14, 2002), PHR-I Archive.

111 Letter from Ms. Hadas Ziv, PHR-I Project Director, to the Chief Military Advocate, Major-General Menachem Finkelstein , (May 12, 2002), PHR-I Archives.

112 According to the PHR-I website's "Mission & History"' section, "PHR-I was founded in 1988 with the goal of struggling for human rights, in particular the right to health, in Israel and the Occupied Territories". PHR-I website, available at <http://www.phr.org.il/default.asp?PageID=145>.

113 For the exchange of letters *see*, letter from Associate Director of the Health Ministry, Dr. Yitzhak Berlovich, to the NGO, (Feb. 19, 2002); letter from the Chief Military Advocate for Advice and Legislation, Hani Kaspi, to the NGO (June 26, 2002).

114 On Feb. 18, 2002 Mr. Assaf Friedman, Assistant to the Knesset FADC's Chairman, responded that the committee had yet to conclude its investigation into the matter and therefore did not have "all the answers to some of the queries". Letter from Assaf Friedman to Dr. Ruchama Marton, the President of PHR-I, (Feb. 18, 2002), PHR-I Archive.

115 Letter from PHR (Project Manager, Ms. Hadas Ziv) to Chief Military Advocate for Advice and Legislation, Colonel Hani Kaspi, (July 22, 2002), PHR-I Archive.

responded: "To the best of our understanding, the use of 'flechette' rounds is not forbidden by international law."[116]

### *CIII. The al Hajan family – (August 2002)*

Although the IDF may have assumed (or hoped) that Lt.-Col. Baruch Mani's response might close the inquiry, just a week later, on August 28, 2002, the weapons once again returned to the public agenda in the context of the severe harm caused to innocent civilians – this time with both sides agreeing over the circumstances.

On that night, the al Hajan family had slept in its fig orchard in the Sheikh Ijleen area of Gaza. The family's permanent residence was in the Jabalya refugee camp but during the harvest season for figs (August-October), the family slept in a small shed on the orchard's perimeter in order to bring the fruit to market at dawn. The orchard was located about 250 meters north of an IDF outpost guarding the Israeli settlement of Netzarim.

Due to its proximity to the settlement, the entire area was designated by the IDF as a "Special Security Zone" (SSZ), in which movement at night was prohibited. A tank was stationed in a nearby guard post facing the direction of Gaza, with the soldiers ordered to open fire on any suspicious figures (i.e., having the means to or showing intent to harm IDF soldiers or the settlements' residents). Work in the orchard during the day was permitted.

Due to the typically heavy August heat, most of the family slept in huts *outside* their small permanent shed, a shed that could have provided protection against the flechettes. This habit, in conjunction with suspicious movement in the orchard (the IDF later claimed that crawl marks and a cellular phone were found), sealed their fate. The shower of darts left four family members dead and eight others wounded (including a three-year old child), some in critical condition.[117]

On the Palestinian side, the outcry was severe. Palestinian President Yasser Arafat called the incident a crime aimed at sabotaging peace efforts while the militant Hamas movement threatened revenge.[118] On the Israeli side, the Minister of Defense, Ben-Eliezer, rushed to apologize and instructed the IDF to "conduct a comprehensive investigation on the event, and present him with its findings."[119] IDF

116 Letter from Southern Command Advocate, Lt-Col. Baruch Mani (to whom Colonel Kaspi had referred the question over the weapons' legality) to PHR-I, (Project Manager, Ms. Hadas Ziv), Aug. 21, 2002, PHR-I Archive.

117 For a detailed report *see*, e.g., Danny Salomon *et al.*, "Hamas: We Will Take Revenge on Israel for the Death of the Family in Gaza", *Yedioth Ahronoth*, (Aug. 30, 2002), at 2; Eli Bohadana *et al.*, "A Palestinian Woman and her Two Sons Killed by an IDF Shell", *Ma'ariv*, (Aug. 30, 2002) at A6.

118 Mark Lavie, "Israeli Regret Does Nothing to Ease Tension", *The Advertiser*, (Aug. 31, 2002), at 53.

119 Amira Hass, "Hamas Threatens to Take Revenge for the Death of Four Family Members in the Gaza Strip", *Haaretz*, (Aug. 30, 2002), at A6.

officers in Gaza expressed their sorrow at the "harm to innocent civilians" while emphasizing that its forces had acted appropriately.[120]

While the incident's timing, three months after Operation *Defensive Shield*, during which Israel had re-taken most West Bank's cities, when the shadow of continued suicide bombing prevented widespread local public criticism,[121] Israel found itself in rather unpleasant straits within the international arena.[122] In Britain, criticism was especially harsh.[123] The British Foreign and Commonwealth Office hurriedly voiced "serious concern" over the incident, while Members of Parliament and human rights NGOs called for this weapon's ban on the grounds that it was indiscriminate in its effects.[124] The BBC – fully aware of the curiosity these weapons aroused – made sure to present a detailed diagram of the weapon and its mode of operation.[125] Some newspapers declared that Israel's use of such weapons made its complaints about Palestinian suicide bombers, with their practice of packing their bomb-belts with nuts and bolts to maximize injuries among Israeli civilians, appear hypocritical.[126] "It was a terrible weapon that took the lives of four fruit-pickers from a Palestinian family yesterday" reported The *Daily Telegraph*. The newspaper went on to clarify for its readers that: "The flechette is not a banned weapon, but it is not used by NATO forces. The Israeli military has assured foreign governments that the weapon is employed only when there is clear evidence that the target is armed men."[127]

In the U.S., a feminist NGO, *Partners for Peace* (PFP), began a campaign to stop Israel's use of the weapon by persuading members of Congress to pressure the Bush Administration.[128] U.S. citizens were called upon to contact their Congressional

---

120 *See,* Ali Waked and Felix Frisch, "Fuad Apologizes for Death of the Palestians; IDF Soldiers Acted Appropriately", *Ynet,* (Aug. 29, 2002), available at <www.ynet.co.il/articles/1,2506,L-2089453,00.htm>.

121 For criticism within Israel, albeit of a limited nature, *see, e.g.*, Gideon Levy, "A Passion for the Grave", *Haaretz,* (Sept. 16, 2002), at 5; Gideon Levy, "Festivals of Indifference", *Haaretz,* (Sept. 23, 2002), at 5. "Under cover of the darkness that has descended on us, in which everything is permitted" – wrote this columnist – "and for a tank to attack a family working in a vineyard with flechette shells…"*Id.*

122 *See, e.g.*, Alan Philps, "Israel used Flechettes in Killing of 4 in Family", *The Gazette* (Quebec), (Aug. 30, 2002), at B1.

123 As written in the *Daily Mail,* "Britain has attacked Israel's use of a controversial 'killer dart' weapon system…". *See,* "Britain Condemns Israel's 'killer dart'", *Daily Mail* (London), (Aug. 31, 2002), at 2.

124 *See,* "Britain Condemns...", *Id.* Labor MP Gerald Kaufman even announced that "[t]o think of the Israeli Army carrying out this kind of murderous attack makes one want to vomit". *See,* Michelle Stanistreet, "So Was Rabbi Sacks Right to Hit Out at Israel's Iron Fist Policy", *Sunday Express,* (Sept. 1, 2002), at 13.

125 For the BBC report, *see,* "Dart Bombs Killed Four Palestinians", *BBC World News,* (Aug. 30, 2002), available at <http://news.bbc.co.uk/2/hi/middle_east/2225706.stm>.

126 Phil Reeves, "Israeli Dart Shells Mirror Suicide Bomber Tactics", *The Independent* (London), (Aug. 30, 2002), at 11.

127 Alan Philps, "Israel's Deadly Darts Wipe out Family – Inquiry Ordered by Minister After Tank Opens Fire on Palestinian Fruit-Pickers sleeping in Orchard", *The Daily Telegraph* (London), (Aug. 30, 2002), at A14.

128 The NGO, defined by right-wing Jewish and Israeli NGOs as a "Palestinian-allied nonprofit group", known for lobbing Washington in favor of an Israeli withdrawal from

representatives while agreeing with the NGO's view, flechette rounds were "clearly a grisly weapon, shooting out thousands of metal darts, obliterating all life in their path. Use in one of the most densely populated places on the face of the planet, Gaza, is grossly irresponsible."[129]

After Israel's political and military officials apologized and expressed sorrow at the death of the al Hajan family members, the issue faded from public view Yet, for human rights NGOs and especially the PHR-I, the Military Advocate's August 2 response, was no longer acceptable. Furthermore, given the Israeli public's indifference over this issue, the legal arena seems to offer the only feasible way to avoid additional "collateral damage" – as detailed in the following chapter.

---

the Occupied Territories, identified itself as providing "a public platform for the voices of Israeli and Palestinian women working for a peaceful and just resolution to the Israeli-Palestinian conflict". *See,* <http://www.partnersforpeace.org/. For its classification see http://www.discoverthenetworks.org/groupProfile.asp?grpid=7236>.

129 For the NGO's call, dated Sept. 6, 2002, *see* <http://partnersforpeace.org/actionalerts/db20020960/> (no longer available, on file with the author).

# Chapter 5

## The Internal Legal Battle

This chapter, devoted to the internal legal dispute in Israel, relies heavily on the general background provided in Chapters 1 and 4. The discussion begins with a survey of the Petitioners' claims regarding international law that were raised before the HCJ in October 2002 and continues with claims that were raised by the Respondent, the State of Israel. Given its length, the survey is divided into sections while comments – legal and factual alike – are spread throughout the corresponding sections. In the first, an overview and comments are presented regarding the Petitioners' claims that the use of flechette weapons in the Gaza Strip is prohibited under international law; since the Petitioners resorted to Israeli national law, a brief overview and comments are provided as well. A similar overview and comments follow regarding the State's attempt to demonstrate that neither conventional rule, nor international community practices, support the Petitioners' claims about the weapon's illegality. In other words, the existing treaty law concerning means of warfare as well as the practice of other states in the matter refute the Petitioners' claim in these respects. A summary of the April 13, 2003 judgment is then briefly provided followed by an in- depth discussion of the HCJ ruling handed down a fortnight later. Two substantive comments on this short ruling, one addressing a factual flaw and the other addressing a legal flaw, conclude this chapter.

### A. Prelude

Within two days after the al Hajan incident (Aug. 28, 2002) – in which four civilians were killed by flechette rounds – Physicians for Human Rights-Israel (PHR-I) sent an urgent letter to Elyakim Rubinstein, then Israel's Attorney General (AG), stating the following:

> As is known the use of the flechette missiles under such circumstances is prohibited by international law, as it inflicts unnecessary suffering (the rounds contain thousands of deadly metal darts), and its employment is not precise and harms innocent civilians.[1]

1 Letter from PHR-I (Tomer Feffer, Director General and Ruchama Marton, President) to Attorney General Elyakim Rubinstein, (Aug. 30, 2002), PHR-I Archive. The appearance of the incorrect term "missile" instead of "round" when describing the weapon used indicates the magnitude of PHR-I ignorance regarding flechette weapons. A copy of the letter was attached to the Petition, *See, Physicians for Human Rights & The Palestinian Centre for Human Rights v. O.C. Southern Command & The State of*

However, PHR-I then reconsidered its former correspondence with the military establishment (i.e., December 2001 incident, *supra* Chapter 4) in which it cast doubts, *inter alia*, over the weapons' legality as hard evidence that the establishment's leadership, the Defense Minister and the Commander in Chief, were aware of the use of "illegal weapons and even authorized its continuous use". Hence, it requested that the AG instigate an investigation into the IDF use of flechette rounds and charged the Defense Minister and other IDF commanders with the use of an illegal weapon.[2]

Coincidentally, the next day (August 31), PHR-I finally received Baruch Mani's letter (dated August 21), in which he noted that the IDF understood flechette rounds to be legal weapons (*supra* Chapter 4, CII). In response, an urgent letter was sent on September 1 to Defense Minister, Ben-Eliezer in which it contested this "understanding," claiming that the weapon's use "inside the Occupied Territories" constitutes a "grave violation" of a cardinal principle of IHL: "the prohibition to employ means which have an indiscriminate effect."[3] The three incidents ("all known cases of flechette use"), in which civilians were killed by the dispersion of darts over a wide area, proved that the IDF used flechette rounds "in area in which it has no possibility of distinguishing between civil population and enemy forces."[4] Furthermore, claiming that the use of flechette rounds had violated the second cardinal principle governing weapon use – avoidance of unnecessary suffering or superfluous injury – PHR-I noted their skepticism over the IDF's choice of weapons for "targeting terrorists," as the IDF had previously explained. PHR-I concluded its harsh letter by demanding that Ben-Eliezer instruct the IDF to stop using the weapon and threatened to consider petitioning the appropriate legal instances if no reply was received.[5]

Not only did Ben-Eliezer ignore the PHR-I's request, a confirmation letter was sent only two months later by one of his assistants.[6] Given such a disparaging response, PHR-I could not back down from its threat to take legal action.

---

*Israel* – Ministry of Defense, Petition for an Interim Decree, Annex F. available at <http://www.btselem.org/hebrew/Legal_Documents/8990_02_Flechette_Appeal.doc> [hereinafter, *The Flechette Petition*].

2 *Id.*

3 Interestingly, although this was later changed in the Petition, PHR-I focused on the issue of the weapon's indiscriminate effects rather than unnecessary suffering. *See,* also a 2001 paper which a PHR-I International Legal Adviser and a Project Director presented a few days before the Petition was submitted. Lubell Noam, "Who's the Target? The Blurring of Distinction Between Civilians and Combatants during IDF Ground Operations in Urban Environments", at 5. *Working Papers, Ground Operations Workshop* (Oct. 17-18, 2002), Carr Center for Human Rights, Harvard University, available at <http://www.hks.harvard.edu/cchrp/ programareas/conferences/october2002.php>.

4 Letter from PHR-I (Hadas Ziv) to the Minister of Defense, Ben-Eliezer, Sept. 1, 2002, PHR-I Archives. A copy of the letter was attached to the Petition. *See, The Flechette Petition*, Annex G.

5 *Id*, at 2. Copies were sent to then IDF Chief of Staff, Moshe Yaalon, O.C. Southern Command, Doron Almog, and Southern Command Judge Advocate, Lt.-Col. Baruch Mani.

6 *See,* letter from Assistant Minister of Defense Adv. Ruth Bar, to PHR-I (Hadas Ziv), (Oct. 3, 2002), see PHR-I Archive.

On October 23, 2002, the PHR-I, along with the Palestinian Centre for Human Rights (PCHR), a Gaza city-based NGO[7] petitioned the Israeli Supreme Court, sitting as a High Court of Justice (HCJ), against then Southern Command's GOC, Doron Almog (Respondent I) and the State of Israel, Minister of Defense (Respondent II).[8] The Petition demanded to know why the use of flechette rounds "*in the context of IDF operations in the Gaza Strip area*" should not be prohibited.[9]

## B. Petitioners' Claims – Overview and Comments

### *BI. The Concept of "Three Layers"*

The Petition[10] relied on what the Petitioners perceived as three layers of IHL:

(a) The first, most basic and extensive layer prohibits attacks against "civilian targets, including civilian [persons].";[11]

(b) The second layer, articulated in Article 3 and common to the Geneva Conventions (1949),[12] notes the obligation that all non-combatants ("persons taking no active part in hostilities") shall be treated humanely in all circumstances. Moreover, "violence to life and person" shall be prohibited at any time and in any place.[13]

---

7 PCHR was founded in 1995 by a group of Palestinian lawyers and human rights activists. PHR-I decided to jointly petition with PCHR for two reasons: one ideological and one practical. Ideologically, PHR-I had given much importance to solidarity "between NGOs working toward similar aims" and felt that "having the petition include those who are most directly affected by it" was the correct manner to work. Practically, PCHR was better equipped to collect testimonies and obtain relevant data given its location in the Gaza Strip. *See,* email correspondence with Noam Lubell, former International Law Adviser and Project Director, Prisoners and Detainees, PHR-I, to the Author (Dec. 21, 2006) (on file with author). For the PCHR's practical advantage *see*, e.g., local medical reports which were sent from Adv. Iyad Alami, Legal Unit, PCHR, To Hadas Ziv, Director of the Occupied Territories Project, PHR-I, (Feb. 23, 2003) (on file with the author).

8 *The Flechette Petition.*

9 Emphasis added. *Id*, at 1.

10 Brought before HCJ Justice, Dorit Beinisch (Supreme Court president as of Sept. 2006).

11 *The Flechette Petition*, para. 22. Since the relevant Petitioners' arguments (i.e., concerning the cardinal principles which govern the weapon's legality in international law) were raised within this layer of IHL, this layer is discussed in detail below.

12 *See,* Convention (I) for the Amelioration of the Condition of the Wounded and Sick in Armed Forces in the Field. Geneva, 12 August 1949, Dietrich Schindler and Jiri Toman, (eds.) *The Laws of Armed Conflicts*, (Leiden: Martinus Nijhoff, 2004), 459, at 461-62 [hereinafter, *Laws of Armed Conflicts]*; Convention (II) for the Amelioration of the Condition of Wounded, Sick and Shipwrecked Members of Armed Forces at Sea. Geneva, 12 August 1949, *Laws of Armed Conflicts*, 485, at 487-88; Convention (III) relative to the Treatment of Prisoners of War. Geneva, 12 August 1949, *Laws of Armed Conflicts, at 507*; Convention (IV) relative to the Protection of Civilian Persons in Time of War (hereinafter, Geneva Convention (IV)). Geneva, 12 August 1949, *Laws of Armed Conflicts, 575 at 580-81.*

13 *The Flechette Petition*, paras. 25-27. Common Article 3 (which appears with identical language in each of the four 1949 Geneva Conventions) provides, *inter alia*, that: "In

(c) The third layer, which comprises "the rules for the control of occupied territory", grants the occupied population – as the Petitioners made sure to stress – special protection *in addition* to the rights and protections that are accrued from the general rules mentioned above.[14] The Petitioners relied mainly on the 1949 Fourth Geneva Convention[15] while pointing out Israel's "commitment to respect the humanitarian rules of the Convention".[16] In addition, they cited Article 46 of the *1907 Hague Regulations,*[17] concerning "[m]ilitary authority over the territory of the hostile state," according to which "*the lives of persons*, and private property ... must be respected".[18]

However, before discussing the first layer of IHL, the Petitioners interestingly cited Israeli national law as well. Under this rubric the Petitioners sought to review the legality of flechette use in two branches of law: (1) Israeli constitutional law, which they supported with an analysis of the Basic Law: Human Dignity and Liberty,[19] noting that the law denotes Israel's "active obligation to protect the life of any person..."

---

the case of armed conflict not of an international character occurring in the territory of one of the High Contracting Parties, each Party to the conflict shall be bound to apply, as a minimum, the following provisions...".

14 *Id.*, para. 28.

15 Geneva Convention (IV), *Laws of Armed Conflicts, 575.*

16 *The Flechette Petition*, para. 32. Geneva Convention (IV) – along with the other three conventions – was already signed and ratified by Israel in 1951. Israel has never explicitly recognized the applicability *de jure* of the Fourth Geneva Convention to the territories it has occupied and, at some point, formally denied such applicability. Yet, Israel has committed on a "de facto basis" to respect those rules which are of a specifically humanitarian nature. *See,* Meir Shamgar, "The Observance of International Law in the Administered Territories" *Israel Yearbook on Human Rights* 1 (1971) 262-266. Shamgar was an Attorney General and later a Supreme Court president (1983-1995). This commitment has been reaffirmed over the years. In reference to a 2004 petition over Israel's Separation Fence and orders to seize plots of land for the purpose of its erection, Aharon Barak, Shamgar's successor as Supreme Court president (1995-2006), noted as follows: "The question of the application of the Fourth Geneva Convention has come up more than once in this Court...(t)he question is not before us now, since the parties agree that the humanitarian rules of the Fourth Geneva Convention apply to the issue under review". *See,* HCJ 2056/04 *Beit Sourik Village Council v. Government of Israel* [2004] P.D. 58(5) 807, at 827, para. 23; [2004] IsrLR *264*. In 1996, however, Geneva Convention (IV) was considered by the ICJ in the *Nuclear Weapons Advisory Opinion* as part of customary international law. *See, Legality of the Threat or Use of Nuclear Weapons, Advisory Opinion* (Int'l Ct. Justice July 8, 1996), *International Legal Materials (ILM)* 35 (1996), 809 & 1343, para. 75 [hereinafter, *Nuclear Weapons Advisory Opinion*].

17 *Laws of Armed Conflicts*, 55, at 78.

18 *See, The Flechette Petition*, paras. 29-30 [emphasis added in the Petition].

19 1992, S.H. 1391. An official English translation is available at <http://www.knesset.gov.il/laws/special/ eng/basic3_eng.htm>. The declared purpose of the basic law (Art.1) "is to protect human dignity and liberty" and it is one of two new basic laws which comprised Israel's 1992 so-called "constitutional revolution." (the second one is Basic Law: The Freedom of Occupation, 1454 LSI 90 (1994) (Is)) For a review of this law and its various effects *see,* Aharon Barak, "Human Rights in Israel" *Israel Law Review* 39(2) (2006) 12-34.

(i.e., Israelis *and Palestinians alike*)";[20] (2) Israeli administrative law using the reasonability requirement. After all, throughout more than 40 years of occupation, the HCJ has adopted a policy allowing petitions against the IDF military commander[21] and usually questions whether "the action taken by the Military Commander was an action that a reasonable military commander, acting proportionally, would have taken under the circumstances".[22] Therefore, the Petitioners argued that the use of flechette rounds "in the heart of a civilian population when those injured sustain unnecessary suffering, and the person firing-shooting the round cannot, even if he so wishes, aim the weapon properly and in a reasonable manner...," fails to meet this requirement.[23]

However, both the State and the HCJ *completely* overlooked the Petitioners' first argument under Israel's national law (i.e., the Basic Law) in their response and judgment, respectively. As to the second argument (i.e., the reasonability requirement) the HCJ refrained from any serious discussion of it – once it accepted the State's argument of non justiciability.[24] Given this book's focus, it is sufficient to note that the Basic Law's relevance to the Petition is doubtful given that the international law on belligerent occupation – and not Israeli law – has been applied to Arab residents of the Occupied Territories.[25] Indeed, in 2005 the HCJ noted that the basic laws grant rights to Israelis even outside Israel proper.[26] Nevertheless, the question of

20 *The Flechette Petition*, para. 43.

21 *See*, HCJ 7015/02 *Ajuri v. IDF Commander in West Bank* [2002] P.D. 56(6) 352, at 376; [2002-3] IsrLR 83; Aharon Barak, "The Role of a Supreme Court in a Democracy, and the Fight against Terrorism" *University of Miami Law Review* 58 (2003) 125, at 130. As Barak stressed, objections based on extraterritoriality, political questions, and non-justiciability – were not heard. Barak, "Human Rights in Israel", *supra* note 19, at 23.

22 Barak, *Id*, at 24. *See*, also HCJ 3114/02 *Barakeh v. The Minister of Defense* [2002] P.D. 56(3) 11, at 16. An English translation is available at: http://elyon1.court.gov.il/files_eng/02/140/031/A02/02031140.a02.htm

23 Barak, "Human Rights in Israel", *supra* note 19, at 54, and *see in this context*, HCJ 156/75, *Daka v Minister of Transport*, P.D. 30(2) 94, at 105.

24 For the State's claim *see*, HCJ 8990/02, *A Response by the State Attorney's Office*, para. 51. The State's Response in its original text is available at <www.btselem.org/hebrew/Legal_Documents/8990_ 02_Flechette_Appeal_State_Response.pdf > [hereinafter, *State Response*]. For the HCJ Decision *see*, HCJ 8990/02, *Physicians for Human Rights v. O.C. Southern Command*, P. D. 57(4) 193, 194 [hereinafter *The Flechette Decision*].

25 *See*, Barak, "Human Rights in Israel" *supra* note 19, at 23.

26 HCJ 1661/05 *Gaza Beach Regional Authority v. Israeli Knesset* [2006] P.D. 59(2) 481. In this case the Court faced petitions from settlers in the Gaza Strip who challenged the constitutionality of the disengagement initiative that ended Israeli settlement. In fact, already in 2002 the HCJ noted that actions by the Israeli military commander are to be examined in light of the constitutional norms that he (the commander) is obligated to uphold as an Israeli government official, in addition to his general obligations to the welfare of the local population. *See*, HCJ 10356/02, *Hass v. Commander of the IDF forces in the West Bank,* 58(3) P.D. 443, at pars. 78-80. An English translation is available at <http://elyon1.court.gov.il/fi les_eng/02/970/104/r15/02104970.r15.pdf>.

whether non-Israelis in the Occupied Territories were to be granted the same rights was left undecided.[27]

### *BII. The First Layer: The Law of Weaponry*

Within this layer – the only one among the three which both the State and the HCJ addressed in their Response and judgment respectively – the Petitioners discussed the two cardinal principles which constitute "the fabric of humanitarian law"[28] and regulate the distinction between legal and illegal weapons: a) The Principal of Distinction and b) The Principle Prohibiting Unnecessary Suffering.

#### (a) The Principal of Distinction

Crowned by AP-I as the "basic rule," customary international law regarding armed conflicts distinguishes between combatants and non-combatants as well as between military and civilian targets. As such, parties to armed conflict shall direct their operations only against military objectives.[29]

When applied to the context of the law of weaponry this principle demands that states – as noted by the ICJ in the *Nuclear Weapons case* – never "use weapons that are incapable [by itself – as opposed to the mode of use and targeting] of distinguishing between civilian and military targets".[30] However, with respect to anti-personnel weapons such as flechettes, the distinction lies between combatants and non-combatants.

Following citing Article 22 of the 1907 Hague Regulations ("The right of belligerents to adopt means of injuring the enemy is not unlimited"[31]), the Petitioners

27 As a prominent Israeli expert in constitutional law remarked, "[a]lthough this holding is understandable on its own, it raises broader questions regarding the scope of the Basic Laws." *See*, Daphne Barak-Erez, "Israel: The Security Barrier – Between International Law, Constitutional Law, and Domestic Judicial Review", *International Journal of Constitutional Law* 4 (2006), 540, at 550; *See*, also Aeyal M. Gross, "Human Proportions: Are Human Rights the Emperor's New Clothes of the International Law of Occupation?" *The European Journal of international Law* 18 (2007) (*EJIL*), 1–35. Gross points out that while in similar cases the HCJ did cite the Basic Laws as the source of the rights of Palestinians and Israelis in the Occupied Territories, in other cases, mostly dealing with the separation wall (*Id.*, at 23, fn.98), it generally made reference to the rights of the local Palestinian population recognized in international law without detailing their source. *Id.*, at 12.

28 *See, Nuclear Weapons Advisory Opinion*, para. 78.

29 *See,* AP-I, Art. 48. *See,* also, *Id.*, para. 78. For the importance that Aharon Barak ascribed to this principal in reference to the sensitive practice of "targeted killings," *see*, HCJ 769/02, *The Public Committee against Torture in Israel v. The Government of Israel*, 53(4), P.D. 817, para. 23. An official English translation is available at <http://elyon1.court.gov.il/files_eng/02/690/007/A34/02007690.a34.htm>

30 *Nuclear Weapons Advisory Opinion*, para. 78.

31 *Laws of Armed Conflicts*, 55, at 72.

pointed[32] to the *apparently*[33] explicit prohibition on the use of a weapon which has indiscriminate effects, as they defined the flechette weapon:[34] Article 51(4) of AP-I.[35] They summarized by declaring that the employment of flechette rounds is prohibited, "particularly in civilian population centers" due to the darts' dispersion over "an enormous area" as well as the fact that this weapon "is very difficult to use precisely".[36]

### (b) The Principle Prohibiting Unnecessary Suffering

This principle prohibits the use of weapons that cause "unnecessary suffering" or "superfluous injury" (to combatants, let alone non-combatants).[37] The Petitioners' main argument (in the first layer) emphasized that the large number of darts that strike the victim constitute in and of itself unnecessary suffering, a claim which they supported by the autopsy reports on the three Gazan youths.[38] Noting that one victim *absorbed more than 71 darts*, the Petitioners maintained that these reports "speak for themselves!"[39] Their legal argument notes that Article 23(e) of the 1907 Hague Regulation forbids "employ[ing] arms, projectiles, or material *calculated* to cause *unnecessary suffering*"[40] as well as Article 35(2) of AP-I[41] which prohibits "employ[ing] weapons, projectiles, and material, and methods of warfare *of a nature*

32 *The Flechette Petition*, para. 35.

33 For a detailed discussion on this issue *see, infra*, Chapter 8.

34 Given the flechette weapon's wide coverage area, it cannot be properly directed. *See*, the *Flechette Petition*, para. 38.

35 As noted (*supra* Chapter 4), indiscriminate attacks are, *inter alia*, "(b) those which employ a method or means of combat which cannot be directed at a specific military objective". *AP-I, supra* note 13. Israel is not a signatory of AP-I but many of its articles –including this one – are now considered part of customary international law. *See, e.g.*, the definition of Articles 57 and 58 by the ICTY Trial Chamber in Kupreskic *et al.* Judgment, (2000), Case No. IT-95-16-T (ICTY; Trial Chamber), para. 524. *See*, also *Nuclear Weapons Advisory Opinion*, para. 75 (ascribing great importance to AP-I & II).

36 *See*, *The Flechette Petition*, para. 7.

37 In various sources the principle prohibiting the causing of unnecessary suffering is referred to as the 'Principle of Humanity'. *See*, for example, *Existing Rules of International Law Concerning the Prohibition or Restriction of Use of Specific Weapons*, UN Doc. A/9215 (Vol. I), Nov. 7, 1973, at 208, para. 203(1). On this principle in general *see e.g.*, Henri Meyrowitz, "The Principle of Superfluous Injury or Unnecessary Suffering" *International Review of the Red Cross* (IRRC) 299 (1994) 98-122. The principle has been reaffirmed in the ICC Statute, which defined its violation as a war crime (*see*, *Rome Statute of the International Criminal Court*, 17 July 1998 [hereinafter, *Rome Status*] Art. 8(2)(b)(xx), *Laws of Armed Conflicts*, 1309, at 1319) and is reiterated in most military manuals. For a partial–yet lengthy–list of these manuals *see*, Jean-Marie Henckaerts and Louise Doswald-Beck, *Customary International Humanitarian Law*, Vol. I., (Cambridge: Cambridge University Press, 2005) at at 242, fn. 35 [hereinafter, ICRC Study Report].

38 *The Flechette Petition*, Annex D, 1-3.

39 *Id.*, para. 38.

40 Emphasis added. *See, id.*, para. 24.

41 *Id.*, para. 35.

*to cause superfluous injury or unnecessary suffering*."[42] They summarized by declaring that the employment of flechette rounds violated this principle due to the unnecessary suffering caused by "the enormous number of darts, which hit the victim's body (similarly to a nail-filled roadside bomb)."[43]

*Comments*: Before proceeding with a discussion of the State's Response, the following comments merit attention: (a) the Petitioners' erroneous decision to cite the English translation of Article 23(e) of the 1907 revised Hague Regulations before the Court and (b) the inaccuracies in reporting the facts, for which both the Petitioners and the State were at fault.

Erroneous Citing: The authentic French text of Article 23(e) of the 1899 Hague Regulations and the 1907 revised version[44] are identical (*Il est… interdit… d'employer des arms, des projectiles ou des matieres propres à causer des maux superflus*) and should be translated into English – as was properly done in the non-binding yet commonly used 1899 English translation – as follows: "it is especially forbidden … (e) To employ arms, projectiles, or material *of a nature* to cause superfluous injury".[45]

Yet, the 1907 English translation was different. "Without any apparent ground" as Prof. Frits Kalshoven stressed,[46] the 1907 English text reads: "it is especially forbidden …(e) To employ arms, projectiles, or material *calculated to cause unnecessary suffering*".[47] That is, the new translation narrowed the prohibition's scope by adding a *subjective element* to the original rule and also replaced the French phrase of *superfluous injury* with the phrase *unnecessary suffering*. Such wording – warned prominent scholars before the 1977 conclusion of AP-I – may lead one to incorrectly conclude that only weapons that were intentionally designed to cause unnecessary suffering were banned.[48]

The Petitioners erroneously chose the 1907 English translation to support their claim. Hence, had the HCJ discussed the claim regarding unnecessary suffering – as it should have done – the Petitioners could have faced a problem. As the detailed overview regarding the weapon's development indicates, causing "unnecessary suffering" is unrelated to the U.S. motives behind the development of any type of fle-

42 Emphasis added. *Laws of Armed conflicts*, 711, at 730.

43 *See*, *The Flechette Petition*, para. 7.

44 For the 1899 French text of Hague Convention see Pearce Higgins *The Hague Peace Conferences and Other International Conferences Concerning the Laws and Usages of War*, (Cambridge: Cambridge University Press, 1909), 206, 234. For the 1907 French Text *see*, *Id.*

45 For the 1899 English text of the Hague Convention, *see*, *Laws of Armed Conflicts*, 55, at 85.

46 *See*, Frits Kalshoven, "Arms, Armaments and International Law", *Recueil Des Cours* 19, (1985[II]) 181, at 217.

47 Emphasis added. *Laws of Armed Conflicts* at 73.

48 *See*, *e.g.*, Hans Blix, "Current Efforts to Prohibit the Use of Certain Conventional Weapons", 4 (1) *Instant Research on Peace and Violence*, (1974), 21, at 24.

chette-fired weapons.[49] Eventually, however, the Petitioners did cite Article 35(2) of AP-I, [50]which incorporated the phrase "unnecessary suffering", within the French text: "as *cause superfluous injury or unnecessary suffering.*"[51] It originated from the fact that the phrase "unnecessary suffering" had acquired its own relevance in states' practices – as reflected in various military manuals and the ICRC model, which chose to use both phrases.[52] Furthermore, while it was accepted that the French expression '*maux superflus*' includes both physical and psychological injury,[53] it was not the case with the English expression and both phrases, which the Canadian Military Manual defined as "closely related concepts,"[54] were required for assessing a weapon's legality during the 1970s discussions on the legality of conventional weapons.[55] As such, it was claimed that the expression unnecessary suffering "has never fully covered the meaning of the French expression (*maux superflus)*" and that "injuries" caused by a weapon may be measured while the suffering inflicted by it "is hard if not impossible to quantify".[56]

49 *See, supra*, Chapter 1, CII. Indeed, according to a U. S Army document, based on experience gained at the Vietnam War "weapons designers developed a wide-fin. (1 cm.) 6.8 grain flechette intended to produce greater damage and incapacitation – P(IH) [Probability of Incapacitation of an individual by a random Hit- E.B.)". *See,* Joint Technical Coordinating Group for Munitions Effectiveness, *Evaluation of Wound Data and Munitions Effectiveness as based upon Battlefield surgeries in Vietnam (WDMEV)*, (Joseph Sperrazza and Joseph R. Blair, Co-Chairman), Alexandria, VA, Defense Technical Information Center (AD511909), at 14. Yet, it is unclear if darts of this type have become operational. In any case, as this document details, the motive was strictly for increasing the darts' military effectiveness (i.e. a legitimate military consideration).

50 *The Flechette Petition*, para. 35.

51 Emphasis added. The French text of AP-I is identical to that of the 1899 Hague Regulations and the 1907 revised version except for the substitution of "de nature" for *"propres".* *See,* Bothe, Michael, K. J. Partsch and W. A. Solf (eds.) *New Rules for Victims of Armed Conflicts: Commentary on the Two 1977 Protocols Additional to the Geneva Conventions of 1949* (Leiden: Martinus Nijhoff 1982) 195, [hereinafter, *New Rules*].

52 *See,* for instance, The Australian Defence Force Publication – Operations Series, Law of Armed Conflict, Art. 402; Humanitarian Law in Armed Conflicts, Manual (German), Art, 402; FM 27-10 the Law of Land Warfare (USA), Art. 34(a). For the ICRC Model see Rogers, A. P. V, *et al.*, *Fight It Right* (Model Manual on the Law of Armed Conflict for Armed Forces), (Geneva: ICRC, 1999), Art. 803(1).

53 *See, New Rules*, at 195-6.

54 *See, The Law Of Armed Conflict At The Operational And Tactical Level* (Canada, Office of the Judge Advocate General, Aug. 2001), Art 4 and Art. 2.

55 *See*, for instance, the use of both concepts in the 1973 expert meeting which the ICRC convened in Geneva. ICRC, *Weapons that may Cause Unnecessary Suffering or Have Indiscriminate Effects* (Report of the Work of Expert) Geneva, (1973), 12 (hereinafter, *1973 Geneva Report*).

56 Hans Blix, "Means and Methods of Combat" in: *International Dimensions of Humanitarian Law*, (Geneva: UNESCO, 1988) 135, at 138. As Kalshoven noted, the experts in the 1974 Lucerne Conference (for which he was the principle rapporteur) generally agreed that "injury" has a more objective connotation than "suffering". *See,* Kalshoven, *Reflections on the Law of War: Collected Essays* (Leiden: Martinus Nijhoff 2007) 159.

Therefore, the incorporation of both phrases in AP-I was not only a Solomonic decision[57] due to the difficulties caused by the 1907 English wording, but a positive although inherently problematic attempt to cover both presumably measurable-objective injury and subjective-psychological suffering and pain.[58]

Inaccurate Reporting: As mentioned briefly in the introduction, the Petitioners only sought to review the legality of flechette use in the Gaza Strip due to the area's endemic features, mainly its dense population, Israel's overwhelming military superiority, and the weapon's known effects. While the State chose to blur legal facts and distort legal details throughout its Response, the Petitioners share this blame as well. First, they deleted the words "in Gaza Strip" when referring to "the legality of use" throughout the Petition.[59] Second, they drew upon an extensive quotation from the ICJ's *Nuclear Weapons Advisory Opinion*, in which the legality of nuclear weapons was discussed *in the abstract*, and placed it within the Petition's main text as well as attaching the full summary as an appendix.[60]

More importantly, aside from matters of form, once the Petitioners raised the argument – in fact their main one – that flechette weapons cause unnecessary suffering – the discussion could not be contained within the boundaries of the Gaza Strip. If flechette weapons indeed inflict unnecessary suffering, they are to be banned universally. Even the question of proportionality between the injury or suffering and the military advantage anticipated – as some scholars have attempted to argue – is redundant.[61] As Professor Dinstein rightly stressed, the principle of proportionality is linked to the issue of collateral damage to civilians resulting from attacks against military objectives and has nothing to do with injury or suffering by combatants.[62] Likewise, the Petitioners claimed that the flechette "is to be considered an 'indis-

---

57 Blix, "Current Efforts to Prohibit..." *supra* note 48, at 24.

58 The AP-I text is adopted in the ICRC Customary Law study Report, Rule 70. *See, ICRC Study Report*, Vol. I, at 237.

59 *See*, for instance, *The Flechette Petition*, para. 19, as follows: "The petitioners shall argue that the use of the flechette round is illegal, constitutes..., *inter alia*, a violation of accepted and/or customary international humanitarian law, since the round causes indiscriminate damage and unnecessary suffering".

60 *Id.*, para. 37 & Annex H.

61 For such a stance *see, e.g.*, Gail Judith Gardam, "Proportionality and Force in International Law" *The American Journal of International Law* 87 (*AJIL*) (1993) 391, at 406. However, as Meyrowitz rightly clarified "the rule adopted by international law-making bodies… absolutely prohibits any recourse to that means or method, and hence excludes any evaluation of the proportional relationship between the suffering, injury or damage that would be caused if it were used and 'the concrete and direct military advantage' that might be 'anticipated.'" Henri Meyrowitz, "Principle of Superfluous Injury…" *supra* note 38, at 110. Indeed, ten years after, Gardam herself retreated from her position stating as follows: "Technically it is more accurate today to talk in terms of superfluous injury or unnecessary suffering in the context of combatant, rather than proportionality". *See*, Gail Judith Gardam, *Necessity, Proportionality and the Use of Force by States*, (Cambridge: Cambridge University Press, 2004), 15.

62 *See*, Yoram Dinstein, *The Conduct of Hostilities under the Law of International Armed Conflict*, (Cambridge: Cambridge University Press, 2004), 59.

criminate weapon' since it disperses over an enormous area, and is very difficult to use precisely, and accordingly...its use is prohibited, particularly in civilian population centers".[63] The principle of proportionality in reference to the principle of distinction means - in the words of Judge Rosalyn Higgins – that "even a legitimate target may not be attacked if the collateral civilian casualties would be disproportionate to the specific military gain from the attack". Yet, it was mentioned *only in* the context of Israeli national law.[64] As with the argument on unnecessary suffering, if the flechette weapon is indeed to be considered "an indiscriminate weapon" it should be banned universally.

### C. The State's Response – Overview and Comments

On January 22, 2003, after its request for a one month extension was granted by Justice Dorit Beinisch, the State Attorney's Office submitted its response on behalf of the State.[65]As expected, after the State argued that the use of flechette weapons was legal, it argued that the issue was non justiciable.[66] This argument was based on two HCJ precedents,[67] in one of which, two of the three sitting justices (Mishael Cheshin and Eliyahu Mazz) stated as follows: "[T]he choice of weapons used by the respondents in order to prevent murderous terror attacks is not an issue that this

63 *The Flechette Petition*, para. 7.

64 *Id.*, paras. 44, 47-55. For the quotation *see*, Dissenting Opinion of Judge Higgins, the *Nuclear Weapons Advisory Opinion*, at para. 20. All Declarations, Separate and Dissenting Opinions of the Judges are available at <http://www.icjcij.org/docket/index.php?p1=3&p2=4&k=e1&p3=4&case=95>. ] For a concise discussion of the application of this important principle to Israel's use of flechettes in the Gaza Strip *see*, *infra* Chapter 6 (D).

65 For Justice Beinisch's decision as to the extension request (Dec. 12, 2002) *see*, HCJ 8990/02 Request by the Respondents, Dec. 19, 2002, available at <http://elyon1.court.gov.il/files/02/900/089/N02/02089900.n02.htm> (in Heb.)

66 Once flechette rounds are to be found legal "under both conventional and customary international law as well as the international practice" – the argument goes – "the case is not justiciable." *See*, State's Response, para. 51.

67 *See*, HCJ 5872/01 *Barakeh v. Prime Minister*, P.D. 56 (3)1 [hereinafter, the *Barakeh Case*]; HCJ 3022/02 *Canon (Law) The Palestinian Society for the Protection of Human Rights and the Environment v. The IDF Commander in the West Bank*, P.D. 56(3) 9. In the latter precedent, for instance, the Court was asked during intense combat in the Jenin refugee camp during *Operation Defensive Shield* (Apr. 2002) to order the IDF to "refrain from shelling and from striking civilian targets in the Jenin refugee camp, by means of tanks, combat helicopters and/or all other types of shooting."(HCJ 3022/02, *Canon (Law)*, at 9). The petitioners in this case (Israeli and Palestinian human rights NGOs) argued that *proportional means* shall be used against the armed Palestinians inside the camp in order to minimize collateral damage (i.e., "kill ... civilians and demolish their property"). The State, however, began its response by stating that "during [the] fighting no judicial review and effective remedies concerning operational considerations are possible" – a stance which the Court fully embraced, explaining that "institutionally and as a matter of course, it is not possible to grant the remedies requested by the petitioners." *Id.*, at 10.

Court will intervene with."[68] As such, although a response regarding weapons use under the specific circumstances (i.e., in the Gaza Strip) would be provided in the event that the preliminary claim was rejected,[69] it was clear that once the weapon was deemed legal – the Petition would be rejected.

In stating that the use of flechette rounds "is legal and does not violate the principles of international law, including the laws of war",[70] the State attempted to avoid confronting the Petitioners' customary IHL claims and thus deliberately overlooked the issue of whether flechette rounds cause *unnecessary suffering* (or are *excessively injurious*) or have *indiscriminate effects*. In fact, these commonplace phrases are mentioned (only once) in the entire Response and even then in the course of a general overview of the two cardinal principals under customary international law governing the use of weapons.[71] Alternatively, following an introduction into the law of weaponry and the IDF's' compliance with customary IHL,[72] a thorough survey of weaponry treaties was provided. A short overview on the alleged relevant practice of the international community concluded this part of the State's Response.

### *CI. Treaty Law Concerning Means of Warfare*

The State pointed out that proposals to prohibit fragmentation weapons, "including flechette-fired weapons" were first raised in the 1970s as part of the international community's effort to reach a consensus on the nature of weapons to be prohibited

68 This citation comprises almost the entire decision. *See* the *Barakeh Case*, and *see* State's Response, para. 53 in which it was cited. For a critique of this Case *see*, Orna Ben-Naftali & Keren R. Michaeli, "Justice-Ability: A Critique of the Alleged Non-Justiciability of Israel's Policy of Targeted Killings" *Journal of International Criminal Justice* 1 (2003) 368-405.

69 In a nutshell, this part of the State's Response was designated to convince the Court that the special circumstances found in the Gaza Strip meet the requirements of the reasonability and the proportionality tests under Israeli administrative law. The state's Response began with an extensive three-part description, as follows: (I) a general overview of the intense conflict in the Gaza Strip since the eruption of the Second *Intifada* (Sept. 2001) in which it was admitted that Palestinian non-combatants were killed. Nonetheless, the State stressed that "it is mainly due to Palestinian practices, which commit gross violations of any legal and moral norm". (Emphasis in original, *State's Response*, para. 27); (II) The IDF's perception vis-à-vis the military necessity of resorting to such weapons even in low intensity conflict is presented (*Id.*, para. 33) with details of its rules of engagement concerning use of flechette rounds included (*Id.*, paras, 36-40). This is of course the important element for us, as the IDF's perception of the weapon's military necessity is required for the general evaluation of the weapons' legality in Chapter 7; (III) The three specific incidents which provided the impetus for the petition are detailed while claiming that the harm to civilians stemmed from a mistake in identification, irrespective of the specific weapons employed. The Gazan youths' incident is excluded as the IDF claimed that they were terrorists. Furthermore, these incidents do not demonstrate the weapon's illegality, as the use of alternative means could certainly have led to the same, if not worse, results. (*Id.*, paras. 45-49).

70 *Id.*, para. 1.

71 *Id*, para 2.

72 *Id*, paras. 2-8.

or restricted by legal instruments.[73] Yet, all such attempts have failed and none of the three original Protocols of the *1980 Convention on Certain Conventional Weapons* (CCW) [74] – the outcome of these efforts – discusses flechette weapons.[75]

The history of the CCW is then presented. While pointing to both the 1995 Protocol IV on Blinding Laser Weapons[76] and the 1996 amended Protocol II on the Use of Mines, Booby-traps and Other Devices,[77] the State noted that proposals to restrict the use of flechette weapons during both the 1980s and the 1990s had failed to attract support.[78] Furthermore, it stressed that no such proposal was then (2002-2003, the time of the legal debate), on the CCW agenda.[79]

In sum, in the State's view, its survey made clear that the international community had reviewed the flechette weapons' legality *over an extended period of time* and had not found a reason to impose restrictions, let alone to prohibit its use. Therefore, it concluded, "the claim that the use of flechettes is prohibited by the customary international law is baseless".[80]

*Comments:*

(a) The fact that the State and, consequently, the Court discussed the CCW in the context of the Israeli-Palestinian conflict is interesting as it was more than two years before Israel's September 2005 Disengagement from the Gaza Strip and the Court December 2006 decision to accept the IDF position that the armed conflict waged inside the Strip should be perceived as an IAC.[81] Indeed, the 2001 Second Review Conference amended the CCW so that it would apply to non-IACs as

73 *Id.*, para. 10. For a detailed overview of these proposals *see*, *supra* Chapter 2 (C-D).

74 Convention on Prohibitions or Restrictions on Use of Certain Conventional Weapons which may be Deemed to be Excessively Injurious or to Have Indiscriminate Effects, Geneva, 10 Oct. 1980. *Laws of Armed Conflicts*, 181.

75 State's Response, para. 13.

76 Protocol on Blinding Laser Weapons (Protocol IV to the 1980 Convention), 13 October 1995, *Laws of Armed Conflicts*, 212.

77 Protocol on Prohibitions or Restrictions on the Use of Mines, Booby-Traps and Other Devices as amended on 3 May 1996 (Protocol II to the 1980 Convention as amended on 3 May 1996), Id., at 196.

78 *State's Response*, para. 14.

79 *Id.*, para. 16.

80 *Id.*, para. 14.

81 In the December 2005 discussion over *the Targeted Killing* case, while the petitioners' initial stance was that the laws regarding police and law-enforcement apply to the conflict with the Palestinians, the State claimed that Israel is involved in an armed conflict with international characteristics – a claim which the HCJ accepted in its decision a year later. For the petitioners' initial positions *see*, HCJ 769/02, *supra note*, para. 4. For the State's position *see*, *Id.* para. 17. As we focus on the Gaza Strip exclusively, an interesting question is the effect of the September 2005 IDF withdrawal on Gaza's legal status. On this point *see*, Yuval Shany, "Faraway, So Close: The Legal Status of Gaza after Israel's Disengagement" *Yearbook of International Humanitarian Law* 8 (2005) 369-383. For opposing positions as to Gaza's legal status before the withdrawal *see*, *id.*, 371.

well.[82] However, not only has Israel yet to accept this amendment,[83] at the time of the Petition (2002-2003), the amendment was not even in force.[84] Therefore, the State's decision to bring the CCW before the Court in this context has rightly attracted interest. As one commentator, who focused solely on the HCJ willingness to discuss the CCW, noted: "Perhaps of even greater interest, the Court tacitly accepted without comment that the CCW in fact applies to the activities of the IDF in the Occupied Territories…".[85] It should be noted, however, that two years after the HCJ Decision and shortly before the conflict was classified as an IAC by the HCJ in the *Targeted Killing* case, the International Criminal Tribunal for the former Yugoslavia (ICTY), established to adjudicate crimes in the former Yugoslavia, has made the conflict's legal classification (IAC/Non-IAC) meaningless by stating the following in reference to weaponry:

> Indeed, elementary considerations of humanity and common sense make it preposterous that the use by States of weapons prohibited in armed conflicts between themselves be allowed when States try to put down rebellion by their own nationals on their own territory.[86]

(b) The State rightly focused on the CCW as the source of treaty law. As already detailed,[87] this "umbrella treaty" is the sole mechanism at the international community's disposal to regulate prohibitions or restrictions on conventional weapons since the early 1980s. In fact, until the emergence of the 1997 Anti-Personnel Mine Ban Convention (Ottawa Convention),[88] the CCW was both the sole treaty concerning conventional weapons in the post WWII era as well as the sole mechanism applied in practice when drafting additional prohibitions or restrictions on conventional means of warfare.[89]

---

82 *See, supra* Chapter 2, EIII.

83 In fact, Israel even made sure to make declarations to the treaty on accession, and to every individual protocol, to the effect that it does not deem the treaty applicable to internal conflicts. For Israel's declarations *see, Laws of Armed conflicts*, 218, 227, 235.

84 This amendment only came into force on May 18, 2004, six months after the deposit of the twentieth instrument of acceptance.

85 *See,* Jeffery K. Walker, 'Introductory Note', ILM 42 (2003) 880, at 881. Walker erroneously thought that the Petitioners had pointed to the CCW and therefore focused only on the HCJ discussion.

86 *Prosecutor v. Dusko Tadic*, IT-94-1-AR72, ICTY Appeals Chamber, Oct. 2, 1995 (Jurisdiction), para. 119. 105 ILR 419, at 516.

87 *See, supra* Chapter 2.

88 The 1997 Convention on the Prohibition of the Use, Stockpiling, Production and Transfer of Anti-Personnel Mines and on their Destruction, (entered into force Mar. 1, 1999), *Laws of Armed Conflicts*, at 285. For a personal account of the emergence of the Ottawa Convention by the 1997 Nobel Peace Prize laureate Jody Williams; *see,* Williams Jody and Stephen Goose, "The International Campaign to Ban Landmines" in: Maxwell A. Cameron *et al.*, (eds.) *To Walk Without Fear: The Global Movement To Ban Landmines*, (Oxford: Oxford University Press, 1999) 20-47.

89 Nine years later, in December 2008, a second legal instrument which had emerged outside the CCW track was opened to signature. *See,* the *Convention on Cluster*

However, the CCW was not mentioned *even once* in the entire lengthy Petition, and therefore, it was not from the Petition, but from the State's Response, that the Court learned that "the international community had reviewed the subject [flechette legality] throughout many years."[90] In other words, the flechette weapon is all but another "regular" conventional weapon. The State's claim that the international community has debated the weapon's legality over the years should have invited the Court to hold serious discussions on the matter, regardless of the CCW's current status and importance.

(c) As already detailed,[91] the failure of the proposal to ban flechette weapons within the CCW's framework was neither exclusively related to the weapon itself nor indicative that its proponents had changed their position.[92] Once the second preparatory session of the UN Conference on the Use of Certain Weapons (September 1979) decided to recommend that the Conference adopt a consensual voting procedure, it was clear that very few of the proposed restrictions or prohibitions would actually be implemented since every delegation enjoyed veto rights.[93] Therefore, the prohibitions which were finally drafted into protocols are few and mostly marginal.[94]

(d) As the State stressed regarding the First Review Conference (1995-1996), "(i) n these discussion as well, proposals to impose restrictions on the use of flechettes have failed to attract support and have not resulted in any restrictions or legally obliging prohibitions".[95] However, as already mentioned,[96] *no proposal was ever sub-*

*Munitions* (CCM), *available* (in all official UN languages) *at* http://treaties.un.org/ doc/ Publication/CTC/26-6.pdf. For a comprehensive and updated overview of the emergence of the CCM *see* John Borrie, *Unacceptable Harm: A History of How the Treaty to Ban Cluster Munitions Was Won* (New-York: United Nations, 2009).

90 *State's Response*, para. 14.

91 *See, supra* Chapter 3.

92 The report of the Second Preparatory Session (1979) reads: "Concerning the question of…, and *flechettes*, brief discussions took place at plenary meetings. *Time did not allow for their discussion in more detail*, therefore agreement thereon could not be reached." (Emphasis added). Rep. of the Preparatory Conferences, UN Doc A/Conf.95/3, (May 25, 1979), at 13, para. 37. This state of affairs was repeated in the September 1980 *UN Conference on the Use of Certain Weapons*, Second Session. *See*, UN Conference on Prohibitions or Restrictions of Use of Certain Weapons which May be Deemed to be Excessively Injurious or to Have Indiscriminate Effects, *Geneva, 10-28 Sept. 1979; 15 Sept.-10 Oct. 1980*, Final Rep. of the Conference to the General Assembly, UN Doc. A/ Conf. 95/15, Oct. 27, 1980, at 7, para. 23.

93 For the recommended rules of procedure, *see*, UN Doc. A/conf. 95/2 (1979). The Conference adopted the recommendation at its first plenary meeting. *See*, UN Conference on Prohibitions or Restrictions of Use of Certain Weapons which May be Deemed to be Excessively Injurious or to Have Indiscriminate Effects, *Geneva, 10-28 Sept. 1979*, Rep. of the Conference to the General Assembly, UN Doc. A/Conf. 95/8, Oct. 8, 1979, at 6, para. 16.

94 To the extent that they were ineffective *see, supra* Chapter 3.

95 *State's Response*, para. 14 (emphasized in original).

96 *Supra* Chapter 2.

*mitted* regarding flechette weapons.[97] In fact, Mexico's delegate Antonio de Icaza complained on the second day of the conference that "[i]t was regrettable that except in the case of laser weapons, no proposal had been submitted concerning prohibitions on the use of small-calibre weapons, cluster bombs, fléchettes ...".[98]

(e) The State rightly claimed that no proposal regarding flechette weapons was on the "current UN discussions on various aspects of the CCW expansion" agenda.[99] However, discussions were then being held (2002-2003) by two working groups of government experts (GGE), which addressed humanitarian concerns more pressing to the CCW State Parties at that time: Anti-Vehicle Mines (AVM) and Explosive Remnants of War (ERW).[100] This is far from implying that the weapon was by then perceived as a legitimate one. By then flechette was simply a "dubious" weapon, forgotten to the point that even the delegate (Gustabo Albin) of the only state to raise the issue in 1995–Mexico–overlooked it.[101]

### *CII. The International Practice*

Despite the apparently universal caption which the State attached to this section in its Response ("*The International Community's Practice*"), it quickly clarified that "it is sufficient for us to relate to the practices of the US Army".[102] Two instances for such U.S. practices were provided as follows: (a) the flechette rounds appearance in the *US Marine Corps Field Manual* (FM 6-50), according to which the M546 flechette rounds (each loaded with 8,000 flechettes) are "intended primarily for anti-

97 For the reports of the Preparatory Committee's (PrepCom) Sessions *see*, The Group of Government Experts to Prepare the Review Conference of the States Parties to the Convention on Prohibitions or Restrictions on the Use of Certain Conventional Weapons Which May Be Deemed to Be Excessively Injurious or to Have Indiscriminate Effects, First Sess. (28 Feb. – 9 Mar. 1994): CCW/CONF.I/GE/4 and Corr.1; Second Sess. (16 – 27 May 1994): CCW/CONF.I/GE/8; Third Sess. (8 – 19 Aug. 1994): CCW/CONF.I/GE/21; Fourth Sess. (9-20 Jan. 1995): CCW/CONF.I/GE/23.

98 Summary Rec. of the 2nd Meeting (26 Sep.1995) CCW/CONF.I/SR.2, Sep. 29, 1995, at 14, para. 69. *Final Report of the First Review Conference of the States Parties to the Convention on Prohibitions or Restrictions on the Use of Certain Conventional Weapons Which May Be Deemed to Be Excessively Injurious or to Have Indiscriminate Effects*, CCW/Conf.i/16 (part II), Documents and Summary Records of the Conference, Geneva 1996.

99 State's Response, para. 16.

100 Eventually, only the latter GGE successfully concluded a fifth protocol in 2003: *Protocol V on Explosive Remnants of War (entered into force* Nov. 12, 2006), UN Doc. CCW/MSP/003/2.

101 For his speech during the third session on December 12, 2001, *see,* Rep. of the Second Review Conference of the States Parties to the Convention on Prohibitions or Restrictions on the Use of Certain Conventional Weapons which May Be Deemed to Be Excessively Injurious or to Have Indiscriminate Effects, CCW/CONF.II/, *Sum. Rec. of the 3rd Plenary Mtg., Geneva*, 12 Dec. 2001, at 82, para. 7.

102 *State's Response*, para. 18.

personnel use at close range";[103] (b) the alleged employment of flechette rounds by the U.S. army "during the war in Korea, the Vietnam War, and the U.S. engagement in Somalia in the 1990s".[104]

Furthermore, in order to bolster its argument over the widespread practice of using flechette rounds, the State argued that they are included in the so-called "Sub-Munitions Category", of which flechette rounds are one of the less destructive weapons to be included. "Cluster bombs – the State provided as an example…, have a widespread footprint and many of them remain 'active' (i.e., duds)."[105] Relying on 2002 HRW's survey on "explosive sub-munitions", the State stressed that "*sub-munitions, under which flechette is included, is in the arsenal of many states and their use is quite widespread, including recent years and including Western countries' armies such as the U.S.*".[106]

Hence, the State's argument is concluded as follows: "It is evident that the international community has not solidified a prohibition norm on the use of sub-munitions, and in effect the opposite is true".[107]

*Comments*:

(a) First, as opposed to the State's presentation, the historical survey suggests that unlike other types of sub-munitions, very few states still maintain flechette *rounds* in their arsenals.[108] Furthermore, Israel's use excluded, the U.S. Army is the only armed force whose use of flechette projectiles was verified (as opposed to flechette-filled air-to-ground rocket warheads) and certainly the only one to be involved in their large scale use. However, the U.S. Army's in-service tanks are equipped with 120mm smoothbore main-gun,[109] whilst the U.S. has produced neither 120mm flechette rounds nor – unlike Israel – adjusted the 105mm flechettes rounds in stock for use by those

103 *Id.*, *US Marine Corps Field Manual* No. 6-50, Marine Corps War fighting Publication no. 3-1.6.23 (Washington, D.C., Dec. 23, 1996), Appendix F., Art. F-10(J).

104 *State's Response*, para. 18. No references were provided to support these claims.

105 *Id.*, para. 20. Given Israel's extensive use of cluster bombs a few years later, in the 2006 Second Lebanon War, this statement seems a bit odd. On that use *see,* Eitan Barak, "None to be Trusted: Israel's Use of Cluster Munitions in the Second Lebanon War and the Case for the Convention on Cluster Munitions" *American University International Law Review* 25 (2010, 423-483).

106 Emphasis in origin. *See, State's Response*, para. 22. For the Report which was attached to the Response *see,* Human Rights Watch, *Memorandum to CCW Delegates a Global Overview of Explosive Submunitions Prepared for the Convention on Conventional Weapons* (CCW) Group of Governmental Experts on the Explosive Remnants of War (ERW) May 21-24, 2002. Indeed, according to the memo this category comprises over 200 known types of sub-munitions that are produced by 33 states.

107 *State's Response*, para. 20.

108 *Supra* Chapter 1.

109 While the M1 Abrams tanks (U.S. main battle tank) were equipped with a 105mm M68 gun, newer versions produced since 1985, the M1A1, M1A2, and M1A2 SEP tanks employ a 120 mm smoothbore main gun (M256). *See,* "Army Accepts First M1A1 Tank", *Marine Corps Gazette* 69(11) (Nov. 1985), 10; Steve Zaloga and Peter Sarson, *M1 Abrams Main Battle Tank 1982-1992*. (Oxford: Osprey 1993) 10-11.

same 120mm guns. In other words, the U.S. Army has virtually renounced the use of flechette rounds in its armor forces (as opposed to flechette shells for its 105mm or 155mm Artillery guns).[110] As to the question of the weapons' legality, flechette rounds are preferable to flechette shells given that tanks are able to direct their fire more accurately. Furthermore, the absence of a particular weapon from an army's arsenal does not necessarily bear on its legality as it may be the result of merely tactical considerations. Yet, it is interesting to note that the IDF it is quite probable the only armed force to continue use of flechette *rounds* in the twenty-first century.

(b) Only one out of the three examples of U.S. flechette rounds use listed by the State in its Response actually took place (the Vietnam War). Prior to 1958, five years *after* the Korean War ended, flechette rounds had not even been tested. In fact, it was the Korean War that provided the main impetus for *developing* such weapons. [111] A careful inquiry indicates that flechette rounds were not employed by the U.S. armor force in Somalia (1992-1994), and it is quite doubtful that flechette shells were used by the small artillery force.[112] In sum, the use by the U.S. during the Vietnam War was the first time the weapon was employed as well as its last large-scale use – Israel excluded.

(c) The State's attempt to bolster the claim that the weapon was widely used in practice by pointing to the general use of sub-munitions is highly problematic. First, flechettes are quite distinct weapons – i.e., *non-explosive sub-munitions* – as reflected, for instance, in their absence from the long list of weapons contained in the HRW report. Second, excluding some types of flechette-filled rocket warheads, they are

---

110 This fact, however, can even be learned from the State's source (*US Marine Corps Field Manual*). As the relevant paragraph reads: "The APERS-T round is available in only 105-mm caliber." *See US Marine Corps Field Manua*l No. 6-50.

111 *See, supra* Chapter 1, C(II). Embarrassingly, in addition to raising these three cases, the State detailed a few types of flechette rounds and provided two references for this, one of which is a brief explanation on the APERS-T 546, issued by the *Federation of American Scientists* (*FAS*), that points explicitly to the Korean War *as the trigger* for developing such weapons". *See*< www.fas.org/man/dod-101/sys/land/m546.htm>.

112 Only after 18 U.S. soldiers were killed and 84 wounded in the "Battle of Mogadishu" on October 3, 1993, was Defense Secretary Les Aspin ready to comply with the request of the Commander of U.S. Forces in Somalia, Gen. Thomas Montgomery, for armored reinforcements. Thereafter, 30 M1A1 tanks as well as self-propelled howitzers and 48 Bradley armored vehicles reached Somalia toward the end of October 1993. Not only did the U.S. troops withdraw from Somalia several months later (March 1994) but, as noted, the U.S. had only 105mm flechette rounds available at its disposal while the M1A1 tanks were equipped with 120 mm smooth-bore gun. *See,* Mark Bowden, *Black Hawk Down: a Story of Modern War,* (New-York: Atlantic Monthly Press 1999) 340 (discussing the rejection of Gen. Thomas Montgomery's request for tanks and armored vehicles at the end of September). For the armor's arrival *see,* Dele Olojede, "U.S. Sends Reinforcements As Somali Withdrawal Nears", *Newsday,* (Nov. 16, 1993), at A2. The author's request for data from the Pentagon (Mar.-Nov. 2005) regarding the possibility that flechette shells were used by the self-propelled howitzers was denied since the Pentagon's Press Office only releases such data to the press. Email correspondences (Mar. 6, 2005, Oct. 28 & 31, 2005); Telephone interview, Nov. 9, 2005.

*anti personnel weapons par excellence* – an important distinguishing feature given the successful prohibition against anti-personnel mines as opposed to the continuous failed attempts to impose even weaker restrictions on the use of AVM.[113]

(d) Finally, the State's argument that the flechette weapon is included in the "submunitions family" as a formal legal argument is problematic from another aspect. Given the worldwide uproar against the use of cluster munitions in the 2006 Lebanon War and the fact that as of July 2010, 108 states have signed the 2008 Convention on Cluster Munitions (CCM)[114] that bans these weapons (it will enter into force on August 1, 2010[115]), one could ask whether this State's claim is in fact another indication of flechette's problematic nature.[116]

## D. The HCJ's Legal Review and Ruling: Overview and Comments

Once the State Attorney's Office submitted its Response on January 22, 2003, it was referred to an HCJ three-judge panel.[117] As already mentioned, the Panel's composition was all but favorable from the Petitioners' point of view since two among the three nominated justices had already rejected a similar petition in 2001 (i.e., the *Barakeh* case) by resorting to the doctrine of non justiciability. The third Judge

---

113 In fact, after five years of discussions within the CCW in Geneva, and despite the great majority of states involved in the process acknowledging that action was needed, by the end of 2006 the international community failed to develop any rules for controlling these weapons. For an analysis of the direct impact of AVMs on civilians, *see e.g.*, Greg Crowther and Richard Moyes, *Anti-vehicle mines: Understanding the Impact and Managing The Risk, A Resource for Security Management*, (Landmine Action Rep.) (Nov. 2007).

114 The Convention's text in all official UN languages is *available at* <http://treaties.un.org/doc/ Publication/CTC/26-6.pdf>. Separate articles refer to assistance to victims, clearance of contaminated areas, and destruction of stockpiles. For a brief overview of its provisions *see, e.g.*, Priya Pillai, "Adoption of the Convention on Cluster Munitions" 12(20) *American Society of International Law Insight* (Oct. 1, 2008) available at: <http://www.asil.org/insights081001.cfm>.

115 On Feb. 16, 2010, the 30th country (Moldova) deposited its instrument of ratification with the UN, a short while after Burkina Faso had done so on the same day. *See*, "UN-backed Pact against Cluster Munitions to take Effect from August" UN Press Release, (Feb. 16, 2010), available at < http://www.un.org/apps/news/story.asp?NewsID=33790&Cr=disarmament&Cr1=# >. An updated list of Ratifications and Signatures is available at: <http://www.icrc.org/ihl.nsf/WebSign?ReadForm&id=620&ps=P>.

116 In the *very month* in which the Flechette Decision was rendered (April 2003), the Dublin Conference on ERW convened. The latter had led to the establishment of the Cluster Munition Coalition (CMC), a group of about 300 NGOs from more than 80 countries, which was an important factor in the successful conclusion of the CCM. For more on the CMC, *see, Cluster Munition Coalition*, <http://www.stopclustermunitions.org/>.

117 For Justice Dorit Beinisch's decision *see* HCJ 8990/02 decision, (Jan. 23, 2003), available at <http://elyon1.court.gov.il/files/02/900/089/N03/02089900.n03.htm>.

(Esther Hayut), a temporary nomination at the time, kept a low profile during the April 13 discussions.[118]

Meanwhile, as the continued discussion was scheduled to April 13 and as no interim order had been requested, the use of flechette rounds in the Gaza Strip went unhindered with all that this implied regarding potential collateral damage and the attendant media uproar.[119] However, for the first time, the HCJ became a target for criticism in this context as well, on the grounds that it intentionally dragged its feet to avoid making a decision about the Petition.[120]

On April 13, 2003, the HCJ finally began its deliberations.[121] The discussion was headed by Justices Mazza and Cheshin; their comments left no doubt as to their likely decision. In fact, Mazza attempted to persuade the Petitioners to withdraw their Petition, commenting that "tomorrow they'll say that it's forbidden to use a stun grenade and tear gas. What authority do we have to decide for the Army which weapon it should use?"[122] Not surprisingly, although no decision was yet rendered, news agencies were already reporting the next day that: "Israel's Supreme Court has given the army the green light to use controversial flechette tanks shells which

118 Telephone Interview with Yossi Zur, the Petitioner's attorney (May 12, 2003).

119 On January 31, 2003, for instance, nine Palestinians, including two children, were wounded by flechette rounds in the Jabalia area. As expected, this event attracted the media's attention. A photograph of one of the children in a local hospital was disseminated worldwide by Reuters, accompanied by the Palestinians' claim that "hundreds of metal spurs from the shells had struck a nearby residence". Following an initial failed attempt to deny their use, the IDF explained that it targeted "squads which were about to fire *kassam* rockets" and emphasized that "it posses a legal opinion allowing the flechette's employment under certain circumstances." *See,* Arnon Regular, "IDF fires outlawed Flachette shells in Gaza", *Haaretz,* (Feb. 7, 2002) at A2. For detailed medical reports on the injuries sustained in the incident, *see* Letter from Adv. Iyad Alami, Legal Unit, PCHR, To Hadas Ziv, Director of the Occupied Territories Project, PHR-I, (Feb. 23, 2003) (PHR-I Archives, Tel-Aviv). Additional incidents involving flechette use accompanied by allegations of collateral damage occurred, for instance, on Feb. 12, 2003 and on Mar. 6, 2003. For a discussion of the first incident *see, e.g.,* Ibrahim Hazboun (AP), "Israelis Clamp Down on Bethlehem", *The Salt Lake Tribune,* (Feb. 13, 2003), at A2. For news reports on the second incident *see, e.g.,* Alan Philps, "Israeli Army Kills 15 in 'Revenge Attacks' on Palestinians", *The Daily Telegraph* (London) (Mar. 7, 2003), at 17; Justin Huggler, "Eleven Palestinians Killed As Israel Hits Back Over Bus Bomb", *The Independent* (London) (Mar. 7, 2003) at 12. As Philips reported: "Designed to kill infantry in large numbers, they [flechette rounds] are banned from use in urban warfare by NATO armies, but are routinely used by the Israeli army in the Gaza Strip." *Id.*

120 Gideon Levy, "The IDF's 'Permissiveness' in the Territories", *Haaretz,* (Feb. 9, 2003), at B1. "In Israel" – he claimed – "this weapon, which is no different from the appalling devices used by terrorists who pack nails into their explosives, is legal": While the HCJ "obviously has time on its hands", the IDF again targeted children with flechettes." *Id.*

121 *The Flechette Decision*, at 195, para. B.

122 Telephone Interview with Yossi Zur, *supra* note 118. For a critique on this specific comment *see,* for instance, Dan Yahav, "HCJ is in Favor of all Weaponry", *Maariv, Deot* (in Heb.), (Apr. 27, 2003), available at <www.nrg.co.il/online/archive/ART/469/522.html>.

spray thousands of darts over hundreds of meters, ripping apart anyone in the killing zone".[123]

On April 27, 2003, two weeks later, the HCJ issued its Decision. As the Petitioners expected, the Petition was rejected in a short Decision rendered by Judge Mazza. After a short introduction noting the Petitioners' claims, the HCJ adopted the State's position stating as follows:

> the question of whether to prohibit the use of flechette shells…[in the CCW context] has been raised several times before various international forums. However, suggestions to prohibit or limit the use of flechette rounds were never really supported and were removed from the agenda.[124]

Consequently, the HCJ concluded that "the claim of the petitioners, that the laws of war prohibit use of flechette rounds, has no basis and should be rejected",[125] resorting to *Barakeh* case (i.e., the 2001 precedent)[126] they rejected the Petition.[127]

Yet, although the State's position on non-justiciability regarding the Petitioners' claims of proportionality and reasonability was accepted, the HCJ made sure to add the following: "We further note that we think the IDF has properly set out the conditions under which the use of the flechette is authorized."[128] In fact, the Decision includes an apparently redundant section given the acceptance of the non-justiciability claim. It opened by repeating the State's position regarding the military necessity of the weapons, the specific incidents, and the IDF rules of engagement concerning the use of flechette rounds. Then, the State's position that no rule of engagement was breached in those incidents was embraced. It concluded by stating that those civilians injured [i.e., the Bedouin women and the Gazan youths] were likely to have been injured by the firing of ordinary rounds – meaning that "their injuries cannot necessarily be attributed to use of flechette".[129]

*Comments:* A careful review of the HCJ's ruling identifies two fundamental flaws concerning: (a) The HCJ's reporting of the facts of the case; (b) The HCJ's interpretation of the legal aspects.

---

123 South African news agency report, "Israel to Use Flechette Shells", *News24*, (Apr. 14, 2004) available at <www.news24.com/News24/World/News/0,2-10-1462_1347525,00.html>. One of the Petitioners, the PCHR, even announced that "there were indications that the judges would avoid issuing a decision on types of weaponry used by the IDF, as they consider it interference into the work of the army". *See,* PCHR Ref: 45/2003, (Apr. 15, 2003), available at <http://www.pchrgaza.org/files/PressR/English/2003/45-2003.htm>.

124 *The Flechette Decision*, at 195, para. B.

125 *Id.*, para. D.

126 *See,* The *Barakeh Case.*

127 *The Flechette Decision*, at 196, para. A.

128 *Id.*

129 *Id.*, at 195, paras. F- G.

### *DI. The Factual Flaw: Ignoring the Petitioner's Main Case*

The incident that acted as an impetus for the filing of the Petition by PHR-I – the killing of the al-Hajan family by flechette rounds[130] – was not mentioned in the HCJ's short judgment. Certainly this incident, in which almost an entire family, sleeping at the time they suffered from a deadly barrage of metal rain, was a clear indication of the high danger posed by the employment of area weapons to non-combatants within the fighting zones. Moreover, given the circumstances in this case (further discussed below but mainly the fact that the victims were lying close to the ground) it seems that the use of other, available targeted weapons would have prevented the heavy collateral damage that was inflicted in this instance.

Accordingly, the State, in its lengthy Response, made sure to address the incident as well, concluding as follows:

> … the consequences of the incident are tragic, but it should be remembered that this was a case of mistaken identification, during fighting, and cannot attest to the illegality of the means [of warfare] and their mode of use.[131]

The absence of this incident from the HCJ ruling is salient; Justice Mazza, who rendered the Decision, *explicitly* counted the incidents and then cited the State's narrative as follows: "In the petition, two cases were mentioned in which the use of flechette rounds resulted in fatal injuries to civilians."[132] One may doubt whether the incident's absence, albeit the wide media attention it received, was indeed a result of a simple oversight.

This disregard, intentionally or mistakenly, was meaningful vis-à-vis the Court's apparently redundant section regarding the IDF's compliance with the requirements of proportionality and reasonability in employing flechette rounds in the Gaza Strip. Once this incident was missing, the HCJ was forced to formulate its opinion according to the remaining two. However, in reference to one of them (of the three Gazan youths), the HCJ paid too much attention to the State's claim that there was no collateral damage involved given the nature of the victims ("a terrorist squad"); hence, "the petitioner's claim regarding this incident is unclear".[133] Therefore, the absence of any discussion of the al-Hajan incident seemed to allow the Court the following comment: "…as far as one can realize these civilians [apparently only the three

---

130 *The Flechette Petition*, para. 14. The incident occurred on Aug. 28, 2002 and listed in detail *in* Chapter 4(c).

131 *State's Response*, para. 48.

132 As the text reads "In the first incident…, (t)he shell landed near a Bedouin encampment and caused the deaths of three women. In the second incident… flechette shells were fired upon individuals suspected of being terrorists… three youths were killed". *The Flechette Decision*, at 194-5, para. 1.

133 For the quotation *see*, the State's Response, at para. 46. For an indication that the HCJ paid considerable attention to the State's position regarding their status as combatants *see*, *Flechette Decision*, 195, at para. 1.

Bedouin women-E.B.[134]] were likely to have been struck by the firing of ordinary rounds."[135]

Yet, both the exact position of the family's sheds, which were located in a fig orchard, and the fact that all but one person was asleep during the fire, is significant. Excluding a precise hit at the sheds, if either HESH rounds or "hollow" rounds – if the specific IDF tank was Merkava Mark III or IV as no HESH rounds were available for 120mm main guns – had hit the orchard, their lethal range would have been significantly lessened, implying that fewer family members would have been injured or killed. This is particularly true for the majority of the family members who were injured in their sleep (i.e., close to the ground).[136] Moreover, the State's description of this incident may suggest that it was aware of this problematic connection.[137] After all, it is in precisely such field conditions, dense tangled shrubbery, let alone a terrain populated by an orchard with multiple trees with large trunks, which gave the flechette weapon its reputation as an effective weapon, especially in particular when aimed at a living target.

### *DII. Legal Flaws: Ignoring Customary International Law*

Following its acceptance of the State's argument – that the CCW does not prohibit the use of sub-munitions, let alone flechette weapons – the HCJ made the following dubious statement: "the claim of the petitioners, that the *laws of war* [i.e., treaty law and customary law alike] prohibit use of flechette rounds, has no basis and should be rejected".[138]

The question of whether it is permissible to use any means of warfare in the absence of an explicit prohibition was already adjudicated in the *Shimoda* case (1963). In this case, Hiroshima and Nagasaki survivors brought an action against the Japanese government for damages that they and their family members suffered as a result of the 1945 U.S. atomic bombing alleging, *inter alia*, that it was an unlawful act.[139] The plaintiffs, well aware of the lack of any agreed prohibition on the use of

134 The incident occurred on June 9, 2001 and described in detail *supra* chapter 4,C(ii).

135 *Id.*, at 195, para. 7. The conditions under which the family was killed – detailed in *supra* chapter 4, C, (III) – seems to negate the factual basis for this statement. However, this sentence in the decision is missing from the formal English translation, available at <http://elyon1.court.gov.il/files_eng/02/900/089/f04/02089900.f04.htm>.

136 Hence, once the IDF used light weapons in this incident, as detailed in chapter four, only the mother who remained indoors, and went outside only after the round was split, was indeed injured by small arms. *See, e.g.*, Amira Hass, "Hamas Threatens to Take Revenge for the Death of Four Family Members in the Gaza Strip", *Haaretz*, (Aug. 30, 2002), at A6.

137 *See, State's Response*, para. 48.

138 Emphasis added. *Flechette Decision*, at 195, para. 4.

139 For an English translation of the decision *(Ryuichi Shimoda et al. vs. The State*, Tokyo, 7 Dec. 1963) *see*, *Japanese Annual of International Law* 8 (*JAIL*) (1964), 213 [hereinafter, the *Shimoda* case]. An English summary is available at UN Doc. A/9215 (Nov. 7, 1973), Vol. II, at 14-21. For a review and in-depth analysis *see*, Richard A. Falk, "The Shimoda Case: A Legal Appraisal of the Atomic Attacks upon Hiroshima and Nagasaki" *AJIL* 59 (1965) 759-793.

nuclear weapons, claimed that if a weapon is permissible unless explicitly prohibited, then a belligerent is entitled to act as "a Merchant of death, or a Politician of Death".[140] The Japanese government contended in response (as does the Israeli court in our case), that in the absence of an explicit prohibition on the use of nuclear weapons, there is no legal basis upon which to object to their use by belligerents. In a 1964 speech to the Japanese parliament, the Japanese foreign minister noted that he could not "state conclusively as a matter of strict law that the bombing [of Hiroshima and Nagasaki] had been in violation of international law, inasmuch as no rule of positive international law would appear to have existed on this point."[141] Nonetheless, the Japanese Government conceded that it is *possible* that the legality of the use of nuclear weapons could be determined by an appeal to more general principles of international law governing belligerent conduct, but during a war a country "is allowed to exercise all means deemed necessary to cause the enemy to surrender".[142]

However, nearly two decades before, the day after the Nagasaki bombings (August 10, 1945), the Government of Japan had filed a diplomatic protest against the U.S. (via the Swiss Government) over the use of nuclear weapons against Hiroshima based *on customary international law,* specifically relying on Articles 22 and 23(e) of the 1907 Hague Regulations.[143] The protest stated as follows:

> It is an elementary principle of international public law that in time of war the belligerents do not have unlimited right in the choice of the means of attack and that they cannot resort to... means capable of causing the enemy needless suffering.[144]

Therefore, the Government was forced in 1963 to reconcile this contradiction by claiming that "it is not possible to draw the same conclusion today."[145] The *Shimoda* Court accepted the Japanese Government's stance and stated (in the context of nuclear weapons) that so long as international law has not prohibited the use of a specific weapon its use is permitted – a statement which attracted harsh criticism. As Prof. Falk stressed: "a prohibition need not be direct or express to be applicable...and, *if the use of the new weapon violates these principles, it violates international law without requiring any specific rule*".[146]

More than thirty years later, in 1995, the conflict between IHL and nuclear weapons arose again before the ICJ in the *Legality of the Threat or Use of Nuclear Weapons*

---

140 The *Shimoda* case, at 217 as reprinted in Falk, *Id.*, at 761.

141 House of Representatives, 46th Sess. of the Diet, Minutes of Standing Committee for Budget, 16th mtg. (28 Feb.1964), 29-32, as cited in UN Doc. A/9215 (Vol. I), at 166, para. 115.

142 The *Shimoda* case, at 225 as reprinted in Falk, "The Shimoda Case..." *supra* note 139, at 764.

143 1907 Hague Regulations, *Laws of Armed Conflicts*, 72-3.

144 *See Foreign Relations of the United States*, 1945, Vol. VI. (The British Commonwealth, The Far East), 1969, at 472-473. The protest was supplemented by a general appeal to what Falk called the "elementary standards of civilization". Falk, "The Shimoda Case..." *supra* note 139, at 765.

145 *See* Falk, *Id.*, at 765.

146 Emphasis added. *Id.*, at 771.

case.[147] However, this time the discussion was abstract and general as opposed to concrete, as in *Shimoda*.[148] Various legal instruments stretching from all-embracing conventions[149] to those that eliminate nuclear weapons in certain areas (e.g., treaties on Nuclear-Weapon-Free Zones)[150] were used to provide the basis for the argument that nuclear weapons are illegal. Yet, the ICJ found that no convention has amounted "to a comprehensive and universal conventional prohibition on the use, or the threat of use, of those weapons as such".[151]

However, the ICJ then addressed the question according to the principles of customary international law.[152] Furthermore, an examination of the declarations that were submitted by four (the U.S., Russia, England, and France) out of the five declared Nuclear Weapons States (NWS) confirms that these four NWS consider nuclear weapons to be subject to the principles of IHL.[153] The fifth state, China, in line with its consistent and persistent policy of no-first use – or threat – of nuclear weapons[154], avoided the December 1994 UN General Assembly vote referring the question to the ICJ for an advisory opinion.[155] In fact, Russia, England, and the U.S. explicitly declared that the use of nuclear weapons must stand up to the principles

---

147 *Nuclear Weapons Advisory Opinion*. For the UNGA's resolution to request an Advisory Opinion from the ICJ *see*, UNGA Res 49/75K (Dec. 15, 1994).

148 For overview and analysis of this historical legal encounter, or – as labeled by Mohammed Bedjaou, then ICJ President,, – between the "ultimate evil" and "the law of less evil [IHL]" *see* , e.g., Richard A. Falk, "Nuclear Weapons International Law and the World Court: A Historic Encounter" *AJIL* 91 (1997) 64-75; Judith Hippler Bello and Peter H. F. Bekker, "Legality of the Threat or Use of Nuclear Weapons" *AJIL* 91 (1997) 126-133; Chazournes Boisson Laurence & Philippe Sands (eds.), Chazournes Boisson Laurence & Philippe Sands (eds.) *International Law, the International Court of Justice and Nuclear Weapons.* (Cambridge: Cambridge University Press, 1999). For Bedjaou's phrase *see, Declaration of president Bedjaou*, July 8, 1996, para. 20, available at <http://www.icj-cij.org/ docket/index.php?p1=3&p2=4&k=e1&p3=4&case=95>.

149 *See*, e.g., The International Covenant of Civil and Political Rights, U.N. Doc. A/6316 (1966), 999 UNTS 171 (entered into force Mar. 23, 1976); Convention on the Prevention and Punishment of the Crime of Genocide, (1951), 78 UNTS 277.

150 *See, e.g.*, Treaty of Tlatelolco of 14 February 1967 for the Prohibition of Nuclear Weapons in Latin America, and its Additional Protocols, [hereinafter, *Treaty of Tlatelolco*] (entered into force Apr. 22, 1968), *ILM* 28, at 1400.

151 *Nuclear Weapons Advisory Opinion*, para. 63.

152 In the words of the ICJ: "The Court will now turn to an examination of customary international law to determine whether a prohibition…" *Id.*, para. 64.

153 *Russian Federation Written Statement* 10 (June 1995). *UK Written Statement* 40 (June 1995); *US Written Statement* 21 (June 1995). *French Written Statement* (June 1995). All Written Statements are available at <http://www.icj-cij.org/docket/index.php?p1=3&p2=4&k=e1&case=95&code=unan&p3=1>.

154 For Chinese policy, *see, e.g.*, its declaration on the day of its first nuclear test, Oct. 16, 1964, as follows: "The Chinese Government hereby solemnly declares that China will never at any time and under any circumstances be the first to use nuclear weapons." Statement of the Government of the People's Republic of China, Oct. 16, 1964, reproduced in John W. Lewis and X. Litai, *China Builds the Bomb*, (Stanford: Stanford University Press, 1988) 241-3. A similar declaration was made upon signing the Second Protocol annexed to the *Treaty of Tlatelolco*, *ILM* 28, at 1414.

155 *See*, UNGA Res 49/75K (Dec. 15, 1994).

of IHL.[156] The U.S. further noted that the applicability of "various principles of the international law of armed conflict" to the use of nuclear weapons was long ago its declared position "as well as to other means and methods of warfare".[157]

In other words, it seems that once the ICJ provided a clear precedent in reviewing weapons' legality (i.e., customary IHL should be followed once no explicit prohibition was found), the *HCJ should have followed suit.*[158] After all, as the HCJ itself ruled, in regard to the incorporation of international law into Israel's municipal law, "customary international law is part of the law of the land, subject to any contradictory provision in Israeli legislation".[159]

Indeed, the Rome Statute has a closed list[160] of weapons whose use constitutes war crimes (bullets which expand or flatten easily [the so-called "dum-dum" bullets[161]], poison and poisoned weapons,[162] and chemical weapons[163]).[164] Even if the International

---

156 The Russian response, for example, reads: "Restrictions set by the rules applicable to armed conflicts in respect of means and methods of warfare also extend to nuclear weapons. However, we are convinced that there is no general prohibition on the use of nuclear weapons as such in treaties codifying those rules". Russian Federation Written Statement 10 (June 1995). *See,* also the UK's response: UK Written Statement 40 (June 1995).

157 US Written Statement 21 (June 1995).

158 As HCJ President Barak stated in *Mara'abe Case* in which the effect of the ICJ Advisory Opinion on Israel's Separation Fence (*Legal Consequences of the Construction of a Wall in the Occupied Palestinian Territory, Advisory Opinion* [Int'l Ct. Justice, July 9, 2004], *ILM* 43 (2004) 1009) upon the HCJ rulings in the Beit Sourik Case (HCJ 2056/04 *supra* note 16) was discussed, "the Supreme Court of Israel shall give the full appropriate weight to the norms of international law, as developed and interpreted by the ICJ in its Advisory Opinion". *See,* HCJ 7957/04 *Mara'abe v. The Prime Minister of Israel,* P.D. 60 (2) 477, para. 74 [English translation is available at <elyon1.court.gov.il/files_eng/04/570/079/A14/04079570.a14.pdf>].

159 HCJ 785/87 *Affo et. al. v. Commander of IDF Forces in the Gaza Strip* P.D. 42(2) 6. *See also* Cr. A. 174/54 *Shtampfer v. Attorney General* P.D. 10(5), (Deputy President, Justice Cheshin held that customary international law is part of the state's laws, "by force of the State of Israel's existence as a sovereign and independent state" (*id,* at 15). For more on incorporating public international law into Israel's municipal law *see,* Ruth Lapidoth "International Law within the Israel Legal System" *ILR* 24 (1990) 451-484. Specifically, the HCJ had recognized that the 1907 Hague Convention, by virtue of the customs which underlie the Convention, is an established component of domestic Israeli law already during Israel's infancy. *See,* C. A. 1/48 *Sylvester V. Attorney General,* 1 Pesakim 513.

160 *See, Rome Statute,* Art. 8(2)(b), *Laws of Armed Conflicts,* 1309, at 1318-19.

161 Using the wording of the Hague Declaration of 1899. *See,* Declaration (IV, 3) concerning Expanding Bullets, The Hague, July 29, 1899, *Laws of Armed Conflicts,* at 99-100.

162 As stated in 1907 Hague Regulations, art. 23(a), *Id.,* at 73.

163 In this respect it repeated the prohibition in the Geneva Protocol, though without the expansion found in this Protocol regarding biological weapons. *See, Protocol for the Prohibition of the Use of Asphyxiating, Poisonous or Other Gases, and of Bacteriological Methods of Warfare. Geneva, 17 June 1925, Id.,* 105.

164 These prohibitions appeared in the U.S. "short-list proposal" for the diplomatic conference in Rome, which included a finalized list, without any option for further modification, of weapons whose use would be prohibited. For the proposal, *see,* Cherif Bassiouni,

Criminal Court (ICC) found that other weapons "cause superfluous injury or unnecessary suffering" or are "inherently indiscriminate" their use may be defined as a war crime *only* if they were already explicitly prohibited and are included in a would-be annex to the ICC Statute.[165] In contrast to this Statute, Israel's HCJ has no such limitation, which in itself has attracted much critique.[166]

In that sense, the HCJ is similar to the ICTY, whose Statute likewise has no such limitation in relation to "poisonous weapons or other weapons calculated to cause unnecessary suffering."[167] Therefore, a similar critique to the one which we

---

*The Status of the International Criminal Court: A Documentary History* (Ardsley, N.Y., Transnational Publishers, 1998) 124.

165 *See, Rome Statute*, Art. 8(2)(b)(xx). Article 8(2)(b) without the provision regarding the future annex to be added, appeared in both the Canadian proposals and the NGO Coalition for the ICC's proposals for the Statute. *See, Id.*, at 124. It should be noted that both France and Russia have stated a similar position (i.e., explicit prohibition required for a weapon's prohibition by virtue of the principle of humanity) *see, France Written statement* 40-41 (June 1995), *Russian Federation Written statement* 12-14 (June 1995).

166 Some experts noted the possibility of expanding the list at an early Review Conference to include weapons like anti-personnel mines, chemical and biological weapons. Other experts, however, suggested that such an appendix is unlikely to be accepted. For the first position *see,* Timothy McCormack & Sue Robertson, "Jurisdictional Aspects of the Rome Statute for the New International Criminal Court", *Melbourne University Law Review* 23 (1999) 635, at 666. For the second position *see, e.g.* Antonio Cassese, "The Statute of the International Criminal Court: Some Preliminary Reflections" *EJIL* 10 (1999) 144, at 152. In any case, for any amendment which could be proposed only from 2009 onwards (i.e., seven years after the Statute entered into force), a two-thirds majority is required; moreover, an amendment related to this article binds only those states that agree to the amendment. *See, Rome Statute*, Art. 121. Indeed, given the 2010 First Review Conference (May 31–June 11), Belgium has initially proposed various far-reaching amendments in this regard, such as defining the use of cluster munitions in both IAC and non-IAC, prohibited in the CCM, as a war crime, or removal of the Annex Amendment to Article 8(2)(b)(xx). Yet, its proposals in the final revised versions, besides adding provisions regarding the use of poison, gases, and certain bullets in non-IAC (proposal No. 1), the Biological Weapons Convention (BWC), the Chemical Weapons Convention (CWC) and the Ottawa Convention were added to Article 8 itself, with the wording changed from "as prohibited by" the respective conventions to "as defined by and in violation of" the respective conventions (proposal No. 2 & 3). In addition, "(e)mploying weapons as defined by" CCW Protocols I (on non-detectable weapons) and IV (on blinding weapons) has been defined as "war crimes". Eventually, only proposal no. 1 (i.e., adding poisonous weapons and expanding or flattening bullets to the ICC war crimes jurisdiction in non-IAC) was accepted. For the Belgian Revised amendments see *Official Records of the Assembly of States Parties to the Rome Statute of the International Criminal Court, Eighth session, The Hague, 18-26 Nov. 2009 (International Criminal Court publication, ICC-ASP/8/20),vol. I, annex II,* Appendix I, at 59-63. For the final outcome see *Draft Resolution Amending Article 8 of the Rome Statute, Rev. Conf. of the Rome Statute*, May 31-June 11, 2010, I.C.C. Doc. RC/DC/1/Add.1 (June 10, 2010).

167 Statute of the International Criminal Tribunal for the Former Yugoslavia, Art. 3(a) (Violations of the laws or customs of war) adopted May 23, 1993 by UNSC Res. 827, *Laws of Armed conflicts*, 1285, at 1288. This specific article remains unchanged despite the various updates. The Statute's updated version (as amended by UNSC Res. 1877, July 7, 2009) is available at <http://www.icty.org/x/file/Legal%20Library/Statute/statute_sept09_en.pdf>. Interestingly, while the restrictive 1907 English wording was adopted,

present against the HCJ's reluctance to review the weapons legality under IHL, was raised at the time against the *Committee Established to Review the NATO Bombing Campaign against the Federal Republic of Yugoslavia.*[168] The Review Committee was established by the ICTY's Prosecutor Carla Del Ponte in May 1999 following claims from human rights NGOs (e.g., Amnesty International) that NATO's high ranking (military and political) officials were also responsible for grave violations of IHL, including, for the purpose of our case, the use of allegedly prohibited means of warfare, cluster bombs and depleted uranium (DU) projectiles.[169] However, the Review Committee simply *refrained* from examining this use under customary international law.[170] Instead, as the HCJ did in the flechette case, the Committee declared that there is no explicit prohibition on the use of these weapons.[171] One can only wonder whether the political circumstances of the two cases led these leading jurists in two different fora to deviate from the ICJ's precedent and stop short of a proper legal review of the weapon's legality.

the Statute's French text parallels the English one ("*l'emploi d'armes toxiques ou d'autres armes conçues pour causer des souffrances inutiles.*").

168 Final Rep. to the Prosecutor by the Committee Established to Review the NATO Bombing Campaign Against the Federal Republic of Yugoslavia, *ILM* 39 (2000), 1257 [available also at <www.un.org/icty/pressreal/nato061300.htm>].

169 For accusations regarding the use of these two means, *see, e.g.*, NATO/Federal Republic of Yugoslavia – "collateral damage" or unlawful killings? Violations of the Laws of War by NATO during Operation Allied Force amnesty international EUR 70/018/2000 (June 6, 2000), Sec. 3.5. For a discussion on the legality of the use of depleted uranium, *see*, e.g., Jason A. Beckett, "Interim Legality: A Mistaken Assumption? An Analysis of Depleted Uranium Munitions Under Contemporary International Humanitarian Law" *Chinese Journal of International Law* 3 (2004) 43-86.

170 For a comprehensive analysis on the use of cluster munitions in this campaign, *see*, Thomas Michael McDonnell, "Cluster Bombs over Kosovo: A Violation of International Law" *Arizona Law Review* 44 (2002) 31-130. As to depleted uranium, after an initial investigation, on March 26, 2001 the ICRC urged states to review the appropriateness of such a weapon in light of the principles of IHL and in accordance with AP-I, Art. 36, as follows: "all States which study, develop, acquire or adopt munitions containing depleted uranium to carry out such legal reviews if they have not already done so." *Depleted Uranium Munitions*, Comments of the ICRC, Geneva, Mar. 26, 2001. To be reproduced in ICRC, "Depleted Uranium Munitions" IRRC 842 (2001), 543-545.

171 For the discussions of depleted uranium and cluster bombs, *see, ILM* 39 (2000), 1257, at 1264-65 (Sec. A(ii), A(iii)). For a critique of the committee in this context, *see, e.g.*, Paolo Benvenuti, "The ICTY Prosecutor And The Review Of The NATO Bombing Campaign Against The Federal Republic of Yugoslavia", *EJIL* 12 (2001) 503, at 512.

# Chapter 6

## Five Reflections

The detailed overview of Israel's use of flechette weapons and the ensued internal legal debate raise few important questions, mainly concerning the weapon's use in the Gaza Strip. As the overview focuses on the period preceding the Petition to the HCJ (2001-2003), one of the first questions arises as to the use of flechettes afterwards. Given the weapon's high profile in the international media following the 2009 Goldstone Report, its disappearance from the headlines during this long interim period (April 2003-December 2009) should be explored. Then, a survey of its use during that period, mainly at the *Operation Cast Lead* is provided to be followed by some thoughts on the renewed debate the Operation generated over the weapon's legality. The third reflection discusses and explores the reasons for the disturbing gap between the Court's seemingly superficial approach to the Petition and the broader context of it, which should have encouraged a distinctly profound discussion. A fourth reflection is devoted to the applicability of the principles of proportionality and precautions in reference to the local use of flechettes in the Gaza Strip. However, while this discursion is done with reference to two international notable cases in which the principle of proportionality was applied to the issue of weapons selection , the last five reflection focus solely on the Israeli case by discussing the ability of the military's legal advisers to ensure a proper balance between the armed forces' inherent tendency to employ all available means at their disposal and the need to protect civilians in conflict zones, given the experience of IDF MAG and the case of flechette use in the Gaza Strip.

### A. The Low Profile Policy after the 2003 HCJ Decision: Origins and Implications

Shortly after the HCJ rendered its Decision (April 2003) on flechette weapons, the issue disappeared from public consciousness. In contrast, flechette use has continued as Israel recently admitted.[1] There was a short cessation of the weapon's use between the IDF's withdrawal from the Gaza Strip in September 2005 and the date when Hamas took full control of it (June 2007), which precipitated the eruption of vio-

1 As Israel's July 2009 report – prepared in anticipation of harsh *Goldstone Report* findings – states: "Subsequent to this decision [HCJ Ruling], IDF forces continued to make limited use of flechette munitions in the Gaza Strip, including during the Gaza Operation". *See*, Government of Israel, *The Operation in Gaza: 27 December 2008 – 18 January 2009, Factual and Legal Aspects*, (July 2009), at 153, para. 435 [hereinafter, *Operation in Gaza*].

lent clashes with Israel. Once the clashes were renewed, flechettes were returned to the arsenal for use by the IDF, along the security fence demarcating the Gaza Strip's boundaries.

During that interim period (April 2003-June 2007), the media – Israeli and international alike – lost interest, a situation supported by the IDF's almost traditional policy of maintaining a low profile regarding these weapons. Repeated – yet unverified – Palestinian claims associated with flechette thus received scant attention.[2] Indeed, as the IDF had already obtained the HCJ's legal approval for the use of flechette weapons in the Gaza Strip, no further legal intervention was forthcoming.[3] Nevertheless, two factors dictated IDF caution: (a) the problematic image accompanying use of the weapon; and (b) the weapon's origin, the U.S.

As to the first factor, the IDF was fully aware of the salience of the Israeli-Palestinian propaganda war and the need to maintain Israel's image – as well as its own – as a highly moral army.[4] Given the prevailing perception – even among military journalists[5] – that flechette use was highly questionable, the need to maintain a low profile was clear.[6] As to the second factor, any use of U.S.-made weapons by Israel must comply with (1) the 1952 U.S.-Israeli Mutual Defense Assistance Agreement (hereinafter, the 1952 Agreement), according to which Israel may employ these weapons if they are "used solely to maintain its internal security, its legitimate

2 *See,* for instance, *Two Palestinians killed during Israeli shelling of Khan Yunis*, BBC Monitoring Middle East, Dec. 23, 2004, at 1; Amos Harel *et al.*, "Exiled Hamas Leader says Bush 'incited the Zionists' to raid Gaza", *Haaretz,* (Jan. 15, 2008), available at <http://www.haaretz.com/hasen/spages /944967.html>. In this incident the IDF refused to confirm – when questioned by Israeli journalists – whether flechette munitions were used.

3 This factor should not be underestimated, as may be indicated by the mysterious appearance of an Israeli press item reporting that the weapon was to be decommissioned, two days after the flechette Petition was submitted. *See, e.g.*, Felix Friesh, "IDF to decommission 'flechette'", *YNET* (Oct. 24, 2002) (in Heb.), available at <http://www. ynet. co.il/articles/1,7340,L-2200793,00.html>.

4 As Col. (Res.) Dr. Roni Bart wrote following criticism of Israel due to the *Cast Lead Operation*: "… Israel's leaders and IDF commanders revel in the claim that 'the IDF is the most moral army in the world.' *This assertion appears to be justified – or more precisely, perhaps, the IDF may be the most moral fighting army in the world.* Compared with what has happened and occurs in the Third World, with the Soviet/Russian army, with the former colonial powers, and even with the US army from Vietnam to Afghanistan, the IDF comes out on top." Emphasis added. Roni Bart, "Warfare – Morality – Public Relations: Proposals for Improvement", *Strategic Assessment* 12(1) (June 2009) 19, at 20. The original paper, in Hebrew, has a more unequivocal title as follows: "The Most Moral Army – Proposal for Improvement".

5 As Rodan concluded: "The use of flechette rounds in war is not proscribed by the Geneva Convention but their use in internal security operations is more problematic" Steve Rodan, "Israel's Military Debates Use of Flechette Round", *Jane's Defence Weekly* 35, (May 23, 2001), at 4.

6 Any further discussion of the implications of the use of flechettes in the Gaza Strip for Israel's image is beyond the scope of this book. However, given the expected international uproar, in the event of collateral damage occurring, prudence dictates that the use of this weapon should be very carefully considered.

self-defense…";[7] and (2) the 1976 U.S. Arms Export Control Act (AECA), which governs use of U.S.-made defense equipment and services by foreign nations while stipulating that the U.S. may stop aid to countries using U.S. military assistance for purposes other than "legitimate self defense".[8]

While Israel's entanglement with the US over the use of cluster munitions (CMs) during the 1982 Lebanon War has already been mentioned,[9] entanglements have also occurred regarding other U.S.-made weapons,[10] such as aircraft and artillery.[11] As one such incident led to the 1983 U.S. embargo on the sale of CMs to Israel – a ban revoked only five years later[12] – Israel's caution is self-explanatory. In fact, in May 2001, two months after Israel introduced flechette weapons into its roster of munitions for use in the Gaza Strip, a State Department official was asked about the possible review of this use. While confirming that Israel's use of U.S.-made weapons was under constant review, the official refused to confirm a targeted review of flechettes.[13] Such confirmation was provided a year later by Israeli officials who stressed the "informality" of the review.[14]

---

7 Emphasis added. Exchange of Notes Constituting an Agreement between the United States of America and Israel Relating to Mutual Defense Assistance, art. 2, July 1 and 23, 1952, 1953 U.N.T.S. 2363, 140. A third allowed purpose is participation in U.N. collective security arrangements. It is important to note that none of these terms (i.e., internal security, legitimate self-defense, or act of aggression) are defined within the agreement.

8 The Arms Export Control Act, 22 U.S.C. chap. 39(1) § 2754 (1976). For the Presidential and Congressional optional responses to violations under this act *see,* 2753 Sec. 3(c) (c). For a brief overview of the act *see,* Richard F. Grimmett, *U.S. Defense Articles and Services Supplied to Foreign Recipients: Restrictions on their Use*, Congress Research Services (CRS) Rep. (updated Mar. 14, 2005), 3-5.

9 *See, supra* chapter 4, B.

10 It should be noted, however, that CMs enjoy a special status in this respect. A classified Bilateral End-User Agreement detailing Israel's use of U.S.-made CMs, signed on December 16, 1976, was reinforced through an exchange of notes dated April 10 and 11, 1978, following Israeli use of CM in the March 1978 Litani Operation. For more on Israel's diplomatic entanglements over U.S.-made CMs *see,* Eitan Barak, "Doomed to be Violated? The U.S.-Israeli Clandestine End-User Agreement and the Second Lebanon War: Lessons for the Convention on Cluster Munitions", *Denver Journal of International Law and Policy* 38 (2009) 1-40.

11 In August 1979, for instance, the Carter Administration reported to Congress that Israel may have violated U.S. law by using U.S.-supplied arms during its raids into Lebanon. In this case he was referring to U.S.-made aircraft and artillery, which were deployed inside South Lebanon. Israel was forced to announce (despite its self-defense claim) that future use of U.S-made weapons would be more carefully supervised. *See,* e.g., Bernard Gwertzman, "Vance Asserts Israel May Violate Pact by Using US Jets in Air Raids", *N.Y. Times*, (Aug. 8, 1979), at A1; Ron Ben-Yeshi, "Vance, We May Take Measures If Israel Keeps Operating Weapons from U.S. in Lebanon", *Yediot Aharonot*, Aug. 8, 1979, at 3. *See,* also Kevin Danaher, "Israel's Use of Cluster Bombs in Lebanon", *Journal of Palestine Studies 11/12 (4/1)* (1982), 53.

12 *See, supra* Chapter 4, BI.

13 *See,* Rodan, "Israel's Military Debates…" *supra* note 5, at 4.

14 Interestingly, according to one report, in the past the US Administration has requested that Israel refrain from using the weapon in the Israeli–Palestinian conflict. *See,* Eli

The low-profile policy failed once notable collateral damage was inflicted or mistakes occurred. The latter was the case when Fadel Shana, a Reuters TV cameraman was the victim of flechette rounds in a well-documented case.[15] Three bystanders were likewise hurt due to the weapon's wide dispersion (labeled indiscriminate effects by the weapon's opponents).[16] As expected, flechette returned to the headlines, with various human rights NGOs – Israeli and international alike – reiterating their call for the immediate cessation of the use of "this prohibited weapon in the Gaza Strip".[17]

### *B. Flechette Use in Operation Cast Lead: The Goldstone Report and Israel's Position*

Notable collateral damage was apparently frequent during *Operation Cast Lead* (Dec. 27-Jan. 17) [hereinafter, the Gaza Operation]. Indeed, during the Operation, limited press coverage together with the spotlight directed at Israel's use of white phosphorous (WP) completely overshadowed the firing of flechette rounds. Ten days after the Operation's end, on January 27, *Amnesty International* issued the following statement: "Apart from white phosphorus," its fact-finding team had found "hard evidence" of the use of flechettes during the Gaza Operation. "An anti-personnel weapon designed to penetrate dense vegetation... flechettes should never be used in

---

Bohadana, Yoav Limor and Eitan Rabin, "A Palestinian Woman and Her Two Sons Killed by an IDF Shell", *Maariv,* (Aug. 30, 2002) at A6.

15 *See*, for instance, Rory McCarthy, "Palestinian Journalist Killed by Metal Darts from Israeli Shell", *The Guardian* (London), (Apr. 18, 2008), at 24. The fact that Shana himself was the target is explained in a letter sent by IDF Chief MAG, Brig. Gen. Avihai Mandelblit to Reuters upon his August 2008 decision to close the investigation into Shana's death. As Mandelblit explained: "The tank crew was unable to determine the nature of the object mounted on the tripod and positively identify it as an anti-tank missile, a mortar or a television camera". For excerpts of the letter, which Reuters made public almost immediately, *see*, e.g., News Agencies, "IDF probe clears tank crew in death of Gaza journalist last April", *Haaretz*, (Aug. 13, 2008), available at <http://www.haaretz.com/hasen/spages/1011266.html>. Reuters expressed dismay over this decision and, as its Editor-in-Chief David Schlesinger stated: "I'm extremely disappointed that this report condones a disproportionate use of deadly force in a situation the army itself admitted had not been analyzed clearly". *See*, e.g., AP, "Israel clears troops in journalist death", *The Sydney Morning Herald*, (Aug. 14, 2008), <http://www.smh.com.au/articles/2008/08/14/1218307050417.html>.

16 Two children, Ahmad Farajallah and Ghassan Khaled Abu 'Ataiwi, as well as Khalil Isma'il Dughmosh, (22). HRW Press Release, *Israel Investigates Death of Gaza Civilians*, (Apr. 18, 2008), <http://www.hrw.org/en/news/2008/04/18/israel-investigate-death-gaza-civilians>.

17 *See*, *Firing of Flechette Shell that Killed Reuters Cameraman Violates Laws of War, B'Tselem* Press Release (17 Apr. 2008), available at <http://www.btselem.org/english/Press_Releases/20080417.asp>.; *See,* also *Israel: Investigate Death of Gaza Civilians*, HRW Press Release (20 Apr. 2008) available at <http://hrw.org/english/docs/2008/04/19/isrlpa18600.htm.>; *Call for investigation into killing of cameraman and other civilians*, Amnesty International Press Release (18 Apr. 2008) available at <http://www.amnesty.org/en /news-and-updates/news/armys-so-called-inquiry-cameramans-killing-gaza-scandal-20080815>.

built-up civilian areas".[18] The press release was followed the next month by a 45-page brief on the munitions used in the Operation, in which a few cases of civilian casualties resulting from flechette were first mentioned.[19] In the following months, the reawakened Israeli press began publishing highly critical commentaries and extensive articles blaming the IDF for targeting "residential neighborhoods"; details were also provided on the victims.[20]

In July 2009, in a rare moment, Israel presented its position on the use of flechettes in the Gaza Strip, especially during the Gaza Operation. The context was the aforementioned detailed document (169 pages), prepared to rebut the report the upcoming *Goldstone Report*.[21] The short section on weapons (i.e., WP and flechettes) opens with the statement that "[t]he IDF uses only weapons and munitions defined as legal under international law and authorised as such by the relevant IDF authorities, including MAG officers".[22] However, unlike other organized forces,[23] such as

---

18 *Israeli Army Used Flechettes against Gaza Civilians*, Amnesty International Press Release (Jan. 27, 2009) <http://www.amnesty.org/en/news-and-updates/news/israeli-army-used-flechettes-against-gaza-civilians-20090127>.

19 *Israel-OPT–Fuelling conflict: Foreign arms supplies to Israel/Gaza* (Amnesty International MDE 15/012/2009, Feb. 23, 2009), 12-13.

20 One victim, doomed to live with two darts embedded in his body given the complications in removing them from his chest and right thigh, described his feelings: "I feel all the time as though needles are stuck in my body". Amira Hass, "Penetrating the Laws of War", *Haaretz*, (Apr. 17, 2009), at 3. *See*, also Amira Hass, "'The Island of Abroad' in Gaza was not Immune from IDF Bombing", *Haaretz*, (Apr. 24, 2009), at 8 (in Heb.). For acerbic commentaries *see*, Gideon Levy, "Ask Your Sons" *Haaretz*, (May 24, 2009), at 5; Gideon Levy, "Man Responsible for Israel's Cruelest Wars Only Attracts Praise", *Haaretz*, (July 16, 2009), at 5. As Levy wrote: "He [Maj. Gen. Yoav Gallant, GOC Southern Command] orchestrated a war... inculcating an awful attitude within the IDF, one that says it is permissible for us to do anything: to drop thousands of bombs, shells, and missiles in addition to phosphorus shells and flechettes..." *Id.*

21 *See*, *Operation in Gaza*. As Israel boycotted the mission, it was published on its Ministry of Foreign Affairs' website. As noted in the *Goldstone Report*: "On 29 July, it received, through UN Watch, a paper on the military operations in Gaza that sets out the Government of Israel's position on many issues investigated by the Mission", *The Goldstone Report*, at 45, para. 173.

22 *Operation in Gaza*, at 145, para. 405.

23 According to a CCW conference study prepared by Prof. McCormack in 2006, out of the 33 States Parties (from among 100 CCW States Parties) that responded to a questionnaire dealing with explosive remnants of war, "[h]alf of the Respondent States indicated that they conduct some form of legal review (ad hoc or otherwise) of the legality of weapons ... The other half of the Respondent States either had no specific mechanism for weapons review, or made no mention of any review mechanism in their responses". *See*, Timothy .L.N. McCormack, *et al.*, Report on States Parties' Responses to the Questionnaire: International Humanitarian Law & Explosive Remnants of War, UN Doc. CCW/GGE/XIII/WG.1/WP.12 (2006), at 30. As of 2006, the states known to have "in place national mechanisms to review legality of weapons and provide the ICRC with information on the established procedure" are Australia, Belgium, Norway, Sweden, and the UK. *See*, Lawand, Kathleen (principle author), *A Guide to the Legal Review of New Weapons, Means and Methods of Warfare: Measures to Implement Article 36 of Additional Protocol I of 1977*, (Geneva: ICRC revised version, Nov. 2006), at 5, fn.7. According to a non-ICRC publication, Greece and South Africa also have an

the U.S. Army,[24] no mandatory procedure for *reviewing the legality of new* weapons, let alone those already in service, has been adopted in Israel. However, as the case of flechette weapons use in the Gaza Strip indicates,[25] once a weapon is perceived as sensitive by military authorities, they tend to request a legal opinion and non-mandatory consultations.

Although most attention was devoted to the use of WP and its legality, Israel opened its single-page position statement on flechette by repeating the argument raised before the HCJ in 2003, specifically, the weapon's "widespread use around the world for decades".[26] Yet, as in the State's Response to the Flechette Petition[27], the reference given for such a far-reaching argument hardly supports it.[28] After stating that "the use of flechette munitions is regulated both by the IDF standing rules of engagement, as well as by specific professional instructions", the 2003 HCJ Decision is summarized. Interestingly, the 2003 State Response and the subsequent HCJ Decision both *completely overlooked* the fact that much of the Petition was devoted to the legality of flechette weapons under the unnecessary suffering prohibition, a shortcoming to be repeated in the 2009 position paper as well.[29] As to the use of fle-

---

established procedure for legal review of new weapons; see Hays Parks, "Conventional Weapons and Weapons Reviews" *Yearbook of International Humanitarian Law 8* (2005) 55 at 136, fn. 264.

24 The U.S. had established such a procedure by 1974, partly due to the international controversy over the M16 rifle. *See,* DoD Directive 5500.15 (16 Oct. 1974), *Subject: Review of Legality of Weapons under International Law.* However, the Judge Advocate General of each of the military arms comprising the Department of Defense (DoD), i.e., the Army, Navy, Marine Corp and Air Force, promulgated its own directive for executing legal reviews for any new weapon developed internally. For the role played by the international controversy over the M16 rifle *see,* Parks, id, at 69, fn. 49. For a detailed survey on the U.S. process *see, id.*, at 110-114.

25 For the IDF's consultations with Israel's MAG regarding flechettes use in the Gaza Strip *see,* Meron Rapoport, "They Walked in the Fields", *Yedioth Ahronoth,* (Mar. 8, 2002), (Weekend Magazine) at 2 (in Heb.).

26 As the text reads: "Many armies in the world have employed them in a variety of armed conflict situations. Despite the fact that these munitions have been in widespread use around the world for decades…". *Operation in Gaza,* at 152, para. 431.

27 *See, supra* Chapter 5, c.

28 Flechettes are mentioned in the reference (W. Hays Parks, "Means and Methods of Warfare" *The George Washington International Law Review* 38 (*GWILR*) (2006) 511-539), only once as part of the writer's (a retired U.S. army Colonel and senior official at the International Affairs Division, Office of General Counsel, U.S. Department of Defense since 2003) attempt to show that the flechette should not be considered a novel weapon, with reference to its appearance in Vietnam, by pointing to the pre-WWI aerial use of flechettes. *See, Id.*, at 513. For a criticism of this very attempt *see, supra* Chapter 3.,

29 The 2009 position paper merely stated: "The allegation before the Court was that flechette munitions were by nature indiscriminate and therefore illegal under international law. The Court squarely rejected this argument, finding that flechettes were not covered by the CCW" (*Operation in Gaza,* at 152, para. 433), without mentioning unnecessary suffering.

chette during the Gaza Operation and the NGOs' protests,[30] Israel noted as follows: "These claims are currently being examined by the relevant IDF authorities, and it is therefore premature to comment on those incidents".[31]

Release of the *Goldstone Report* in mid September 2009 revealed that it included a section dedicated to "The Use of Certain Weapons" during the Gaza Operation.[32] The report in effect targeted these weapons (flechettes rounds but mainly WP)[33] "in relation to the obligation to take all feasible precautions in the choice of the means and methods of warfare".[34] After finding that "in all cases those hit by these devices were civilians",[35] the Mission stated that flechettes,

> ... are an area weapon incapable of discriminating between objectives after detonation. They are, therefore, particularly unsuitable for use in urban settings where there is reason to believe civilians may be present.[36]

However, unlike the July 2009 Israeli document, the *Goldstone Report* not only attends to the question of whether flechettes inflict much suffering, it also appears to take the position that the suffering is, a priori, unnecessary. The *Report* opens by stating that "(f)lechettes... penetrate straight through human bone and can cause

30 Shortly before this document was published, on July 2, 2009, Amnesty International published its 118-page comprehensive report on the Gaza Operation of which three pages are dedicated to the use of flechettes. *See, Israel/Gaza: Operation "Cast Lead": 22 Days of Death and Destruction* (Amnesty International MDE 15/015/2009, July 2, 2009), 38-40. After noting that "(f)lechette rounds are designed to be used against massed infantry attacks or squads of troops in the open and obviously pose a very high risk to civilians when fired in densely populated civilian residential areas" (Id, at 39), the report detailed three incidents, and concluded that "(t)he manner in which shells containing flechettes were used by Israeli forces in Gaza – fired in densely populated civilian areas – violates the international prohibition on indiscriminate attacks". *Id.*, at 40. Israel's 2009 document detailed three incidents in which the use of flechettes caused the death of six civilians, two of which appeared in Amnesty's report. *See, Operation in Gaza*, at 121.

31 *Operation in Gaza*, at 153, para. 435. The section ends by stating that "(a)s with all other IDF investigations, however, it is certain that any findings shall be subject to review by the MAG, the Attorney General, and possibly an ultimate review by the Israeli courts". *Id.*

32 *The Goldstone Report*, Chap. XII. For a short general overview of its findings and some comments on Israel's reason for denouncing this report *see*, David Kay, "The Goldstone Report" *American Society of International Law Insight* 13(16) (Oct. 2009) available at: <http://www.asil.org/insights091001.cfm>.

33 In addition, the Mission cited allegations regarding the IDF's use of Dense Inert Metal Explosive (DIME) in its "factual findings" section and noted that "further clarifications" were required (at 198, para. 909); as to "depleted and non- depleted uranium munitions" it stated that: "While it cannot be excluded that such weapons were used, on the basis of the information received the Mission decided not to investigate the matter further". *Id.* at 198, para. 911.

34 *Id*, at 194, para. 886.

35 *Id*, at 197, para. 904.

36 *Id.* at 21, para. 48.

serious, often fatal, injuries".[37] Then, perhaps somewhat hastily, it mentions that: "Flechettes are known to bend, break or 'tumble' on impact with human flesh. Such performances are often part of the flechettes' design characteristic and are marketed as such...".[38] More pointedly, in its evaluation of a tragic incident in which the *Report* accused Israel of purposefully targeting civilians with flechette rounds,[39] the *Report* made special efforts to mention the "... severe and *unnecessary suffering* of the civilians affected".[40] Although admittedly acknowledging that flechette weapons are not "currently proscribed under international law",[41] the *Goldstone Report* did call for urgent UNGA discussions on the weapons mentioned in the report, "in particular white phosphorous, flechettes and heavy metal such as tungsten".[42] As to Israel specifically, it recommended that a moratorium should be imposed "on the use of such weapons in the light of the human suffering and damage they have caused in the Gaza Strip".[43]

It is premature to reach any conclusions over the specific incidents mentioned in the *Goldstone Report*, especially given the fact that some are still under IDF inves-

---

37 *Id.* at 193, para. 880.

38 Yet, it added that "flechettes can be designed to be free of these post-impact characteristics if it is desired that they should do so". *Id.*, at 197, para. 905. The report referred to the 1965 Kokinakis and Sperrazza 's research (*see,* William Kokinakis and Joseph Sperrazza, "Criteria for Incapacitating Soldiers with Fragments and Flechettes", *Ballistic Research Laboratories Report* No.1269, Aberdeen Proving Ground, Maryland [Jan. 1965]). However, it is clear the Mission overlooked the required thresholds (i.e., velocities) specified in this research for having such effects. For more on this issue see *infa* chapter 7, v.

39 The *Report* noted in this context that one of the survivors of this incident, where, allegedly, firing was directed at participants at a condolence ceremony (Abd al-Dayem family, Jan. 4), "has several flechettes embedded in his body...and is unable to move freely without pain". *Id.* at 192, para. 873.

40 Emphasis added. *Id.* at 193, para. 883. In this case, the report accused the IDF of firing flechette rounds with the intent to "kill and maim the victims directly and otherwise to terrify the people in the area rather than to pursue any genuine military objective." Hence, the Mission defined it as "willful killing and the willful infliction of suffering". *Id.*, paras. 881-883. As the Report mentioned, Israel did not respond to questions on the use of flechettes during the Operation, particularly regarding this incident. *See, Id.*, at 196, fn. 485. However, in its initial response to the *Report,* (Sept. 24, 2009) Israel duly announced that "directing flechette munitions towards civilians or civilian targets" is among the causes of the new criminal investigations opened by the Military Police "regarding matters more closely connected to 'operational activities'". *See,* State of Israel, *Initial Response to Report of the Fact Finding Mission on Gaza Established Pursuant to Resolution s-9/1 of the Human Rights Council* at 21, fn.23, available at < http://www.mfa.gov.il/MFA/Terrorism-+Obstacle+to+Peace/ Hamas+war+against+Israel/Initial-response-goldstone-report-24-Sep-2009.htm>. In July 2010 Israel's MAG, Maj. Gen. Avichai Mendelblit decided to close the case against the commander of a tank company who was responsible for this incident. *See,* Anshel Pfeffer, "A Year and a Half after [Operation] Cast Lead, Givati Brigade Soldier is Indicted for Manslaughter", Haaretz, (July 7, 2010), at 1, 4 (in Heb.).

41 For the call *see, The Goldstone Report,* at 414, para. 1924.

42 *Id.*, at 425, para. 1971(d).

43 *Id.*

tigation, a process that could offer a clearer picture of Israel's view of the events. Yet, the relatively heavy use of flechettes and the collateral damage inflicted (even to the extent acknowlelged by Israel) are sufficient to imply that in the long run, the 2002 Petition failed to improve the IDF's awareness of the weapon's inherently problematic character. Moreover, as the 2003 HCJ Decision has become the State's first line of defense against any accusations of its improper use of flechette weapons, we can only assume that the HCJ Decision, unintentionally, serves as encouragement for the weapon's continued use. In other words, from a humanitarian point of view, a question mark hangs over the desirability of directing such appeals to national courts and – given the outcome of the ICJ *Nuclear Weapons Advisory Opinion*[44] – some would add international courts as well.

### C. "From Warfare to Lawfare": The Flechette Decision, the Court and the Role of Politics

There is no doubt that the 2003 HCJ debate over the legality of flechette rounds was a rare occasion. Its singularity is reminiscent of the 1965 *Shimoda* case[45], considered twenty years after the the nuclear bombing of Japan, with respect to which Prof. Falk argued that "[i]t is the one and only attempt by a court to assess the legality of atomic and, by extension, nuclear weapons".[46] Yet, irrespective of the uniqueness of

44 *See*, Legality of the Threat or Use of Nuclear Weapons, Advisory Opinion (International Court of Justice July 8, 1996), *International Legal Materials* 35 (*ILM*) (1996), 809 & 1343, [hereinafter, *Nuclear Weapons Advisory Opinion*].

45 *Ryuichi Shimoda et al. vs. The State*, Tokyo, 7 Dec. 1963. English translation available at *Japanese Annual of International Law* 8 (*JAIL*) (1964), 213.

46 Richard A. Falk, "The Shimoda Case: A Legal Appraisal of the Atomic Attacks upon Hiroshima and Nagasaki" *The American Journal of International Law* 59 (*AJIL*) (1965) 759, at 759. Furthermore, in the *Shimoda* case, the legal instance was the Tokyo District Court while the two sides (the five plaintiffs and the Japanese Government) both avoided executing their right of appeal. *Id.* For U.S. national courts' approach to issues of weaponry, *see,* for instance, *the 1984 Greenham Women Case* in which the plaintiffs (U.K. citizens, a U.S. citizen residing in London and two U.S. congressmen) brought an action before the Southern District Court of New York to enjoin the U.S. from deploying nuclear ground-launched cruise missiles from a U.S. base situated at Greenham Common, the U.K. The Defendants sought dismissal on the grounds of lack of subject matter jurisdiction, ripeness and standing. The court denied the plaintiffs' motion claiming, *inter alia*, that "as the court lacked judicially discoverable and manageable standards for resolving the dispute, the case was rendered non-judiciable under the political question doctrine." *See,* 99 ILR (1994), 44-53, at 44-5. The decision was affirmed by the Court of Appeals. *See,* id., at 53. In the U.K., however, in a similar case known as the *Zelter case*, brought to court a few years later (March 2001), the High Court of Justiciary in Scotland clarified after a through discussion of the 1996 ICJ *nuclear weapons Advisory Opinion* that "the relevant rules of conventional and customary international law, and in particular the *rules of international humanitarian law*, are not concerned with regulating the conduct of States (i.e., deployment of nuclear weapons) in time of peace". Emphasis added. *See,* 122 ILR (2002), 618-659, at 653. The context was a response to a query of the Lord Advocate (Reference No. 1 of 2000) regarding three Trident Ploughshares activists Angela Zelter, Bodil Roder, and Ellen Moxley, who were indicted at Greenock Sheriff Court, for damaging components of

the HCJ's debate, there is a disturbing gap between the Court's seemingly superficial approach to the Petition in its substantive (as the factual and legal flaws suggest) and procedural (i.e., its limited scope) aspects, and the broader context of the issue, which should have encouraged a thorough discussion. The most notable factors to have been wrestled with include: (a) The flechette's problematic nature as an *area weapon* in populated areas under the two *principles* governing the legality of weapons (Prof. Ingrid Detter noted, "(s)uch weapons [flechettes] tend to be indiscriminate in their effect over a large area and they also tend to cause much unnecessary suffering by dispersal of tiny fragments"[47] )(b) The international interest generated by the case; (c) The seriousness with which the Petitioners and the State alike accorded to the legal dispute, as expressed in the volume of the documents submitted; and (d) The Court's traditional *rhetoric* in these matters. The latter was pointedly manifest in the statement made by the then HCJ President, Aharon Barak, in his decision over the *Targeted Killings* petition. Following a discussion on "the central principle of the distinction" while referring to the principle's wording in the 2005 ICRC Study on customary international law ("The parties to the conflict must at all times distinguish between civilians and combatants"[48]), Barak stated:

> This approach – which protects the lives, bodies, and property of civilians who are not taking a direct part in the armed conflict – passes like a thread throughout the case law of the Supreme Court.[49]

This gap may be explained by factors such as the complexity of the topic, the dearth of appropriate information and, above all, the HCJ's reluctance – probably affected by the Court's lack of familiarity with the Law of Weaponry – to review the case under customary international law.

However, given the suboptimal conditions for making a judgment, we can legitimately ask why the Court refrained from simply dismissing the entire Petition – as it *had done in the previous year* with respect to the *Barakeh case*, in which two of the three appointed judges participated as well.[50] Alternatively, the Court could also have dragged its feet – as it had done in many delicate instances (including petitions refer-

this weapons system, and were subsequently charged with malicious damage and theft. For a criticism on this decision *see*, Charles Moxley, "The Unlawfulness of the United Kingdom's Policy of Nuclear Deterrence: The Invalidity of the Scottish High Court's Decision in *Zelter*," *58 Disarmament Diplomacy* (June 2001), available at <http://www.acronym.org.uk/dd/dd58/58moxle.htm>. For a similar October 2000 decision by the Divisional Court in a similar case, where another Trident Ploughshare activist (Ms. Hutchinson) was convicted of criminal damage to a fence at the Atomic Weapons establishment at Aldermaston, *see*, 122 ILR (2002), 499-513.

47 Ingrid Detter, *The Law of War*, (Cambridge: Cambridge University Press, 2000), 217.

48 *See*, Henckaerts Jean-Marie and Louise Doswald-Beck, *Customary International Humanitarian Law*, (Cambridge: Cambridge University Press, 2005) at 237 [hereinafter *ICRC Study Report*] Vol. I., at 3 (Rule No. 1).

49 *See*, H.C.J. 769/02 The Public Committee against Torture in Israel v. Gov't of Israel *et al.* 53(4) P.D. 817, at para. 23.

50 *See*, HCJ 5872/01, *Barakeh v. The Prime Minister*, 56(3) P. D. 1.

ring to Targeted Killings[51]) – in the hope that the quandary would be resolved or dropped from the public agenda in response to political developments.

One possible explanation for this deviation from the court's normal reticence in this area is to apply, *mutatis mutandis*, Prof. Michael Schmitt's explanation for the ICJ's willingness to discuss the legality of nuclear weapons. Schmitt wrote:

> It [the ICJ] certainly demonstrated institutional courage in taking on a highly politicized, extremely difficult issue which it could easily have, in its discretion, turned away... Therefore, in the long term the Court enhanced its credibility by taking on the case....Such courage is essential if the Court is to be considered as more than a prestigious arbitral panel.[52]

After all, the HCJ's policy not to intervene in a weapons-related issue was severely criticized within Israel. As Ariel Bendor, an Israeli Professor of Administrative Law noted with respect to the dismissal of the *Barakeh Case*:

> There is a tendency in some Israeli jurisprudence to imply that certain aspects of the fight against terrorism are not institutionally justiciable. For example, in a number of cases it was held that the examination *of the means of warfare* in which the army acts.. *is not one of the subjects in which the court sees fit to interfere*. Yet *such an approach is erroneous* because a refusal to decide a case acts as a de facto approval of the challenged actions.[53]

However, the Court *went beyond* discussing the flechette's legality according to *treaty* law when rejecting the Petition, an act that seemed sufficient for meeting the institutional interests required for discussing the case while eventually rejecting it. Instead, after accepting the State's claim of non-justiciability, the HCJ expressed *unequivocal* support of the State's position and indirectly praised IDF behavior in the Gaza Strip.[54]

---

51 Whilst the petition was submitted in 2002, a ruling was handed down only on Dec. 14, 2006, four years later.

52 *See,* Michael N. Schmitt, "The International Court of Justice and the Use of Nuclear Weapons", *U.S. Air Force Annual Journal of Legal Studies* 7 (1996-1997), 57, at 81.

53 Emphasis added. *See,* Ariel L. Bendor "Justiciability of The Israeli Fight Against Terrorism" *GWILR* 39 (2007) 149, at 163. Perhaps with such criticism in mind, the Court's President at the time of both petitions (the 2001 *Barakeh Case* and the 2003 *flechette case*), Aharon Barak, may have chosen the weapons issue to demonstrate that from the perspective of the HCJ, everything is justiciable. As Barak stated: "The Court does not determine how to engage in a battle, but when civilians are hurt, it can decide *whether the use of the chosen weapons was in compliance with the rules of international law* to which Israel is subject. The army is expert in war; the Court is expert in determining the proportionality of the war's harm to human rights" (emphasis added). Ari Shavit (interviewer) "Aharon Barak, A Worried Citizen", *Haaretz* (Friday Magazine), (Apr. 11, 2008), at 36 (in Heb.). The context of this statement, made after Barak's retirement, was a challenge to the question of whether "everything is justiciable".

54 As summarized in the *Israeli Yearbook of Human Rights*: "...the Court noted that it had been convinced that the IDF has properly set out the conditions for the use of

Therefore, it seems that the unique circumstances within which Israel's HCJ functions may provide a better explanation for why the HCJ, which was *well aware* of the problematic aspects of the case – as attested to by its *obiter dictum* regarding flechette weapon's indiscriminate effects – rejected the Petition.[55] By the same token, the Court was well aware of the *political role* it was playing in the context of this case. After all, the orthodox approach taken by social scientists and jurists, which views the HCJ as an apolitical, objective institution that resolves disputes on the basis of a neutral paradigm (i.e., the rule of law in its formal and material sense) was undermined by the 1980s.[56] Instead, a new approach that views the Court as a political institution, with all that this entails, has since taken hold primarily among social scientists in Israel. In reference to petitions from residents in the Occupied Territories touching upon security matters , they argue that the HCJ has adopted the basic principles of a political dialogue based on a friend/foe axis.[57] One researcher, Prof. Gad Barzilai, has termed this state of affairs as the "illusion of separation" between the law and politics in democracies in general, and in Israel in particular.[58]

---

flechettes.... Referring to the explanation presented by the respondents in regard to the events in which innocent civilians had been killed, the Court opined that there is no reason to believe that the use of flechettes in both events was contrary to the instructions". Fania Domb, "Judgments of the Supreme Court of Israel Relating to the Administered Territories" *Israeli Yearbook of Human Rights* 34 (2004) 293, at 321.

55 As the Court noted: "[The] Respondents do not dispute that the use of flechettes – which are intended to be used against military targets in a combat zone – entails increased danger towards civilians that inadvertently enter the zone of fire". HCJ 8990/02, *Physicians for Human Rights v. O.C. Southern Command*, P. D. 57(4) 193, at 195 D [hereinafter *the Flechette Decision*].

56 Aharon Barak, an eminent proponent of this approach, had stated "... judicial rulings are made fairly, objectively, neutrally, and without bias. *It is not the identity of the claimants, but the weight of their claims – which decides the verdict* ...the judge is not party to the legal battle and he is not fighting for his own power but for the rule of law" (emphasis added). *See,* Aharon Barak, *Judicial Discretion* (Tel Aviv: Papyrus, 1987), 292 (in Heb.). For further support of the orthodox approach *see,* Eliakim Rubinstein, *Judges of the Country* (Tel Aviv: Shocken, 1981) (in Heb.).

57 In a comprehensive study of HCJ rulings regarding the Occupied Territories in the 1990s, it was found that the HCJ was prepared to accept the claims of the authorities pertaining to security considerations, without thoroughly probing the matter and without examining the claim's legitimacy or the necessity for government action. *See,* Leon Sheleff, "The Green Line is the Border of Judicial Activism: Queries about Supreme Court Judgments in the Territories", *Iyunei Mishpat* (Tel Aviv University Law Review) 17 (1993), 757 (in Heb.).

58 According to Prof. Gad Barzilai, the Court's then exclusively Jewish justices are inclined to serve the aspirations of the Jewish majority. Hence, the Court has legitimated the implementation of the notion of 'rule of law' according to the military's political requirements, preferences and purposes. Barzilai has attempted to demystify the view that on issues of national security, the Court has been an objective arbitrator, acting purely on the merits of objective security considerations. *See,* Gad Brazilai, "Political Institutions and Conflict Resolution – The Israeli Supreme Court and the Peace Process", in: Ilan Peleg (ed.), *The Middle East Peace Process – Interdisciplinary Perspectives*, (Albany NY: State University of New-York Press, 1988), 87-105.

Once non-legal considerations were included, strong considerations – such as the propaganda war with the Palestinians – arose in favor of the Petition's rejection given its high political sensitivity.[59] Furthermore, these non-legal considerations may have motivated the Court to go beyond mere rejection of the Petition (on the basis of no explicit prohibition on the use of flechette) and uphold the State's claims as to flechette use specifically in the Gaza Strip, including claims to the weapon's irrelevance regarding harm to civilians, that had motivated the Petition.[60]

However, a careful reading of the State's Response indicates that no supporting evidence (such as an affidavit issued by a weapons expert) was submitted to justify its decision. A year later, the HCJ rejected a petition presented by residents from Rafah requesting the nullification of an IDF decision to demolish their family homes in a decision that shared a similar structure with the flechette decision (i.e., a short dismissal with favorable comments made by the Court regarding the IDF's actions). In the Rafah case, the HCJ essentially accepted the State's explanation of "immediate military necessity" without reviewing the demolition order's proportionality and reasonableness.[61] In response, Prof. Ze'ev Segal, an expert on Israeli Constitutional and Administrative Law, wrote the following:

> It would have been better for the court to choose one of two other courses of action: Either it should have declared that under the circumstances, this is a nonjusticiable security matter and *the court will therefore not examine the substance of the claims*, or *it should have examined it in depth by requiring affidavits and even cross-questioning those who filed the affidavits*... the public's faith in it [the Court] is liable to be undermined if the justices of the nation's highest court are willing to serve as a "rubber stamp" for the defense establishment. The High Court should not serve as a fig leaf for the government.[62]

The HCJ's informal function as the Israeli government's "fig leaf",[63] especially in cases such as that of flechette use, should not be underestimated given "the rising

59 The fact that the Jerusalem Magistrate's Court was required to address the question of whether an advocate's consent to submit such a petition constitutes evidence for his political identity as "non-Zionist" made this heightened sensitivity eminently clear. The context, however, was of a libel suit filed by one of the Petitioners' advocates, Yosef Arnon, against a right-wing advocate and fellow member of the Israel Bar Association's Central Committee. The judge (Noam Solberg) rejected the claim stating that "in my opinion...I do not believe that the position taken by Y. Arnon evidences a non-Zionist worldview." See C.C. (Jerusalem) 5665/02, Yosef Arnon v. Yaron Pastinger, Pad-Or 04(8) 296, para. 47 (2004).

60 As the Court stated: "The civilians injured by the flechettes would probably have been injured by standard shells as well." *Flechette Decision Id.*, at 195, para. G.

61 *See*, HCJ 4585/04-B *Shakfahat v. IDF Commander in Gaza Strip*, (2) Takdin-Supreme 1284. *See*, also HCJ 4573/04 *Al-Besioni v. Commander of the IDF Forces*, P. D. 2004(2) 1580.

62 Ze'ev Segal, "When Security is Above the High Court", *Haaretz*, (May 24, 2004), at B1.

63 For the empirical basis of this perspective in reference to the Occupied Territories, *see, e.g.*, David Kretzmer, *The Occupation of Justice: the Supreme Court of Israel and the Occupied Territories*. (Albany NY: State University of New York Press 2002) 2, 190.

phenomenon of law as a weapon capable of producing military advantages or disadvantages on the twenty-first century battlefield":[64] "Lawfare",[65] defined as "the use of law as a weapon of war", a concept coined by Charles Dunlap in 2001,[66] reached the point where a recent EJIL editorial entitled "Gaza – From Warfare to Lawfare" was forced to protest that "where conjecture is presented as established fact, where legal submission is presented as legal conclusion, where the arrow is stuck and the target is then drawn around it, brings honour to no one".[67] Given this new environment, the HCJ's Decision can be duly presented by executive authorities as a first response to any protest against the use of any area weapons in populated areas.[68]

Therefore, from a humanitarian perspective, it appears that the Petitioners lost more than they gained. For those who opposed the use of flechettes, the Court – in the words of Israel's former Minister of Education (1992-1993), Shulamit Aloni – "had made the inedible fit to eat".[69] As expected, some of those involved in the Petition's submission have already regretted their action.[70]

---

64 *See,* Troy R. Stone, "War Is Too Important to be Left to the Lawyers", (Naval War College Newport RI Joint Military Operations Dept. Report) (Oct. 2008) available at <http://www.stormingmedia. us/06/0634/ A063494.html>.

65 As Yehuda Ben-Meir, Israel's Deputy Foreign Minister (1981-1984), wrote regarding the HCJ's "tremendous contribution to Israel's diplomatic and information struggle", the Court's "... profound involvement in the activities of the Israel Defense Forces and Israel as a whole in the territories, and its thorough and profound rulings on these issues *have helped Israel rebuff harsh diplomatic and propaganda attacks on it by various actors worldwide*". Emphasis added. Yehuda Ben-Meir, "The Country's Flak Jacket", *Haaretz* (Dec. 19, 2006), at B1.

66 Charles J. Dunlap, "Law and Military Interventions: Preserving Humanitarian Values in 21st Conflicts," Prepared for the Humanitarian Challenges in Military Intervention Conference, Washington D.C., Nov. 29, 2001, at 2, available at <http://www.duke.edu/~pfeaver/dunlap.pdf>.

67 *See,* Editorial, "Gaza – From Warfare to Lawfare", *The European Journal of International Law* 20 *(EJIL)* (2009), *259-262*, at 261. For more on this new aspect of warfare *see,* Elizabeth Samson, "Warfare through Misuse of International Law", *Perspectives* 73, Papers on Current Affairs (Mar. 2009) (Bar Ilan University, BESA Center), available at <http://www.biu.ac.il/SOC/besa/perspectives73.html>. For lawfare against Israel solely *see,* e.g., Anne Herzberg, *"Lawfare": Exploitation of Courts in the Israeli-Arab Conflict* (NGO Monitor Monograph Series, Sept. 2008).

68 A typical IDF spokesman's response to a query about IDF policy regarding the use of cluster bombs and flechettes: "For operational reasons, the IDF does not detail the types of weapons at its disposal, and the circumstances of their use. As for the IDF's use of flechette shells, see the ruling by the High Court... and the State's response to the petition, attached for your perusal." Letter from the head of the Public Affairs Department, Capt. Henrietta Levi, *The Use of Cluster Bombs and Flechette Shells in the IDF* (Dec. 23, 2003) (on file with the author).

69 Literally, "to declare the vermin *kosher*," [*hichshir et hasharatz*]; Shulamit Aloni, "A License to Kill Civilians", *Haaretz*, (May 4, 2003), at B2 (in Heb.).

70 Interview with Yosi Zur (Dec. 13, 2006).

### D. The Use of Flechettes in The Gaza Strip: Some Reflections on the Principles of Proportionality and Precautions in Attack

The main objective of Part Two is to provide a solid basis for a reappraisal of the legality of flechette weapons – a task to be continued in the next two chapters. If the flechette is to be considered a weapon that inflicts unnecessary suffering, it should be prohibited at any place and time; and vice versa: if not, its employment is to be permitted under all circumstances. However, as to the principle of distinction, even if flechette weapons were not to be considered as inherently indiscriminate weapons, this judgment is far from constituting a warrant for their employment in all circumstances. As the prominent jurist Stefan Oeter stated in 1995: "The traditional approach, according to which all collateral damage that is unavoidable in attacking a certain military objective must be accepted, is no longer valid".[71] To reiterate the words of Judge Rosalyn Higgins, application of the *principle of proportionality* in attacks dictate that "even a legitimate target may not be attacked if the collateral civilian casualties would be disproportionate to the specific military gain from the attack".[72]

Therefore, although the use of flechette in the Gaza Strip served as a point of departure for a reappraisal of its legality *per se*, its compliance with the said principle in the *specific* circumstances endemic to that arena, merits some attention. After all, as the *Goldstone Report* stated: "While international humanitarian law does not explicitly prohibit the use of flechettes in all circumstances, the principles of proportionality and precautions necessary in attack render their use illegal".[73]

The proportionality principle, for which Higgins found "no specific mention"[74] in treaty law, is currently "a norm of customary international law applicable in both international and non-international armed conflicts".[75] The principle was first set out in treaty form[76] and codified in Art. 51(5)(b) of AP-I, although the term "excessive " was preferred to that of "disproportionate". According to this provision – iterated in Article 57(2)(a)(III) of AP-I– an attack is disproportionate and thus indiscriminate if it

> ... may be expected to cause incidental loss of civilian life, injury to civilians, damage to civilian objects, or a combination thereof, which would be excessive in relation to the concrete and direct military advantage anticipated.[77]

71 Oeter Stefan, "Methods and Means of Combat" in: Dieter Fleck (ed.) *The Handbook of Humanitarian Law in Armed Conflicts*, (Oxford: Oxford University Press, 1995), at 173.

72 Dissenting Opinion of Judge Higgins, the *Nuclear Weapons Advisory Opinion*, at para. 20. All Declarations, Separate and Dissenting Opinions of the Judges are available at <http://www.icj-cij.org/docket/index.php?p1 =3&p2=4&k=e1&p3=4&case=95>.

73 *The Goldstone Report*, at 193, para. 880.

74 *Id.* Some writers, however, argued that the principle can be inferred from Arts. 15 and 22 of the Lieber Code. *See*, A. P. V. Rogers *Law on the Battlefield*. 2nd ed. (Manchester: Manchester University Press, 2004), at 17. For a brief overview of the principle's origin *see*, Judith Gail Gardam, "Proportionality and Force in International Law," *AJIL* 87 (1993) 391, at 394-403.

75 *ICRC Study Report*, Vol. I., at 237.

76 Rogers, *Law on the Battlefield*, *supra* note 73, at 19.

77 For the history of this important paragraph *see*, *ICRC Commentary*, *paras. 1967 -1981*. For an extensive discussion on the proportionality principle in modern IHL *see*, Gardam Gail

Since then, this provision was reiterated in the 1980 CCW Protocol II on the Use of Mines, Booby-Traps and Other Devices[78] and the 1996 amended Protocol II.[79] Moreover, "intentionally launching" such a disproportionate attack was defined as a war crime in the Rome Statute.[80] Nonetheless, as an in-depth study on this principle noted: "The determination, however, of what attacks will fall within this prohibition is obviously a subjective one and will often be difficult".[81]

The same can be said when this principle is applied to weapon selection. In other words, do the particular weapons selected by belligerents for carrying out their attacks on military targets ensure that the attack will not fall within the AP-I prohibition given the expected effects of these weapons?

Article 57(2)(a)(ii) of AP-I, which deals with "Precautions in attack", specifically imposes an obligation on "those who plan or decide upon an attack", to wit:

> Take all feasible precautions in the choice of *means* and methods of attack with a view to avoiding, and in any event to minimizing, incidental loss of civilian life, injury to civilians and damage to civilian objects...".[82]

However, as the ICRC's Commentary considered this provision – which was also recognized (as well as Article 58) by the trial Chamber of the ICTY as "part of customary international law"[83] – regarding weapon selection, "their precision and range

---

Judith, *Necessity, Proportionality and the Use of Force by States*, (Cambridge: Cambridge University Press 2004), 59-84. For state practices *see*, *ICRC Study Report*, Vol. II, at 297-335.

78 Protocol on Prohibitions or Restrictions on the Use of Mines, Booby-Traps and Other Devices (Protocol II). Geneva, 10 October 1980, Art. 3(3) in: Dietrich Schindler and Jiri Toman, (eds.) *The Laws of Armed Conflicts* (Leiden: Martinus Nijhoff, 2004), 191 [hereinafter, *Laws of Armed Conflicts*]

79 Protocol on Prohibitions or Restrictions on the Use of Mines, Booby-Traps and Other Devices, Amended Protocol II (entered into force Dec. 3, 1998), Art. 3(8), *Laws of Armed Conflicts*, 196.

80 Rome Statute of the International Criminal Court, July 17, 1998, Art. 8(2)(b)(iv). Id, 1309 at 1318.

81 Gardam, "Proportionality and Force...", *supra* note 74, at 407. Indeed, other jurists who have debated this principle admit that: "The precise scope of the rule... is not entirely clear" while noting that "it probably prohibited military acts that were grossly disproportionate to the object to be obtained". Rogers, *Law on the battlefield supra* note 74, at 19.

82 Emphasis added. *Laws of Armed Conflicts*, at 739. However, as the ICRC Commentary noted, in order to avoid all ambiguity, it would have been better to say "with a view to reducing incidental loss [...] to a minimum". *Commentary on the Additional Protocols of 8 June 1977 to the Geneva Conventions of 12 August 1949* (Claude Pilloud *et al.*, eds., 1987), (hereinafter *ICRC Commentary*), 683, para. 2203.

83 *See*, Kupreskic *et al.* Judgment, (2000), Case No. IT-95-16-T (ICTY; Trial Chamber), para. 524. For such a position, declared long before the ICTY statement, *see*, e.g., Christopher Greenwood, "Customary Law Status of the 1977 Geneva Protocols", in Astrid Delissen & Gerard Tanja (eds.), *Humanitarian Law of Armed Conflict, Challenges Ahead, Essays in Honour of Frits Kalshoven*, (Leiden: Martinus Nijhoff, 1991), 100, at 111. For more on the precaution's requirement and its roots in the *1907 Hague Regulations see*, Yoram Dinstein, *The Conduct of Hostilities* under *the* Law *of International Armed Conflict*, (Cambridge: Cambridge University Press, 2004) 125-129.

should be taken into account". The example it provides is the uselessness of firing a series of projectiles or a 10-ton bomb when "a well-placed 500 kg projectile is sufficient to render a military objective useless".[84]

As might have been expected, the application of this principle arose in the *Legality of Nuclear Weapons* case (1996). States opposing their legality argued that it was inconceivable that circumstances could arise where the use of nuclear weapons would comply with the provision's principle requirement; in their argument, they also pointed to the "high probability of an escalation of nuclear exchange".[85] The ICJ, however, refrained from resolving the issue.[86] As one of the U.S. representatives to the discussions, Michael Matheson, later noted:

> The rule of proportionality... necessarily requires a case-by-case assessment of the specific circumstances of the particular situation, and it is not possible in the abstract to conclude that all uses of a weapon will violate the principle".[87]

As for conventional weapons, the principle was applied, for instance, in the case of Milan Martic, the former political leader of Croatian Serbs. Martic was accused of ordering the shelling of Zagreb on May 2-3, 1995, with rockets fitted with cluster munition (CM) warheads ('Orkan M-87'),[88] causing the deaths of seven civilians and wounding over a hundred.[89] In the proceedings conducted to reconfirm the indictment against an accused not in custody (he surrendered in 2002), pursuant to Rule 61 of the ICTY Rules of Procedure and Evidence, the Trial Chamber stated that there is no formal provision forbidding the use of CMs as such.[90] Yet, it noted that the principle of proportionality is part of customary international law. Given the evidence, including the

---

84 *ICRC Commentary*, at 682, para. 2200. It should be noted, as the Commentary stressed, that in and of itself, "this rule does not imply any prohibition of specific weapons". Id., at 682, para. 2201.

85 *Nuclear Weapons Advisory Opinion*, para. 43. As Mexico stated before the Court "Even by actualizing the requirement of absence of deliberation or choice of means, the condition of proportionality would not be respected since in the case of a nuclear attack *the proportionality of a nuclear response would be uncontrollable and would endanger the security of mankind as a whole*" [emphasis in original]. *Mexico Written Statement* 10 (June 1995). *See*, also, e.g., *Egypt Written Statement* 10 (June 1995) and *Sweden Written Statement* 3 (June 1995). The latter stated as follows: "It would be difficult to regard this principle as consistent with the use of nuclear weapons in retaliation against an attack using conventional weapons". All written statements are available at <http://www.icj-cij.org/docket/index.php?p1=3&p2=4&k=e1&case=95&code =unan&p3=1>.

86 *Nuclear Weapons Advisory Opinion*, para. 43.

87 Michael J. Matheson, "The Opinions of the International Court of Justice on the Threat or Use of Nuclear Weapons" *AJIL* 91 (1997) 417, at 428.

88 For technical details on the Orkan 262 mm rocket fitted with 288 bomblets *see*, Anthony G. Williams & Leland Ness, (eds.), *Jane's Ammunition Handbook 2005-2006* 14th ed. (London: Jane's Information Group 2005) 660.

89 *See*, *Prosecutor v. Martić*, Decision (No. IT-95-11-R61, T. Ch. I, 8 Mar. 1996) ("*Martić Rule 61 Decision*"). Note: all indictments, trial decisions and other related material concerning this case are available at <http://www.icty.org/case/martic/4>.

90 *Id.*, para. 18.

fact that the specific type of rocket fired was considered inaccurate,[91] the Chamber concluded that "attacking and terrorising the civilian population was the main reason for using such rockets"[92] and thus its use went counter to the principle of proportionality.

A more comprehensive discussion on the principle's applicability in the context of weapons selection was conducted by the *Committee Established to Review the NATO Bombing Campaign against the Federal Republic of Yugoslavia* (hereinafter, Review Committee).[93] The Review Committee, which discussed, *inter alia*, NATO's use of CMs and depleted uranium projectiles,[94] found that framing the principle was simple in comparison to its application, which must be judged on a case-by-case basis.[95] Hence, it proposed four secondary tests – albeit for assessing military commanders' criminal responsibility as opposed to state responsibility – such as "the relative values to be assigned to the military advantage gained" and the injury to non-combatants and/or the damage to civilian objects or the extent to which "a military commander was obligated to expose his own forces to danger" in order to limit collateral damage.[96] Nevertheless, it admitted:

> The answers to these questions are not simple. It may be necessary to resolve them *on a case by case basis*, and the answers may differ depending on the background and values of the decision maker.[97]

In sum, conclusions as to whether the collateral damage inflicted by the use of flechette weapons in the Gaza Strip is proportional to the military target for which they were employed are troublesome to reach. In all of the cases tested, the State defined the objective in employing flechette weapons as "injury to Palestinian terrorists and preventing attacks carried out by them".[98] With such a general definition, it is clear,

91 Other evidence: (II) the rockets landed in areas with no legitimate military targets; (III) The rockets were used as an anti-personnel weapons; (IV) Martić himself indicated that he intended to attack the city as such. *Id.*, paras. 23-31.

92 *Id.*, para 30. In January 2007 he was sentenced to 35 years in prison (see Prosecutor v. Milan Martić (Judgment)). (No. IT-95-11-T. ICTY, 12 June 2007). In October 2008 the ICTY Appeals Chamber dismissed his appeal and confirmed the sentence while sustaining the Trial Chamber's finding that the M-87 Orkan – as used under the circumstances of the case – was an indiscriminate weapon and that Martić knew of the effects of this weapon when he ordered the shelling. *See*, Prosecutor v. Milan Martić (Appeal Judgment). (No. IT-95-11-A, ICTY. 8 Oct. 2008), at 94.

93 Final Rep. to the Prosecutor by the Committee Established to Review the NATO Bombing Campaign Against the Federal Republic of Yugoslavia, *ILM* 39 (2000), 1257, at 1271 [hereinafter, *NATO Bombing Report*] .

94 An important non-weaponry issue was the environmental damage caused by NATO bombing of "industrial facilities such as chemical plants and oil installations" *Id.*, at 1261, para. 14.

95 *Id., at 1271*, para. 48.

96 *Id., at 1271*, para. 49.

97 Emphasis added. *Id.*, at 1271, para. 50.

98 *See*, HCJ 8990/02, *A Response by the State Attorney's Office*, para. 40, available at <www.btselem.org/hebrew /Legal_Documents/8990_ 02_Flechette_Appeal_State_Response.pdf>.

for instance, that in the employment of flechette in the incident involving the three Gazan youths (December 2001), no collateral damage should have been expected – and indeed no such damage was inflicted – given the specific circumstance (at night and within an exposed area with no residents observed in its vicinity). In contrast, a different conclusion may be reached regarding the al-Hagan family (August 2002), in which the site, in proximity to the Netzarim settlement, was populated by civilians. In other words, a permanent civilian presence was found not only adjacent to the anticipated areas of the weapon's deployment, it literally characterized that area. In fact, in repelling the criticism over this incident, the IDF claimed that "the family members were previously warned not to wander in the area [i.e., their orchard] at night".[99] This very warning bears witness to the persistent presence of civilians and, more importantly, to the fact that the IDF was quite conscious of this when it chose to fire flechette rounds.[100] As one scholar pointed out while discussing U.S. rules of engagement in Vietnam,[101] they were:

> particulary noteworthy for suggesting ways in which some of the indiscriminate effects of the use of *fragmentation weapons* could be reduced…they show that it *is necessary and possible to determine safety zones for specific weapons in order to protect civilian populations.*[102]

However, viewed from another perspective, the Review Committee did cite a statement made by the trial Chamber in the *Kupreskic* Judgment that might contradict the pervious conclusion:

> In case of repeated attacks, all or most of them falling within the grey area between indisputable legality and unlawfulness, it might be warranted to conclude that the cumulative effect of such acts entails that they may not be in keeping with international law.[103]

As the Review Committee itself clarified, the "mere *cumulation*" of individual and legitimate attacks on military objectives "cannot *ipso facto* be said to amount to a crime". Instead, the Review Committee stated that it understands the formulation as referring "to an *overall* assessment of the totality of civilian victims as against the

99 *See,* Eli Bohadana, *et al.*, "A Palestinian Woman and her Two Sons Killed by an IDF Shell", *Maariv,* (Aug. 30, 2002) at A6.

100 One can interpret the State's defense, according to which the main factor responsible for the injury to the family was a case of "mistaken identity", as an explicit admission of knowing that civilians might be present in the area.

101 As one of the rules reads: "Unobserved fires will not be fired closer than 1000 meters to noncombatant or friendly troop locations when engaging known or suspected targets". *See,* Para. *4b* (5) *(e)),* as cited in SIPRI, *Ant-Personnel Weapons (*New-York: Taylor and Francis, 1978*) 26*

102 Emphasis added. Lumsden, Malvern "The Prohibition of Inhumane and Indiscriminate Weapons" *SIPRI Yearbook 1979: World Armaments and Disarmament*, (New-York: Taylor & Francis, 1979) 453-462, at 458.

103 *See, Kupreskic* Judgment (Case No: IT-95-16-T 14 Jan 2000), para. 526.

goals of the military campaign".[104] Essential data required for such a review in the case of flechette use in either South Lebanon or the Gaza Strip are missing, specifically, the number of times in which flechette rounds were safely employed (i.e., no collateral damage was inflicted).[105] Hence, any unequivocal determination that the number of flechette rounds used in the Gaza Strip "amounts to a crime" is precluded.

### E. The Use of Flechettes in the Gaza Strip and Israel's MAG: A Call for Specific Rules

Nevertheless, the Israeli case does provide clear evidence – if there was still need for any within the framework of this book – that despite the lack of explicit prohibitions, flechettes are anything but ordinary weapons. As the review of Israel's use of flechettes in the Gaza Strip has revealed, this view was not solely that of the Petitioners. The IDF seems to share this view as well – as the following four events indicate: (a) *The internal IDF debate over the weapon's introduction*, *before* any collateral damage was inflicted. This debate indicates that the weapon's sensitivity was clear to all involved, especially considering the specific circumstances of its intended use: low intensity conflict in densely populated areas; (b) *The request for the MAG's approval* of its employment. In fact, as already noted, unlike some armies, the IDF has not formulated even one standard operating procedure regarding a new weapon, let alone one 30 years in service. Israel is also not a State Party to AP-I and is therefore exempted from the requirement (Article 36) to conduct such a legal review of its weapons choices;[106] (c) As the Petitioners rightly mentioned, unlike the IDF's Southern Command, in which Maj.-Gen. Doron Almog decided to employ flechette rounds,[107] Maj. Gen. Yitzhak Eitan, then the Commander of the IDF Forces in the West Bank, had "completely prohibited" their use in the West Bank due to the "concern for civilian casualties".[108] In addition to recognition of the weapon's *indis-*

---

104 *NATO Bombing Report*, at 1271, at para. 52.

105 The Committee concluded, for instance, that the NATO Air Campaign was legal despite the large number of casualties (approx. 495) and wounded (820) due to the high number of strike sorties that were conducted (10,484). Id., paras. 53-54. For a critique of this position, *see*, e.g., Orna Ben-Naftali and Yuval Shany, *International Law between War and Peace*, (Tel Aviv University Press, 2006) (in Heb.), 156.

106 As long as this specific article is yet to be considered a customary one. Yet, as the ICRC guide to the legal review of new weapons stated "The requirement that the legality of all new weapons…be systematically assessed is arguably one that applies to *all* States, regardless of whether or not they are party to Additional Protocol I". *See*, Kathleen Lawand, *Guide to the Legal Review…"*, *supra* note 23, at 4.

107 Almog's stubborn struggle against the IDF Chief MAG and other members of the General Staff on both the legality of his imprecise instructions on opening fire in the Gaza Strip, and the necessity of preventing Palestinians from requesting any judicial hearing prior to their homes' destruction, suggests that individual views played a prominent role in the different decisions about the use of the flechette weapon in the two arenas. For Almog's struggle *see*, Harel Amos & Avi Isacharoff, *The Seventh War*, (Tel-Aviv: Miskal, 2004), at 263-64 (in Heb.).

108 *See*, Amos Harel, "Weapons Effective in South Lebanon Becomes Problematic in the Gaza Strip", *Haaretz*, (June 11, 2001) at A1 and A2 (in Heb.). The reporter noted that

*criminate effects*, this claim had great weight at the time of the Petition's rejection given the Israeli estimate of population density in the Gaza Strip as 2,800 per km$^2$, as opposed to 345 per km$^2$ in the West Bank.[109] The State was thus forced to explain in its Response that the reason for the different decisions in each Command stemmed from the "difference between the areas, ... the character of the IDF's operations, the conditions under which they are conducted, the ... terrain, and the operational scenarios";[110] and (d) Following the IDF investigation of the al-Hajan family incident (August 2002), the tank company's deputy commander was suspended for employing flechette rounds without authorization, in addition to the fact that such use was not permitted at targets located beyond the range of 400 meters.[111] Put briefly, all these details confirm the fact that of all the means of warfare available to the IDF's commanders in the Gaza Strip, flechette rounds required special authorization prior to use, a clear indication of senior command's awareness of their problematic nature. Further proof of this conclusion lies in the increased level of required authorization – from the battalion's Deputy Commander to the Brigade Commander[112] – following this incident.

To summarize, the IDF authorities, which had decided on using flechettes, and the political authorities, which had approved that decision, were conscious of the weapon's character and its consequences in terms of potential injury to bystanders. The fact that an armed force that employs flechette weapons has developed a view circumscribing flechette use, one far from the U.S. Army, Vietnam-era view of flechette as a standard weapon ("conventional ironmongery"), is quite meaningful.[113]

However, because, as the incidents cited have shown, awareness is insufficient to halt the use of weapons in the field, the case of flechettes may point to the importance of banning or heavily restricting the distribution, and certainly the use, of dubious weapons. History has shown that presumed military necessity has almost always outweighed humanitarian concerns. One may argue that the military's legal advisers should ensure a balance between this inherent tendency and the protection of civilians in conflict zones. However, as experience in the Gaza Strip points out, such a task has so far been proven beyond their capabilities. As an Israeli military

---

Maj. Gen. Yitzhak Eitan realized that flechette rounds are area weapons, with all that this implies.

109 *See*, Ephraim Kam & Yiftah Shapir, (eds.) *The Middle East Strategic Balance 2002-2003*, (Tel Aviv: Jaffee Center for Strategic Studies, 2003), at 235.

110 *State's Response*, para. 42. In other words – as IDF officials were cited – in the Gaza Strip, "the boundaries between settled Palestinian areas and uninhabited areas are clearer, there is an organized fence, and therefore it is easier to distinguish between combatants and civilians, and also the response time [the time required for the IDF to properly deployed against any immediate threat-E.B.] around the Jewish settlements is very short". *See*, Rapoport, "They Walked in the Fields", *supra* note 25.

111 Amira Hass, "Hamas Threatens to Take Revenge for the Death of Four Family Members in the Gaza Strip", *Haaretz*, (Aug. 30, 2002), at A6.

112 For the orders of the Division Commander, Brig. Gen. Yisrael Ziv *et al.*, "Hamas: We Will Take Revenge on Israel for the Death of the Family in Gaza", *Yedioth Ahronoth*, (Aug. 30, 2002), at 2.

113 *See, supra* chapter 3.

legal adviser commented with respect to the MAG's sanction of such weapons in the Gaza Strip, "this was a decision of the operational field command, and we believe that the person who gave us the information believed in what he said".[114]

It is also difficult to find more favorable circumstances for the responsible use of flechettes than the current case: a professionally trained and modern army with legal advisors knowledgeable in IHL, who are parties of every major decision. Yet, in late 2007, the Southern Command Headquarters' Chief of Staff during February 2000 to 2003, Brig. Gen. (Res.) Zvika Fogel, explained the legal restraints as follows:

> When you want to use something [i.e., munitions-EB], you have no problem finding the justification, especially when we hit those we wanted to hit when we used them at the beginning of the operation. If we could justify our actions operationally at the beginning, then even if personnel from the Advocate General's Office or from the prosecution were present, it was easy to bend their views in our direction in the face of the results.[115]

This officer's comment indicates that even modern armies with active legal frameworks are not immune to "military preferences". The failure of Israel's MAG to withstand the IDF pressures for the purpose of providing some feasible level of protection to Palestinian civilians stresses just how much we need to regulate area weapons' use in general and flechette weapons in particular within the framework of treaty law given these weapons' high potential of indiscriminate effects.[116]

---

114 *See*, Rapoport, "They Walked in the Fields" *supra* note 25.

115 *See*, Yotam Feldman, "Collision Course", *Haaretz*, (Dec. 21, 2007) [Friday Magazine] at 22 (in Heb.). An English (yet incomplete), version is available at <http://www.haaretz.com/hasen/spages/936744.html>.

116 After all, as Prof. Benvenisti, a leading scholar on the law of occupation, stressed "the right to life of enemy civilians is entitled to the same protection granted to that of our own civilians, that all civilians be shown the same concern during combat, regardless of their national affiliation". *See*, Eyal Benvenisti, "Human Dignity in Combat: The Duty to Spare Enemy Civilians" 39 *Israel Law Review* (2006), 81, at 87.

# Part III.

## Old Arguments, Fresh Victims and New Findings: The Legality of Flechette Weapons in International Customary Law – A Reappraisal

# Chapter 7

## Flechette Weapons and the Principle Prohibiting Unnecessary Suffering

As Part II suggests, the legality of the flechette weapon should be reappraised under the parameters of international law. As Israel's HCJ Decision declared, no conventional rule exists that specifically prohibits the use of flechette weapons. However, the HCJ should have reviewed the legality of flechette weapons according to the two cardinal IHL principles governing weapons' legality – identified by the ICJ in the *Nuclear Weapons Advisory Opinion* – specifically, the principle prohibiting the infliction of unnecessary suffering or 'superfluous injury' and the principle of distinction between combatants and civilians. In the absence of such a review, the next two chapters take on the task of assessing the legality of the flechette weapon according to each of these two principles.

The order of the chapters corresponds to the order of the arguments raised in favor of prohibiting flechette weapons: the 1973 Swedish Foreign Ministry Report ("needless suffering"),[1] followed a year later by discussions in the first session of the *Ad Hoc Committee on Conventional Weapons* (March 1974). Dr. Blix, the vocal representative of the states calling for the prohibition of flechette use (the "Swedish Group") was the first to focus on the requirement to prevent unnecessary suffering or superfluous injury.[2] In fact, the *Explanatory Memorandum* to the first proposed ban on the use of flechettes focused *entirely* on the rationale of unnecessary suffering, a theme that has remained unchanged throughout the course of the Ad Hoc Committee's sessions over the years.[3] Furthermore, the Swedish Group's main concern – that the flechette weapon inflicts unnecessary suffering – has likewise retained its position in the NGOs' Petition against the weapon's use.[4] By appending

1 *See*, Conventional Weapons, Their Deployment and Effects from a Humanitarian Aspect: Swedish Working Group Study, *Conventional Weapons, Their Deployment and Effects from Humanitarian Aspects, Recommendations for the Modernization of International Law*, (Stockholm: Almqvist & Wiksell, 1973) 167-8.

2 *See*, *Official Record of the Diplomatic Conference on the Reaffirmation and Development of International Humanitarian Law Applicable in Armed Conflict*, Geneva (1974-1977), [hereinafter: *CDDH O. R.*], Vol. XVI, (1978), at 47.

3 *See*, Doc. CDDH/DT/2, Working Paper, Feb. 21, 1974 (on file with the author) and compare it to CDDH/IV/201 (III) (CDDH, O. R., Vol. XVI, at 614). The original draft proposal was slightly revised and re-submitted following the discussion at the Lucerne Conference of government experts (Sept.-Oct. 1974).

4 *See*, *Physicians for Human Rights & The Palestinian Centre for Human Rights v. O.C. Southern Command & The State of Israel* – Ministry of Defense, Petition for an Interim Decree, at 5, para. 24, available at <http://www.btselem.org/hebrew/Legal_Documents/8990_02_ Flechette_Appeal.doc> (hereinafter, *The Flechette Petition*).

autopsy reports to their Petition,[5] which attested to the fact that one victim *absorbed more than 71 darts,* the Petitioners were able to maintain before the Court that these reports "speak for themselves!"[6]

Following an introduction to the historical-legal sources of the principle that prohibit causing unnecessary suffering, and the difficulties in defining such suffering or its correlate, superfluous injury, we describe the measures traditionally applied when quantifying human suffering. An attempt will then be made to assess flechette weapons according to these measures together with an analogous examination of shotgun use: During WWI, Germany disputed the American Expeditionary Forces' (AEF) use of shotguns fired from the trenches. In a manner resembling flechette weapons, shotguns disperse large numbers of buckshot pellets with every shot, which are simultaneously projected at different trajectories. Also re-examined are the arguments, often repeated by experts, that in addition to their tendency to bend or break at impact, even at low velocities, flechette darts have a so-called "high velocity effect" (i.e., the infliction of extensive damage to the human body once a fragment propelled at a velocity exceeding a certain specific threshold strikes the human body). A brief assessment of the weapon's legality according to the standards proposed by the suspended 1997 ICRC-sponsored Superfluous Injury or Unnecessary Suffering Project (SIrUS) is then conducted while bearing in mind that the project was lauded by its initiators as "an attempt to bring objectivity to the legal notion of superfluous injury or unnecessary suffering."[7]

Nevertheless, as one Jurist rightly noted, "... when the effects of any weapon are described in detail, a case could probably be made for its prohibition."[8] Given these normative conundrums, the remaining crucial questions relate to the weapon's military necessity, addressed in the section D.

## I. Introduction

The "infliction of suffering for the sake of suffering" was explicitly prohibited in 1863, in the *very first* offcial military manual ("Lieber code").[9] This prohibition received its

5 *Id*, appendix D, 1-3.

6 *Id.*, para 38.

7 Robin Coupland and Peter Herby, "The SIrUS Project and Reviewing the Legality of New Weapons" *International Review of the Red Cross* 835 (*IRRC*) (1999), 583, at 586.

8 *See*, Andrew M. Stone, "Conventional Weapons, Their Deployment and Effects from a Humanitarian Aspect: Recommendations for the Modernization of International Law" *The American Journal of International Law* 68 (*AJIL*) (1974) 149-150 (book review). It should likewise be noted that a case can be made for "military necessity" in almost all situations.

9 Dr. Francis Lieber, then a professor at Columbia University, prepared the manual for the U.S. armed forces during the American Civil War; after revision by a board of officers it was promulgated by President Lincoln. For the prohibition *see*, *Instructions for the Government of Armies of the United States in the Field,* General Order No. 100, 24 Apr. 1863, Art. 16, cited in Dietrich Schindler and Jiri Toman, (eds.) *The Laws of Armed Conflicts*, (Leiden: Martinus Nijhoff, 2004), 3, at 6 [hereinafter, *Laws of Armed Conflicts*]. On the Manual's importance *see, e.g.*, Burrus M. Carnahan, "Lincoln, Lieber and the Laws of War: The Origins and Limits of the Principle of Military Necessity", *AJIL* 92 (1998), 213.

ultimate imprimatur five years later, when referred to in the preamble of the 1868 St. Petersburg Declaration.[10] The Declaration, the first formal agreement prohibiting the use of certain weapons (i.e., explosive projectiles weighing less than 400 grams) stated:

> … That the only legitimate object which States should endeavour to accomplish during war is to weaken the military forces of the enemy; that for this purpose it is sufficient to disable the greatest possible number of men; That this object would be exceeded by the employment of arms which *uselessly aggravate the sufferings of disabled men, or render their death inevitable*; That the employment of such arms would, therefore, be contrary to the laws of humanity …. [11]

Six years later, in 1874, during the first modern attempt to codify the laws of war (the "Brussels Declaration"), the "employment of arms, projectiles or material calculated to cause unnecessary suffering" was "especially forbidden."[12] A similar prohibition, with slightly different wording, appeared in 1880, in the draft of the treaty on the laws and customs of war, drawn up by the Institute of International Law (Oxford Manual).[13]

These two instruments constituted the basis for the 1899 successful codification of the laws of war, the *Hague Convention (II) Respect to the Laws and Customs of War on Land*, where this principle appears as Art. 23(e) of the Hague Regulations.[14] As previously mentioned (*supra*, chapter 5B), the principle was repeated in its original wording (i.e., "it is especially prohibited … [t]o employ arms, projectiles, or material of a nature to cause superfluous injury") in the authentic French text in the 1907 Revised Hague Regulations,[15] and finally in the 1977 AP-I (Article 35(2)), but with the dual English expression "superfluous injury or unnecessary suffering."[16] As the text reads: "It is prohibited to employ weapons, projectiles and material and methods of warfare of a nature to cause superfluous injury or unnecessary suffering".[17] In 1996,

10 Declaration Renouncing the Use, in Time of War, of Explosive Projectiles Under 400 Grammes Weight, 1868, *Laws of Armed Conflicts*, 91.

11 Emphasis added. *Id.*, at 92.

12 *See*, Project of an International Declaration concerning the Laws and Customs of War, Brussels, 27 Aug. 1874, Art. 13(e), *Laws of Armed Conflicts* 23, at 24.

13 The text reads: "…calculated to cause superfluous suffering, or to aggravate wounds," *The Laws of War on Land*, Oxford, 9 Sept. 1880, Art. 9(a), *Laws of Armed Conflicts*, 29, at 32.

14 *Id.*, 55, at 73.

15 Pearce Higgins, *The Hague Peace Conferences and Other International Conferences Concerning the Laws And Usages of War*, (Cambridge: Cambridge University Press, 1909), 206, 234.

16 *Laws of Armed Conflict*, 711, at 730.

17 The same wording is repeated in the preamble to the 1980 Convention on of Certain Conventional Weapons. See Convention on Prohibitions or Restrictions on Use of Certain Conventional Weapons which May Be Deemed to be Excessively Injurious or to Have Indiscriminate Effects (hereinafter, the CCW), *Laws of Armed Conflicts*, 181, at 184.

it was inserted into the ICJ *Nuclear Weapons Advisory Opinion* as an independent criterion for distinguishing between legal and illegal weapons use.[18]

Notwithstanding the difficulties created by the English translation of the 1907 Revised Hague Regulations (i.e., the subjective element "calculated to cause unnecessary suffering"), even the widely accepted (1977) wording of AP-I (Art. 35(2)) leaves many unanswered questions regarding the definition of "superfluous" injury and "unnecessary" suffering. After all, any wound inflicted by a weapon induces the victim's suffering. It was therefore suggested that whenever an instrument of war does not offer a military advantage over other available measures, but does inflict greater suffering or injury than that inflicted by the other options, one should refrain from its use.[19] As the government experts in the 1974 Lucerne Conference noted, the implementation of this principle:

> ... involved some sort of equation between, on the one hand, the degree of injury or suffering inflicted (the humanitarian aspect) and, on the other, the degree of necessity underlying the choice of a particular weapon (the military aspect).[20]

The problem, they agreed, rests in the criterion's implementation. Neither side of the equation, the military nor the humanitarian, could be quantified precisely. The two sides were, furthermore, so different from one another that their comparison was difficult.[21]

Indeed, this was the stance taken by the declared nuclear weapons states (excluding China), together with like-minded states, in the *Nuclear Weapons Advisory Opinion* case,[22] in their response to the claim that nuclear weapons inflict great suffering because they release powerful and prolonged radiation in addition to immense

18 *See*, "Legality of the Threat or Use of Nuclear Weapons, Advisory Opinion" (International Court of Justice July 8, 1996), *International Legal Material* 35 *(ILM)* (1996), 809 & 1343, para. 95, [hereinafter, *Nuclear Weapons Advisory Opinion*].

19 As one scholar noted, the ban on dumdum bullets (The 1899 Declaration (IV, 3) concerning Expanding Bullets, *Laws of Armed Conflicts,* 99) demonstrated that the primary underlying consideration was that the type wounds which result from this type of bullet was thought to constitute superfluous injury or causing unnecessary suffering, as ordinary bullets were adequate for putting a man out of action. *See,* Hans Blix, "Current Efforts to Prohibit the Use of Certain Conventional Weapons" *Instant Research on Peace and Violence* 4 (1) (1974) 21, at 25. For an opposing perspective on the 1899 Declaration *see*, Parks Hays, "Conventional Weapons and Weapons Reviews" *Yearbook of International Humanitarian Law 8* (2005) 55, at 69.

20 *Conference of Government Experts on the Use of Certain Conventional Weapons* (Lucerne, 24.9-18.10.1974) ICRC Report (1975) [hereinafter, *1975 Lucerne Report*], at 9, para. 24.

21 *Id.*

22 *See,* for instance, Holland's statement that: "The availability of considerably less harmful means to attain the military advantage or the causing of suffering out of proportion to the military advantage to be gained therefore appear[s] to be the essential yardstick for determining whether the use of certain weapons must be deemed to cause unnecessary 'suffering'." *Netherlands Written Statement* 7-8 (June 1995). *See,* also the Russian response on this matter, *Russia Written Statement* 12-14 (June 1995). All Written Statements are available at <http://www.icj-cij.org/docket/index.php?p1=3&p2=4&k=e1&case=95&code=unan&p3=1>.

quantities of heat and energy.[23] The U.S. – abiding by the English translation of the 1907 Revised Hague Regulations – declared that: (t)his prohibition was intended to preclude weapons *designed* to increase the injury or suffering of the persons attacked beyond that necessary to accomplish the military objective.[24]

While the ICJ itself only mentioned that the test is "… a harm greater than that unavoidable to achieve legitimate military objectives",[25] with no further elaboration, Judge Rosalyn Higgins addressed this issue in greater detail in her *Dissenting Opinion*; she nonetheless eventually came to conclusions similar to those reached by government experts in their 1974 meeting. This balance is the reason why, she noted, states have been able to move to a specific prohibition on the use of dum dum bullets while other weapons "that cause vastly greater suffering" have escaped specific bans and prohibitions in practice.[26] Hence, the effects of any weapon are not to be examined in and of themselves but, rather, in reference to other available weapons as well as the weapon's military necessity.

Determination of whether flechette weapons inflict unnecessary suffering therefore requires balancing between the weapon's military advantages and the degree of suffering and injury it inflicts on its victims. The question is whether the flechette darts themselves cause unnecessary suffering when compared to – if we consider the Israeli case – other "regular" rounds or, as in the three incidents described in the Petition, the use of heavy machine guns or light weapons. Stated differently: Do victims of flechette weapons indeed experience unnecessary suffering in comparison to victims of these alternative means? Moreover – and perhaps more importantly – is it possible to quantify suffering?

### B. Measures for Quantifying Suffering: The Search for a Solid Scientific Basis of Comparison

Despite the numerous discussions that the Conferences of Government Experts held during the 1970s, quantification of suffering was found to be an impossible task. Long before the experts' discussions many writers argued that the "measure of humaneness" (in contemporary terms: the suffering criterion) involves comparisons of (1) the degree of suffering caused at the time of injury by the different weapons (*the pain criterion*); (2) the permanent after-effects resulting from the injuries

---

23 Sweden claimed that "(t)he effects of radioactive radiation as a result of the use of nuclear weapons cause unnecessary suffering, not merely for third parties who are directly affected, but also future generations, for example, as a result of genetic damage". *See, Sweden Written Statement* 5 (June 1995); see also, *Mexico Written Statement*, 13 (June 1995).

24 Emphasis added. *See, U.S. Written Statement*, 28-29 (June 1995).

25 *Nuclear Weapons Advisory Opinion*, para. 78.

26 Dissenting Opinion of Judge Higgins, para. 16. *See, also Id.*, paras. 14-15, 17. Note: all Declarations, Separate and Dissenting Opinions of the Judges are available at <http://www.icj-cij.org/docket/index.php?p1=3&p2= 4&k=e1&p3=4&case=95>.

inflicted by each weapon (*the disability criterion*);[27] and (3) the percentage of deaths among total casualties produced by each weapon (*the mortality criterion*).[28]

However, as the medical experts discovered, all of these criteria are quite problematic. The *pain criterion* was found as impossible to be objectively defined from a strictly medical standpoint because the pain threshold varies between individuals, and is dependant on circumstances. As such *it varies even for the same person* as circumstances change.[29] Pain, it appears, does not depend solely on the extent of the injury but, rather, on the victim's psychological state; that is, the subjective aspect of pain constitutes the principal barrier to its quantification. A psychologist, who has explored the psychophysiology of pain has written that

> [p]ain is a curious sensation. It ... can be modified by opiates, by hypnosis, by the administration of pharmacologically inert sugar pills, by *emotions*, and even by other forms of stimulation, such as acupuncture.[30]

Moreover, as far as pure psychology is concerned, initial pain is in fact felt with the receipt of the wound in the battle due to the victim's state of agitation and increases several hours afterwards due to the change in the injured party's mental state as well as the body's release of pain-producing substances, together with the frequent onset of infection. It has even been argued that surgical intervention can add to the victim's pain.[31]

The multiple factors constituting the *disability criterion* render it incapable of providing solid ground for assessing the degree of suffering imposed by any particular weapon. If they receive medical or surgical care within a few hours of injury, most victims experience disability for only a short duration. However, with respect to those who suffer disability due to injuries to the central nervous system or to non-

---

27 Some writers prefer the term "the incidence of permanent damage or disfigurement". *See*, Burrus M. Carnahan & Marjorie Robertson, "The Protocol on 'Blinding Laser Weapons': A New Direction for International Humanitarian Law" *AJIL* 90 (1996) 484, at 485.

28 *See*, for instance, Augustin M. Prentiss, *Chemicals in War; A Treatise on Chemical Warfare*, (New-York: McGraw-Hill, 1937), 679. Interestingly, the 2005 ICRC Study Report mentioned only the last two factors as relevant "in establishing whether a weapon would cause superfluous injury or unnecessary suffering". *See*, Henckaerts Jean-Marie and Louise Doswald-Beck, *Customary International Humanitarian Law*, Vol. I., (Cambridge: Cambridge University Press, 2005) at 241.

29 *Conference of Government Experts on the Use of Certain Conventional Weapons* (Lugano, 28.1- 26.2.1976) ICRC Report (1976), [hereinafter: *1976 Lugano Report*], 140; *See* also ICRC, *Weapons that may Cause Unnecessary Suffering or Have Indiscriminate Effects* (Report of the Work of Experts) Geneva (1973), [hereinafter: *1973 Geneva Report*], at 27, para. 89.

30 Neil R. Carlson, *Physiology of Behavior* 9th ed. (Boston: Pearson Allyn & Bacon, 2007), at 240. For instance, certain types of injuries cause great psychological suffering, even at the moment of injury, which is not quantifiable: destruction of victims' faces, the loss of various senses, damage to procreative capacities, and so forth. For the experts' unanimous agreement about the variance of suffering, *see*, *Statement Concerning Unnecessary Suffering presented by the Informal Working Group of Medical Experts*, *1976 Lugano Report*, at 140.

31 On this matter, *see*, *1973 Geneva Report* at 27.

vital organs and extremities, factors such as the psychological trauma experienced cannot be properly gauged.[32]

Similar difficulties exist with reference to *the mortality criterion*.[33] Prima facie, the probability of death is an easier measure to wield. However, as the expert reports suggest, mortality is as complex as pain and depends upon several factors, only some of which are determined by either the type of weapon or the wound inflicted. These factors include: (1) location of the wound in the body's different organs; (2) the time lag between injury and adequate medical care – mortality declines if good first aid and adequate surgical treatment are received within six hours, after which the mortality rate rises sharply;[34] (3) the victim's level of physical resistance – pregnant women, children, and the elderly are more vulnerable than soldiers to war-related injuries. However, soldiers' resistance to mortality is significantly lowered by exposure to famine, dehydration, or extreme temperature.

Yet, the experts in the 1973 Geneva meeting pointed to another important factor that affects mortality: *multiplicity of wounds*. It also argued that the number of wounds sustained is important in determining mortality: (a) it increases the likelihood that a vital organ will be injured; and (b) the "additive effect," meaning that "...two wounds that in themselves are not lethal, may together put such stress on the body that the probability of death increases."[35] However, at the 1974 Lucerne Conference where, unlike in the Geneva session, an extensive delegation from the United States (10 delegates) participated, this causal relationship (number of wounded organs and probability of death) was criticized as being devoid of a proper scientific basis.[36]

It is no wonder, then, that the *Informal Working Group of Medical Experts* unanimously stated following the end of the 1976 Lugano Conference:

> From a strictly medical standpoint, it seems impossible *at the present stage of medical knowledge* to objectively define suffering or give absolute values permitting comparisons between human individuals.[37]

---

32 *Id.*, at 29.

33 For a list of military manuals as well as official statements stating that "weapons which render death inevitable are prohibited" see *ICRC Study Report*, at 241, fn. 31.

34 This finding was based on data collected during WWII. *1973 Geneva Report* at 46, para. 146. Yet, the decrease in mortality among U.S. battle casualties, from 17% in WWI to about an average 2% in the Korean and the Vietnam Wars, which was attributed to the decreased amount of time elapsing before casualties receive medical care, highlights its importance. At the same time, this finding demonstrates the problematic nature of the relationship between the type of weapon and mortality. *See, 1973 Geneva Report*, at 28-9, para. 72.

35 It was even claimed that "if more than five different organs, for instance, are injured in the abdomen, death is statistically inevitable even if the victim survives the primary injury". *1973 Geneva Report*, at 28, para. 70.

36 Their argument rested on the grounds that mortality rates depend on many other factors, such as the particular organs or tissue injured, the degree of damage, and the type and quality of available medical care. *See, 1975 Lucerne Report*, at 58, paras. 206-208.

37 Emphasis added. *1976 Lugano Report*, at 140. Given the criticism of the 1997 ICRC-sponsored SIrUS Project, the validity of this statement's applicability nowadays is an open question.

## C. Flechette Weapons and Unnecessary Suffering or Superfluous Injury

### *CI. Measures for Quantifying Suffering*

In spite of the absence of an agreed-upon scientific basis for quantifying suffering, utilization of these measures as a theoretical baseline for evaluating flechette weapons may nevertheless enhance our ability to assess the repeated claims regarding the unnecessary suffering they may impose.

*The Pain Criterion*: It appears that among the three proposed criteria, the pain criterion poses the strongest case against flechette weapons. Perhaps the most salient feature of any type of flechette weapon is the *multiplicity of simultaneous wounds,* which served as the central argument by proponents for banning the flechette weapon during the 1970s intensive experts discussions, whether in the context of the level of pain caused, or the mortality rate. The crux of this argument rests on the presumably causal relationship between the number of tissues injured and the resulting level of pain. As the ICRC report on the first 1973 Geneva experts meeting stated: "As a very general rule, the more tissue that is damaged the more painful will be the wound."[38] However, at the 1974 Lucerne Conference, several experts who opposed any restriction on fragmentation weapons disputed this "very general rule" and questioned the causal relation between degree of pain and multiplicity of wounds.[39]

Given the fact that the *lowest* number of flechette wounds found in the December 2007 incident's victims and noted in their autopsy reports was eight (*supra* chapter 4, CII), this criterion seems valid. Furthermore, the fact that another victim of this incident absorbed no less than 71 multiple wounds indicates that flechette weapons can *potentially* inflict an *enormous* number of multiple wounds and – given the ICRC claim above – cause great suffering to its victims.[40]

*The Disability Criterion*: Apparently, this criterion is the least relevant to any evaluation of flechette weapons, especially when compared to the damage caused by other types of weapons, such as anti-personnel mines or blinding laser weapons. Injuries caused by the first would necessarily cause amputation (at least to the foot that set off the mine),[41] while the second causes permanent blindness. In the case of

38 *1973 Geneva Report*, at 27, para. 67. It should be noted, however, that the amount of tissue damaged by a single dart will not be very great if there is no cavitation or breaking-up of the dart.

39 Their claim was that the degree of harm depends on a great number of factors, such as the type of organs or tissue injured, the degree of damage, and the sensitivity of tissue damaged. *See*, *1975 Lucerne Report*, at 59, para. 207. In principle, the report of the U.S. delegation criticized the 1973 ICRC data regarding fragmentation and blast weapons as 'exaggerated'. *See*, Parks, "Conventional Weapons…", *supra* note 19, at 72, fn. 64.

40 In fact, if PHR-I was right in its claim that the third youth was still alive after suffering eight darts (four of which had entered his right lung) one can only guess at the great suffering he had undergone before being run over by an IDF tank. The other four darts hit his back and left forearm. *See*, National Center of Forensic Medicine, Rep. 3/2294/2001, (Jan. 20, 2002), at 9.

41 For supporting statistical data *see*, Robin M. Coupland, "The Effect of Weapons: Defining Superfluous Injury and Unnecessary Suffering. *Medicine and Global Survival*

laser weapons, it was stressed that the prohibition of its use and transfer (i.e., CCW, Protocol IV on Blinding Laser Weapon[42]) represents an exception from the traditional perspective regarding weapon restrictions. While the traditional approach focused on the military and medical situation at the time of the injury and relatively soon thereafter, in the case of this Protocol, "the most persuasive arguments for its [the Protocol's] adoption focus on the long-term impact on society of an increased numbers of blind veterans".[43] However, as the discussion above suggests, injuries to the central nervous system or to non-vital organs and extremities can potentially inflict long-term disability. Given the multiple wounds that flechette weapons may cause, there is always a possibility that even a single dart could hit the nervous system or an organ and cause permanent disability. As the pathology reports of victims of flechettes weapons clearly indicate, the darts are not targeted at any single organ but rather are spread out throughout the body. As such, it is clear that permanent disability is a possible outcome for any flechette weapon victim.[44]

*The Mortality Criterion*: As mentioned, the experts attending the 1973 Geneva meeting considered the multiplicity of wounds an important factor influencing the mortality rate. As the ICRC report stated: "A hit by a multiplicity of projectiles or fragments is more likely to injure a vital organ than [is] a hit by a single projectile. This is self-evident."[45] Naturally, this seemingly causal relationship was further discussed when the impact of fragmentation weapons was reviewed.[46] In contrast, those who opposed any restriction on fragmentation weapons not only questioned this causal relationship,[47] they also argued that in principle, while multiple small fragmentation increased the likelihood of incapacitation, such weapons inflicted less-severe wounds and consequently less morbidity and mortality.[48] After reviewing the autopsy reports on the three youths from Gaza, it seems almost certain that once 71 darts hit an individual while penetrating and traversing his body, *at least a few darts* would strike vital organs, such as the brain, as in this case.[49] In a second case, it turned

3 (1996), table 3 [online publication as of Jan. 1996], available at <http://www.ippnw.org/MGS/V3Coupland.html>.

42 *Laws of Armed Conflict*, 212.

43 *See*, Burrus M. Carnahan & Marjorie Robertson, "Protocol on 'Blinding Laser'..." *supra* note 27, at 485.

44 According to the autopsy report the death of the victim, who was thrashed by 71 darts, was the result of a single dart which traversed his brain. *See,* National Center of Forensic Medicine, Rep. 3/2295/2001, (Jan. 16, 2002), at 8.

45 *1973 Geneva Report*, at 28, para. 70.

46 Data culled from WWII medical records show that even with adequate medical care within the required time frame (7 hours), "the risk of fatality rose by 15% for each additional abdominal organ injured". *Id.*, at 46, para. 146.

47 As to the validity of the above data, it was claimed that the figure of a 92% fatality rate for those with five wounded abdominal organs in WWII was reduced to only 16.6% in the Korean War for the same number of injured abdominal organs. *See*, *1975 Lucerne Report*, at 59, paras. 205-206.

48 *Id.*, at 59, para. 208.

49 *See*, National Center of Forensic Medicine, Rep. 3/2295/2001, (Jan. 16, 2002), at 9.

out that a few darts, entering the lungs through the back, were sufficient to induce massive loss of blood that caused death.[50]

As the discussions concentrated on the issue of the multiplicity of wounds, natural bias was introduced given the fact that flechette weapons could *potentially* inflict the greatest numbers of multiple wounds among the weapons currently available in military arsenals. It appears that this feature of flechette weapons is crucial concerning the IHL requirement – to avoid inflicting unnecessary suffering. Therefore, before turning to the debate that evolved around the darts' structure, more attention should be given to the multiplicity of wounds and its importance in assessing the weapon's legality. Comparisons with other similar weapons may provide some useful insights into this matter. Given the fact that firearm flechettes originated with flechette shotguns (32 flechettes), successfully produced for the U.S. army (standard combat shotguns) by 1953 (*supra*, Chapter 1, CI), shotguns seem appropriate candidates for such a comparison.

Both weapons inflict multiple wounds, depending on the number of flechettes or buckshot pellets, respectively, contained. Indeed, as mentioned (*supra*, Chapter 1, CI), the use of shotguns (12-gauge shotgun shells) during the Vietnam War received poor media coverage, which prevented any public debate over the weapon's attributes and strategic value. Yet, their use some 50 years before, during WWI, did become the subject of a serious international legal debate, as the following section details.

### *CII. Shotguns: World War I and the 1918 German Protest*

In the summer of 1917, the vanguard of the American Expeditionary Forces (AEF) had reached France.[51] A year later, in the summer of 1918, when the AEF first took a significant part in the war,[52] the Germans noticed that American soldiers were equipped with a weapon that had not been seen in previous WWI trench warfare: shotguns.[53] Nearly 30,000 shotguns (the 12-gauge Winchester Model 97 [repeating] Trench shotgun), which fired a "cloud" of nine buckshot shells (No.00) with every shot, were procured for the AEF, with large numbers of these weapons finding their way into the trenches from 1918 on.[54] Considering the average short firing range between the belligerents' lines that marked European trench warfare, German

50 *See*, National Center of Forensic Medicine, Rep. 3/2294/2001, (Jan. 20, 2002), at 9.

51 For the official history of the AEF's emergence and its operations see, *American Military History*, (Office of the Chief of Military History United States Army, 1989), chaps. 17-18. For a concise overview *see*, Allen R. Millett, "Over Where ? The AEF and the American Strategy for Victory, 1917-1918" in: Kenneth J. Hagan & William R. Roberts (eds.) *Against All Enemies: Interpretations of American Military History from Colonial Times To The Present*. (Westport, CT: Greenwood Press, 1986) 235-256.

52 Even this was "on a small scale." *See*, Cyril B. Falls, *The First World War*, (London: Longmans, 1960) 340.

53 According to Jenkins, both the British and the French High Commands briefly considered the use of double-barreled shotguns in trench defense but decided against them due to their being insufficiently effective. P.B. Jenkins, "Trench Shotguns of the AEF", *American Rifleman* 84 (Nov. 1935), 14-15, 22, at 15.

54 Thomas F. Swearengen, *The World's Fighting Shotguns*, (New-York: T.B.N. Enterprises, 1978), 10. Swearengen, A former Chief Warrant Officer of the US Marine Corps, is considered a highly authoritative source on this affair.

soldiers were very likely to be wounded by nine buckshot shells fired at once.[55] As one writer argued, the shotgun's highly effective use had such a telling effect on the morale of front-line German troops that it reached the point that it became "a real problem for the German General Staff."[56]

Following the capture on September 11, 1918 of a second American soldier armed with a shotgun, the German government attempted to stop use of this weapon.[57] On September 19, 1918, a diplomatic protest against the use of shotguns was issued via the Swiss government, which represented German interests in the U.S. The protest claimed that the use of shotguns was prohibited by the law of war as it constituted a violation of the ban on unnecessary suffering as set out in the Hague Regulation (Art. 23(e)). A reply to the protest was requested by October 1, 1918, with a severe threat concluding the protest:

> The German Government... calls attention to the fact that according to the Laws of War (*Kriegsrecht*), every prisoner found to have in his possession a shotgun or ammunition belonging thereto forfeits his life.[58]

The protest was forwarded by the Department of State to the Department of War, which sought the advice of the U.S. Judge Advocate General, Brig. Gen. Samuel Ansell. On September 26, 1918, the latter responded in a lengthy memorandum in which he noted the shotgun's characteristics and maintained that the multiplicity of bullets is adapted to the military necessity of "putting out of action" more than one enemy soldier with each shot. In this sense, the weapon was similar, in his view, to either a shrapnel shell, which discharged a multitude of fragments, or a machine gun. He therefore concluded his legal opinion by stating that "the protest is without legal merit." Yet, Brig.-Gen. Ansell did acknowledge instances in which a shotgun *causes more injury* than a regular rifle: "... instances where the enemy soldier has approached so close to the shooter *that he is struck by more than one of the nine* ... [buckshot projectiles] contained in the cartridge."[59]

Accordingly, US Secretary of State Robert Lansing rejected the German protest as unjustified on the following day. Notwithstanding the German threat, Lansing announced that it was the opinion of the U.S. that the use of the weapon did not violate Article 23(e) in view of the weapon's history, its well-known effects, and a comparison with other weapons permitted in war. Therefore, Lansing declared, it

55 For U.S. soldiers' eyewitness reports on the effect of these shotguns *see*, Jenkins, "Trench Shotguns..." *supra* note 53, at 22.

56 Swearengen, *World's Fighting Shotguns*, *supra* note 54, at 10.

57 *See*, W. Hays Parks, "Memorandum of Law (Jan. 24, 1997), Subject: Joint Service Combat Shotgun Program; Legal Review", at 7-9; reprinted in Department of the Army Pamphlet (hereinafter, *DA Pam*) 27-50-299, *The Army Lawye*r (Oct. 1997), 16, at 19.

58 For the text of this protest, *see*, U.S. Department of State, Foreign Relations, 1918, Supp. 2 (*The World War*), 785 (1933).

59 Emphasis added. The Opinion of the Judge Advocate General, Brigadier General Samuel T. Ansell is cited in Parks, "Memorandum of Law..." *supra* note 57, at 19-20.

"cannot be the subject of legitimate or reasonable protest..., its use will not be abandoned ...".[60]

The war ended six weeks later, and no accounts exist of either a German response or the implementation of the German threat.[61] The legal dispute, however, remained untouched. While the German 1961 Military Manual[62] and its 1992 revised edition[63] defined the shotgun as an illegal weapon, causing unnecessary suffering and devoid of any real military advantage, the U.S. official position, to this day, has remained unchanged: "The combat shot gun and its lead-and-antimony buckshot (or shot) ammunition are consistent with the law of war obligations of the United States".[64]

To conclude this short section on weapons which, *by their nature,* cause multiple injuries, it seems that the inclusion of flechette weapons in this *relatively rare group* should serve as an important factor in any attempt to impose restrictions on its use. Leaving aside the protracted debate about the shotgun's legality, the 1918 German protest strongly indicates that anti-personnel weapons that inflict multiple wounds are far from typical "conventional weapons."

### *CIII. The Dart's Lethal Features*

In addition to the claims raised against flechettes under the rubric of multiplicity of wounds, two additional arguments, voiced in almost every significant forum against the weapon, focused on the darts' structure and form. As stated in the *Explanatory Memorandum* submitted to the Ad Hoc Committee Second Session: "When the flechettes hit a person at high velocity they bend or break at impact and the wounds resulting are aggravated by the high velocity effect."[65]

In effect, this statement makes two different arguments:

---

60 As for the threat, Lansing stated that "if the German Government should carry out its threat in a single instance, it will be the right and duty of the Government of the United States to make such reprisals as will best protect the American forces...." *See,* Foreign Relations, 1918, Supp. 2, 785. For a summary of the protest and the American response *see* also, Marjorie Millace Whiteman, *Digest of International Law*, Vol. VI (Washington D.C.: U.S. Government Printing Office, 1943), 271-272.

61 *See,* Swearengen, *World's Fighting Shotguns, supra* note 54, at 10. It should be noted that the presumed German retreat from its threat may have been the result of poor documentation regarding the capture of soldiers armed with shotguns.

62 *See,* The German ZDv 15/10, Mar. 1961, para. 76.

63 *Humanitarian Law in Armed Conflicts –Manual* (DSK VV207320067) (Aug. 1992), para. 407. Interestingly, the specific instruction also references the *1899 Declaration concerning Expanding Bullets* in this context.

64 *See,* Parks, "Memorandum of Law..." *supra note 57,* at 24. This 1997 legal review seems to be the last to be issued on this subject. The weapon's legality, however, was conditioned on the type of bullets used (i.e., bullets covered by the above 1899 Declaration would negate the shotgun's legality). As the 1960 legal review stated: "the legality of the use of shotgun depends upon the nature of the shot employed and its effect on a soft target..." *See,* OTJAG, Army, JAGW 1960/1305 (Jan. 4, 1961) as reprinted in Parks, *Id.*, at 22. This statement was reaffirmed a year later (see *Use of Shotguns in Conventional or Unconventional Warfare*, Op. OTJAG, Army, JAGW 1961/1210 (Sept. 11, 1961) and reiterated in DA Pam 27-161-2.

65 *See, e.g., CDDH O. R.* Doc CDDH/IV/201, *Explanatory Memorandum*, CDDH, O. R., Vol. XVI., at 614.

1. The first argument focuses on the *"high velocity effect"*: Once a fragment propelled at a velocity exceeding 700-800 m/sec strikes the human body, it often *tumbles* rapidly inward, causing extensive damage to tissues, adjacent blood vessels, nerves, and organs. In addition, there is a great risk of injury from secondary fragments as bones splinter.[66] There was real concern about this issue regarding flechettes.[67] As observed in classified U.S. army experiments during the mid 1960s, to ensure such an effect, the darts should have a striking velocity exceeding approximately 900m/sec.[68] Yet, the muzzle velocity of flechette tank gun rounds, coupled with the fact that the built-in round's propellant and the centrifugal forces provide additional velocity (*supra* Chapter 1, B), gave rise to concerns that such an effect would occur despite the lower initial velocity.[69] In reference to flechette rounds, it appears that such an effect is unlikely. The maximum muzzle velocity of the powerful 120mm smoothbore main gun – with which modern Israeli tanks (Merkava Mark III, IV) are equipped – does not exceed 900m/sec. As to the propellant and centrifugal forces previously mentioned, flechette rounds, like all other rounds or artillery shells, lose considerable velocity depending on the target's distance from the firing point, two factors that cannot be offset. However, as mentioned, contrary to the limited proliferation of anti-personnel flechette projectiles in tank rounds and artillery shells, deployment of flechette-filled air-to-ground rocket warheads has become quite widespread. Given the rocket's high velocity, they pose the risk of becoming "tumbling flechettes", a fact that has intensified the respective concerns.[70]

---

66 *See,* for example, SIPRI, *The Law of War and Dubious Weapons* (Stockholm: Almqvist & Wiksell, 1976) 71. This argument was voiced mainly in reference to the dispute over the legality of 5.56 mm bullets (the standard bullet's diameter in many contemporary armed forces). *See, e.g., 1975 Lucerne Report*, at 40, para. 129. Yet, according to one of the Conference's participants (Mr. Eric Prokosch, *Friend World Committee*) the 800 m/sec "threshold" was strongly disputed. Email correspondence, Jan. 9, 2009 (on file with the author).

67 *See, e.g., 1975 Lucerne Report*, at 60, para. 211a.

68 *The* 900m/sec" threshold" applied to both the 1.7- and 7.2-grain flechettes. Other types even required a higher "threshold" for tumbling effects. *See*, William Kokinakis and Joseph Sperrazza, *Criteria for Incapacitating Soldiers with Fragments and Flechettes*, Ballistic Research Laboratory (BRL) Report No. 1269, (Jan. 1965), at 9, 25. It seems that the dart's aerodynamic form (i.e., a low drag that maintains flight stability for longer distances than do other fragmentations) is the main factor requiring the higher velocity necessary to produce the tumbling effect. As most experiments were still classified in the 1970s, in most cases rather general statements, not exact figures, were provided during the Experts Conferences on this issue. *See, e.g., 1975 Lucerne Report, 60*, at para. 211a.

69 *1976 Lugano Report*, at 78. The fact that flechettes are fitted with concave-tapered points increased these concerns.

70 *See, e.g.*, Erich Prokosch & Ernst Jan Hogendoorn, "Antipersonnel Weapons" in: Ilkka Taipale *et al.*, (eds.) *War or Health? A Reader*, (New-York: Zed Books, 2002) 65, at 73. Indeed, the HYDRA 70 Rocket has a velocity of 1000m/sec. while the Canadian HYDRA 70 (known as CRV-7) rockets have a velocity of 1034 m/sec. See "Air-Launched 2.75-Inch Rockets" in: Directory of U.S. Military Rockets and Missiles (Appendix 4: Undesignated Vehicles), available at <http://www.designation-systems.net/dusrm/app4/275in-rockets.html>. The CRV-7 velocity was compiled from the Manufacture's (Magellan Aerospace Corp) brochures , available at <http://www.magellanaerospace.com/_bin/rockets/rocket.cfm>.

2. The second argument refers to the possibility that upon impact, *"flechettes could deform in the tissues and could become bent or hooked"*.[71] However, while the available autopsy reports contain no details in this regard (only the darts' number and location are noted), testimony given by Palestinian doctors since 2001 and Lebanese doctors during the 1990s provide no support for this argument. Rather, their complaints focused on the darts' deep penetration capability and the additional damage inflicted by the dart's fin-bearing segment (i.e., the tail).[72]

Yet, the question remains as to the legal meaning of such effects. In the 1980s, two experimental flechette rifles were offered to the U.S. Army as part of the *Advanced Combat Rifle* (ACR) Program (*supra*, Chapter 1, CI). Both, it should be stressed, could fire flechettes at a muzzle velocity of more than 1371m/s (4500f/s), fast enough to ensure the high velocity effect if the darts struck flesh.[73] In accordance with the 1974 U.S. regulations,[74] the U.S. Judge Advocate General conducted a legal review to ensure that all four types of experimental rifles (the other two fired "conventional" jacketed bullets) under development were consistent "with the law of war obligations of the United States".[75] While a legal scholar – given the Lebanese doctors' testimonies – stated that "the effects of which [the darts] clearly cause unnecessary

71 *1975 Lucerne Report*, at 59-60, para. 210. In fact, even at low velocities, there is concern that on impact, the dart's body looses rigidity and tends to break or bend, thus increasing injury.

72 In August 1995, for instance, Lebanese doctors asserted before Human Rights Watch interviewers that the darts have "devastating effect" as: "(t)he nail's rotation and its sharp, pointed head maximize the damage. The spinning is what makes the flechette dangerous... The speed of the nails allows them to penetrate one part of the body and move on to another, doing damage along the way...when the nail enters the body, it is like a drill [as the tail end is spinning]." *See*, HRW, *Civilian Pawns: Laws of War Violations and the Use of Weapons on the Israel-Lebanon Border*, (HWR Rep.), (May 1996), at 127. For Palestinian doctors' reports, *see, e.g.*, medical reports of flechette victims following an incident on Jan. 31 , 2003; *see*, Letter from Adv. Iyad Alami, Legal Unit, PCHR, to Hadas Ziv, Director, the Occupied Territories Project, PHR-I, (Feb. 23, 2003), Appendix (PHR-I Archive). As to the additional damage inflicted by the tail, should it break off inside the victim's body and then act as a separate fragment (see *1975 Lucerne Report*, at 59-60, para. 210), the Lebanese Doctor explained as follows: "... the tail end does additional damage as the nail moves, ripping apart tissue as it spins", *Id.* It may be worth noting that these testimonies were reproduced in the context of an Israeli NGO's protest against flechette use in South Lebanon, *see*, *B'Tselem*, *Israeli Violations of Human Rights of Lebanese Civilians*, (Jan. 2000), at 58.

73 Kevin Dockery, *Future Weapons*, (Berkley: Berkley Books, 2007), 78.

74 *See*, Department of Defense Instructions 5500.15, Subject: Review of Legality of Weapons under International Law (16 Oct. 1974). It should be noted that a year later, in 1991, the requirement for a legal review was incorporated into the overall DoD acquisitions directive. *See*, DoD Instruction 5000.2 (Feb. 23, 1991), Subject: Defense Acquisition Management Policies and Procedures, Part 2, para. 5: Review of the Legality of Weapons Under International Law. For an extensive overview of the legal review of new weaponry in the United State *see*, Parks, "Conventional Weapons..." *supra* note *19*, at 109-113.

75 *See*, W. Hays Parks, "Memorandum of Law: Review of Weapons in the Advanced Combat Rifle Program" reprinted in: *The Army Lawyer*, DA PM 27-50-211 (July 1990) 16, at 18.

suffering,"[76] a different view was presented in a U.S. legal review, dated May 1990, as follows:

> Even if a flechette is designed to or has a tendency to bend or otherwise deform on impact, this would not necessarily constitute a violation of the prohibition on unnecessary suffering; it would be legally permissible if military necessity requires such a characteristic in order to increase the projectile's incapacitation effect or lethality at its maximum effective range, and is not incorporated into the design merely for the purpose of needlessly aggravating the wounding effect of the projectile.[77]

In short, the argument over deformation on impact has yet to be substantiated when it comes to flechette projectiles. Yet, even if it were to be substantiated, in order to assess whether such effects indeed constitute unnecessary suffering or superfluous injury, the military necessity of the weapon's use should be assessed. [78]

### *CIV. The ICRC's Proposed Criteria for Superfluous Injury or Unnecessary Suffering (The "SIrUS Project")*

In 1997, the ICRC initiated the SIrUS project, the purpose of which was to introduce objectivity into the legal notion of "superfluous injury or unnecessary suffering" for the purpose of facilitating review of weapon legality.[79] However, the project, which was based on an analysis of extensive ICRC data on war wounds,[80] was suspended in 2001 given the lack of support by states and the negative assessment it received.[81]

---

76 David Turns, "Some Reflections on the Conflict in Southern Lebanon: The 'Qana Incident' and International Humanitarian Law" *Journal of Conflict and Security Law* 5 (2000), 177, at 205.

77 Parks, "Memorandum of Law..." *supra* note 75, at 21.

78 As Parks himself made sure to stress, "what is prohibited is the design (or modification) and employment of a weapon for the purpose of increasing or causing suffering beyond that required by military necessity." *Id.*, at 19.

79 Robin M. Coupland (ed.), *The SIrUS Project: Towards a Determination of which Weapons Cause Superfluous Injury or Unnecessary Suffering*, (Geneva: ICRC, 1997) 7. *See* also, Robin Coupland, "Abhorrent Weapons and 'Superfluous Injury and Unnecessary Suffering': from Field Surgery to Law" *British Medical Journal* 315 (1997) 1450 The catalyst behind the SIrUS Project was an ICRC symposium on "The Medical Profession and the Effects of Weapons" (Montreux, March 1996), where the major finding was the importance to define in objective terms "which weapons were inherently abhorrent and which weapons caused superfluous injury or unnecessary suffering." Coupland, *The SIrUS Project*, id., at 5.

80 The data were collected from military medical publications and the ICRC wound database of 26,636 weapon injuries ("war-wounded"). *See*, Coupland and Herby, "The SIrUS Project ...", *supra* note 7, at 586.

81 As McClelland noted: "(t)here were several flaws in the solely scientific approach of the proposals...criticism ranged from the medical to the legal. The legal concern [the non-consideration of the factor of *military necessity* in the equation-E.B.] caused the greatest unease ..." Justin McClelland, "The Review of Weapons in Accordance with Article 36 of Additional Protocol *IRRC* 85(850) (2003) at 400. Another writer claimed that the project had not been subjected to a peer review. See Parks, "Conventional Weapons..." *supra* note *19*, at 86, fn.121. For a summary of the legal critique on this project, *see e.g.*,

A brief review of the criteria proposed by the ICRC for determining "superfluous injury and unnecessary suffering" may yet provide some useful insights into the issues. This review was, after all, a genuine effort to provide some common ground for the discussion. The medical experts attending the Lugano Conference, it should be recalled, had sufficed themselves with the declaration that "...instead of 'suffering', *the wound or injury* caused by a weapon offered a better but still very complex way of defining the effect of that particular weapon."[82]

The four single medical criteria proposed for determining which foreseeable weapon design-dependent effects of weapons constitute "superfluous injury or unnecessary suffering" were:[83]

1. Specific disease, abnormal physiological state, permanent disability or disfigurement; or
2. Field mortality of more than 25% or hospital mortality of more than 5%; or
3. Grade 3 wounds as measured by the ICRC wound classification;[84] or
4. Effects for which there are no well-recognized and proven treatment.

However, these criteria (mortality excluded) appeared to have offered nothing to those who categorically opposed the use of flechettes. While it was argued with respect to other fragmentation weapons (i.e., cluster bombs) that they "fall squarely

Andrew Koch, "Should War Be Hell?" *Jane's Defense Weekly* 33 (*JDW*) (May 10, 2000) 23-26; Donna Marie Verchio, "Just Say No! The SIrUS Project: Well-Intentioned, But Unnecessary and Superfluous" *Air Force Law Review* 51 (2001) 203-215. For the ICRC's report on the conference in which the most criticism was expressed (Expert Meeting on Legal Reviews of Weapons and the SIrUS Project, Jan. 29-31, 2001) see Isabelle Daoust, "ICRC Expert Meeting on Legal Reviews of Weapons and the SIrUS Project" *IRRC* 842 (2001) 539-542.

82 *1976 Lugano Report*, at 140.

83 Coupland, "The SIrUS Project ...", *supra* note 7, at 23.

84 The ICRC's wound grading scheme is as follows: Grade 1 denotes skin wounds of less than 10 cm without a cavity; Grade 2 denotes skin wounds of less than 10 cm but with a cavity; Grade 3 denotes skin wounds of 10 cm or more with a cavity. However, as the ICRC admitted, it is impossible "to establish a precise correlation between [the wound's] grade and energy deposit[a close-range shotgun shells, for instance, deposit more than 1500 joules of energy while handgun bullets deposit up to 500], nor between the grade and type of weapon." *See*, Coupland, "The SIrUS Project..." *supra* note 7, at 15. For the ICRC wound grading scheme, *see also* Robin M. Coupland, *The Red Cross Wound Classification*, (Geneva: ICRC, 1997), 4-5.

under the first three criteria",[85] pathology reports indicated that flechettes weapon failed to meet all, including the third criteria.[86]

The fact that flechette injuries failed to meet even the third criterion (i.e., skin wounds of 10cm or more with a cavity) points to a major obstacle with which attempts to ban flechette weapons must contend on the ground that they inflict unnecessary suffering. It was noted during the 1974 Lucerne Conference that fragments with a regular shape cause less tissue damage but retain high velocity, which means that they have a greater strike velocity.[87] However, although it has been empirically shown that flechette darts indeed penetrate deeper into a victim's body, unclassified U.S. experiments have proven that unless "bone were encountered," flechettes could traverse the human body at a velocity as low as 152m/sec.[88] The reason for this level of penetration is the *little* tissue damage that the darts cause given their aerodynamic shape. In other words, the damage the darts cause to any *single organ* is less than the expected damage caused by standard fragmentation weapons.[89] Therefore, to prove unnecessary suffering, the depth of penetration – based on tissue damage – should ordinarily (with rounds that disperse regular fragments) be associated with increased suffering or, alternatively, with decreased military benefits. However, due to the lack of sufficient and reliable medical data on the level of suffering caused by flechette penetration,[90] it is difficult to unequivocally accept the Swedish group's position to ban flechette weapons *solely* on the grounds that it inflicts high levels of suffering.

---

85 For the full quotation and a review of cluster munitions according to these criteria *see*, Thomas M. McDonnell, "Cluster Bombs over Kosovo: A Violation of International Law," *Arizona Law Review* 44 (2002) 31, at 73. As Parks noted "The cluster munitions issue had changed from the original 1978-1980 argument that cluster munitions cause superfluous injury, which CCW delegations did not accept, to one of indiscriminate effect either in delivery or as to their failure rates" (*see*, Parks "Conventional Weapons...", *supra* note 19, at 102, fn. 169). Yet while most contemporary critics of cluster munitions indeed focus on their indiscriminate character during and after the conflict, the argument that cluster munitions cause superfluous injury or unnecessary suffering and should be banned as inhumane is still raised. For more on this debate *see*, *e.g.*, Thomas J. Herthel, "On the Chopping Block: Cluster Munitions and the Law of War – Unexploded Submunitions from Cluster Bombs" *Air Force Law Review* 51 (2001) 229, at 256-39; Karen Hulme, "Of Questionable Legality: The Military Use of Cluster Bombs in Iraq in 2003" *Canadian Yearbook of International Law* 42 (2004), 143, at 162-69.

86 However, as it points to cavitation at a given level of penetration as an indicator, one can relate it to the three "traditional" criteria. In other words, it seems reasonable to suppose that an increase in the probability of cavitation at a low level of penetration can entail increases in pain, disability and mortality.

87 *1975 Lucerne Report*, at 58.

88 *See*, Kokinakis and Sperrazza, *Criteria for Incapacitation*, *supra* note 68, at 26. *See also* in this respect Hans Blix's remarks at the first session of the Ad Hoc Committee, *CDDH, O. R.*, Vol. XVI, at 48.

89 It seems that the multiplicity of wounds makes little difference within the framework of this consideration.

90 For the lack of adequate medical literature on the issue, *see*, Correspondence with Dr. Robin Michael Coupland, Medical Advisor, ICRC Chief Medical Officer (May 12, 2003) (on file with the author).

### D. Indispensability – The Question of "Military Necessity"

The concept of military necessity was recognized as far back as the 1863 Lieber code:

> Military necessity, as understood by modern civilized nations, consists in the necessity of those measures which are indispensable for securing the ends of the war, and which are lawful according to the modern law and usages of war.[91]

The principle which was reaffirmed in the preamble to the St. Petersburg Declaration of 1868 according to which there are "technical limits at which military necessities ought to yield to the requirements of humanity."[92] As Professor Dinstein has noted, the law of international armed conflicts (IACs) "in its entirety is predicated on a subtle equilibrium between two diametrically opposed impulses: military necessity and humanitarian considerations."[93]

While the built-in tension between the demands of military necessity and those of humanitarian considerations is confronted by every IHL norm, this built-in tension increases exponentially when the legality of weaponry is at hand. As Prof. Baxter stated: "A rifle can be used indiscriminately against civilian and soldier. A knife can be used to cause unnecessary suffering".[94] Therefore, as one scholar noted, the more crucial question regarding dubious weapons turn on their military aspects: "*How and in what circumstances are these weapons used*; what are the military advantages sought by their use; could other weapons, or the same weapons with less potency, accomplish these same ends?"[95] As one expert noted:

> Injuries can only be 'superfluous' either if they are not justified by any requirement of military necessity or if [they]...are *manifestly disproportionate* to the military advantage reasonably expected from the use of the weapon. The first will only rarely be the case, … The second condition will only be fulfilled if the weapon is at least relatively superfluous-which requires a *comparative judgment* as to how much suffering various weapons cause and whether *alternative military means could achieve the same results with less suffering*.[96]

91 Instructions for the Government of Armies of the United States in the Field (Lieber Code), 24 April 1863. *Laws of Armed conflicts*, 3, at 6.

92 St. Petersburg Declaration, *id.*, at 91.

93 Yoram Dinstein, *The Conduct of Hostilities under the Law of International Armed Conflict*, (Cambridge: Cambridge University Press, 2004), 16. In fact, he further clarified, "LOIAC [Law of International Armed Conflicts] amounts to a checks-and-balances system, intended to minimize human suffering without undermining the effectiveness of military operations". *Id.*, at 17.

94 Baxter, R. R. "Criteria of the Prohibition of Weapons in International Law", in: Horst Ehmke *et al.* (eds.) *Festschrift fur Ulrich Scheuner – Zum 70. Geburtstag*, (Berlin: Duncker & Humblot, 1973), 41, at 46.

95 Andrew M. Stone, "Conventional Weapons …", *supra* note 8, at 149.

96 Emphasis added. Stefan Oeter. "Methods and Means of Combat", in: Dieter Fleck (ed.), *The Handbook of Humanitarian Law in Armed Conflicts*, (Oxford: Oxford University Press, 1995), 105, at 109.

However, many factors are involved with the determination of military necessity, too. For example, the direct military advantages from a certain weapon's use, the availability of adequate alternatives[97] (which requires exploring each weapon's military and humanitarian implications), logistics, etc. Furthermore, as the 1982 Commentary on the Additional Protocols noted in this context: "(I)nevitably, the assessment will be subjective even when sufficient agreed factual data are available on wound effects and military effectiveness".[98]

The expert conferences' protocols indicate that flechette's military necessity was an intensely debated question. Arguments in favor of its necessity were based primarily on the Vietnam War experience. Regarding use of artillery-fired fragmentation weapons, it was argued during the Lucerne Conference that in defensive contexts, "... there is a military need to use weapons which cover a large area, in such a manner that there will be little chance for the attacking force to go unscathed."[99] When flechette weapons were specifically discussed (again in the context of artillery), it was asserted – as noted in our overview (*supra*, Chapter I) – that "flechette weapons are very efficient in this application, surpassing the effect of a conventional artillery round by up to ten to one."[100] A year later, during the Lugano Conference, it was pointed out that flechette weapons offered military advantages by reaching beyond the artillery's range, as "... anti-personnel weapons for tanks and in the rockets launched from helicopters against enemy troops."[101] Some also argued that its appearance constituted a natural technological development, a response to the greater firepower and broader dispersal of infantry formations.[102] Logistical considerations were likewise raised in support of the flechette's military necessity.[103]

In contrast, those who questioned its military necessity retorted that they never proposed a sweeping ban on fragmentation weapons; nonetheless, recent developments have enabled the production of extremely destructive versions of these weapons. They therefore stressed that the proposed prohibition relate exclusively to anti-personnel weapons[104] and to the "extreme area coverage" of these weapons, with the

97 Indeed, the 2005 ICRC Study Report noted that some states (e.g., the U.S.) refer to this issue as an element that has to be considered in the assessment of weapon legality under this principle. *See*, *ICRC Study Report*, Vol. I, at 240.

98 Michael Bothe, K. J. Partsch and W. A. Solf (eds.) *New Rules for Victims of Armed Conflicts: Commentary on the Two 1977 Protocols Additional to the Geneva Conventions of 1949*, (Leiden: Martinus Nijhoff, 1982) 196.

99 *1975 Lucerne Report*, at 50, para. 162.

100 *Id.*, at 55, para. 178. As Major General William C. Westmoreland, Commander of the U.S. forces in Vietnam's during the war's peak (1964-1968) recalled: "Beehive artillery rounds were markedly effective at close range, sometimes making the difference between life and death when isolated artillery units came under attack". *See*, William C. Westmoreland, *A Soldier Reports*, (New-York: Doubleday, 1976) 281.

101 *1976 Lugano Report*, at 72, para. 6.

102 *Id.*, at 71, para. 6.

103 Relying on the assumption that flechette weapons had also increased the accuracy of artillery, it was claimed that "For that reason, greater fragmentation would not be used to increase the numbers of human losses, but to reduce the number of shots that had to be fired, thus helping to solve the logistic problem", *1976 Lugano Report*, at 72, para. 6.

104 *See*, CDDH/IV/201, *CDDH, O. R.*, Vol. XVI, at 48.

attendant implications for minimizing inadvertent "civilian casualties".[105] The debate merged into a discussion over "substitutes for fragmentation weapons". In line with arguments that flechettes might be replaced with "ordinary fragmentation warheads",[106] heard at the previous Lucerne conference, one expert noted that finding alternate weapons within the existing inventory of arms should not pose any difficulty. After all, only a strictly limited list of weapons was proposed for possible ban or restrictions.[107] In the final analysis, it appears that flechette opponents believed that in the final balance between military necessity and humanitarian considerations, the latter should triumph.[108] As the *Explanatory Memorandum* to the draft proposal on flechette weapons submitted at the Lugano Conference stated:

> It is queried whether the military advantage of these weapons is so great as to outweigh the humanitarian concern which is raised by their use. It is submitted that the answer is in the negative.[109]

However, the Israeli case (*supra*, Part II) is far from being a clear-cut source of comparison regarding the military necessity of flechette weapons. Neither the conflict in South Lebanon nor the one in the Gaza Strip represent a "traditional" conventional war between regular armies, *especially* in regard to the types of warfare employed. Two comments can be made in this respect. First, according to Israel's armored combat doctrine in South Lebanon, three flechette rounds were to be followed by a single High Explosive Squash Head (HESH) (*supra*, Chapter 4). If flechette rounds were truly irreplaceable, then this doctrine should have been proved pointless and destined for revision.

Second, as to the weapon's *military necessity* in the Gaza Strip, the State made sure to address this issue in its Response to the Petition:

> The need to use flechette in Gaza Strip derived from the assessment of the military and professional ranks... that *the use of these munitions is currently necessary in this area and has no appropriate substitute.*[110]

The basis for this assessment, as detailed in the State's Response, was the weapon's effectiveness against "lone terrorists infiltrating in open areas," coupled with the

105 *1976 Lugano Report*, at 73, para. 8.

106 *1975 Lucerne Report*, at 56, para. 184.

107 *Id.*, at 6, para. 5.

108 As Blix concluded in the Explanatory Memorandum to his proposal to ban the weapon in the Ad Hoc Committee from March 22, 1974: "Fechettes caused a high degree of suffering. The proposal that they should be banned was based on the view that their utility was not so great as to outweigh the suffering they were likely to cause." *CDDH, O. R.*, Vol. XVI, at 48.

109 *1976 Lugano Report*, at 205. A similar text appeared in the Explanatory Memorandum to the 1974 proposal to ban flechette weapons that was submitted to the Ad Hoc Committee's first Session (Mar. 22, 1974). *See, CDDH, O. R.*, Vol. XVI at 48.

110 Emphasis in original. *See,* HCJ 8990/02, *A Response by the State Attorney's Office*, para. 33, available at www.btselem.org/hebrew/Legal_Documents/8990_ 02_Flechette_ Appeal_State_Response.pdf.

clear separation – where "exposed areas exist" – between the Israeli presence and the Palestinian population.[111] Yet, a careful review of the three incidents (i.e., the June 2001 Bedouin Camp, the December 2001 three Gazan youths and the August 2002 Al Hajan family incident) suggests that the objectives chosen for the firing of flechette rounds could have been attained by alternative means, mainly machine gun and small arms (personnel rifles) fire. In fact, in the incident where the three Gazan youths were killed, one of the autopsy reports mentioned bullet wounds from the type of machine gun with which IDF tanks are equipped *in addition* to the darts' injuries.[112] The problem, however, is that in contrast to the State's Response, according to which flechette use in the Gaze Strip was considered the last means to be employed,[113] this incident, together with news reports, indicates that flechettes were, in most cases, the first resort.[114]

## E. Concluding Remarks

First and foremost, this chapter demonstrates why such legal reviews are rarely seen in the literature. The absence of universally agreed-upon scientific criteria, the multiplicity of related factors, and the heavy reliance on medical data and insight have deterred most legal scholars from attempting this task. In reference to the flechette weapon, specifically, however, more empirical data is needed in order to form a solid scientific base.[115] Even the ICJ, despite its proclamations to the contrary,[116] has refrained from any serious discussion of the legality of nuclear weapons in respect to this cardinal IHL principle. This state of affair implies that unequivocal conclusions are currently highly problematic.

Yet, three general comments are in order. First, it appears that the multiplicity of wounds criterion suggests that the flechette has rightly been labeled a "dubious weapon".[117] Despite the absence of conclusive scientific evidence, it is highly probable

111 The darts' low penetration capability when aimed at hard substances – "unlike normal rounds" – was also mentioned in reference to the shelter provided by house walls against flechettes. *Id.*, para. 14.

112 *See*, National Center of Forensic Medicine, Rep. 3/2293/2001, (Jan. 16, 2002), at 3-8.

113 According to the State's response, the use of flechette rounds is "reserved for cases in which the operational target… cannot be achieved by less severe means, such as detention, the firing of rubber bullets, small arms fire...". *The State Response*, para. 40.

114 *See e,g.*, Avichai Becker, "Things which Can Be Seen From There", *Haaretz*, (Friday Magazine), (Dec. 5, 2003), 16-23, at 22 (in Heb.). Becker was for a few days attached to the IDF force that secured the then-isolated settlement of Nezarim and documented his stay.

115 As Sellier and Kneubuehl noted in reference to the wound ballistics of flechettes: "Little has been published to date on the behaviour of wing-stabilized projectiles". *See*, Karl G. Sellier and Beat P. Kneubuehl, *Wound Ballistics and the Scientific Background*, (Amsterdam: Elsevier, 1994), 136.

116 As the Court stated: "It is imperative for the Court to take account of the unique characteristics of nuclear weapons, and in particular their destructive capacity, their capacity to cause untold human suffering, and their ability to cause damage to generations to come". *See, Nuclear Weapons Advisory Opinion*, at para. 36.

117 It is worth noting in this context that during the nineteenth century, legal scholars (e.g., the Russian Prof. Friedrich Fromhold de Martens) asserted that the machine gun

that the multiplicity of wounds causes a high level of pain for the wounded person, raises the mortality rate and may even make it difficult to supply the wounded with proper medical care. This last point should not be treated lightly. That flechettes affect the feasibility of receiving adequate medical treatment in the field has been suggested as a factor to be considered when evaluating its legality.[118] Furthermore, it has been argued that the very difficulty in *treating* multiple wounds should be considered a form of unnecessary suffering.[119] As most flechette-injured persons require surgical attention to several wounds simultaneously, concern has been raised as to whether medical treatment for the wounded can be assured in developing states, where several surgical teams working at the same time remains outside the norm.[120] Employment of such weapons likewise increases the medical workload as *many victims are expected to arrive for treatment concurrently.*[121]

In contrast, however, it was argued that while a greater number of victims can be anticipated, the average wound may be less severe; the resulting strain on available medical resources is thus expected to be lower.[122] Second, the fact that *many victims are expected* once flechette weapons are employed was not disputed as these anti personnel weapons were designed to cause mass causalities. The results of the famous December 1966 battle at Landing Zone Bird in which 266 North Vietnamese casualties were counted, mainly due to two beehive shells, speak for themselves.[123] However, recent legal writing on the issue of weapon legality suggests that two aspects should be considered when assessing the level of suffering or injury in consideration of new weapon's legality: one qualitative and one quantitative. While the first, the qualitative aspect, "relates to the nature of the suffering itself"[124] (i.e., whether flechette use does increase the level of suffering caused to the individual), the second, quantitative aspect, *relates to the scale of the suffering*. Here, the question is "will there be greater numbers of victims injured by the weapon".[125] While this aspect is clearly related to the issue of distinction, to be considered in the next chapter, it is clear that in this respect flechette weapons do inflict unnecessary suffering.

Third, there are other types of directional fragmentation devices or charges (e.g., Claymore-type devices) that also inflict multiple wounds.[126] These have provided U.S.

and the firing of multiple projectiles from a single weapon violated the law of war. *See*, SIPRI, *Anti-personnel Weapons*, (London: Taylor & Francis, 1978) 212.

118 *See, e.g.*, the implementation of this measure in the context of blinding laser weapons, Bengt Anderberg *et al.*, "Blinding Laser Weapons and International Humanitarian Law", *Journal of Peace Research* 29 (1992) 287, at 294.

119 *See, e.g., 1976 Lugano Report*, at 75, para. 10.

120 *1973 Geneva Report*, at 46, para. 149.

121 For this claim *see, e.g.*, Blix speech at the Ad Hoc Committee second Session *CDDH, O. R.*, Vol. XVI, at 122.

122 *See, 1975 Lucerne Report*, at 58, para. 202.

123 *See, supra* chapter 1(b)(IV).

124 Justin McClelland, "The Review of Weapons...." *supra* note 81, at 407.

125 Id.

126 The M18 Claymore, for instance, contains 700 steel balls that, on detonation, are dispersed in a 60-degree horizontal arc extending 50 meters at a maximum height of 2 meters in front of the mine. The mine covers a casualty radius of 100 meters but the optimum effective range (the range at which the most desirable balance is achieved

practitioners with a justification to object considering this feature as crucial when weighing a weapons' legality.[127] Proponents of this position (i.e., that weapons producing multiple wounds do not automatically violate the principle of unnecessary suffering) claim, in fact, that the proliferation of these devices in world military arsenals,[128] together with their *de jure* approval in the CCW 1996 Amended Second Protocol,[129] "reconfirm the legality" of any potential multiple-wounding weapon.[130] However, the said Protocol does impose restrictions on the use of "other devices"[131] – similar to the restrictions imposed on anti-personnel mines. Claymore-type devices are therefore permitted for use if placed within fenced, marked and guarded minefields (unless fitted with self-destructing and self-deactivating mechanisms).[132] Such weapons enjoy an exemption from those restrictions but this exemption is *strictly reserved* in terms of time (up to 72 hours), location ("immediate proximity to the military unit that emplaced them") and, finally, presence of active monitoring personnel to ensure the effective exclusion of civilians as required.[133] No doubt these rules were meant to protect civilians rather than combatants. However, while one can view these rules as acknowledgements of the legality of multiple-fragmenting munitions against combatants,[134] the very fact that heavy restrictions have been placed on their use is quite meaningful.

Furthermore, most of the current stocks of Claymore-type devices are covered by Article 2 of the 1997 Anti Personnel Mine Ban Convention (Ottawa Convention).[135]

---

between lethality and area coverage) is limited to 50 meters. Its successor, the M18A1, contains an additional 60 10.5-grain steel balls. For more on this "defensive weapon" *see, e.g.*, <http://www.globalsecurity .org/military/systems/munitions/m18-claymore.htm>. Its Field Manual (FM 23-23, Department of the Army, Washington, DC, 6 Jan, 1966) is available at <http://www.globalsecurity.org/military/library/policy/army/fm/23-23/>.

127 *See e.g.*, Parks, "Review of Weapons…" *supra* note 75, at 22; Parks, "Memorandum of Law…" *supra* note 57, at 20-21.

128 According to a list compiled by the HRW (Feb. 2003), among the States Parties to the Ottawa Convention, 20 states are known to retain operational stocks of Claymore-type mines. *See*, Landmine Monitor Fact Sheet, available at <http://www.icbl.org/content/download/20104/387716/file/claymore_jan_2003.pdf>.

129 *See*, Protocol on Prohibitions or Restrictions on the Use of Mines, Booby-Traps and Other Devices as amended on 3 May 1996 (Protocol II to the 1980 Convention as amended on 3 May 1996), Art. 5(6), *Laws of Armed Conflicts*, 196, at 200.

130 Parks, "Memorandum of Law…" *supra* note 57, at 22.

131 Article 2(5) reads: "'Other devices' means manually emplaced munitions and devices, including improvised explosive devices designed to kill, injure or damage and which are activated manually, by remote control or automatically after a lapse of time", *Laws of Armed Conflicts*, 196, at 197.

132 *See*, Art 5. For the self-destruction and self-deactivation requirements see, Technical Annex, Art. 3, *id.*, at 208.

133 *See*, Art. 5(6), *id.*, at 200.

134 *See*, Parks, "Conventional Weapons…" *supra* note 19, at 79, fn. 92.

135 Claymore-type devices are typically triggered either by remote control or by tripwire. Once they are fitted with tripwires, they are defined as mines by the Ottawa Convention. *See*, the Convention on the Prohibition of the Use, Stockpiling, Production and Transfer of Anti-Personnel Mines and on their Destruction, 18 September 1997, Art. 2. *See*, *Laws of Armed Conflicts*, 285, at 287.

These devices are, therefore, not only *prohibited* but also legally subject to the requirements of their *destruction* under the Convention's article 4.[136]

In sum, these instances appear to reinforce the call for drafting a codex of conventional rules regarding the use of flechette weapons or, at the very least, for conducting an urgent discussion over their legality, similar to that made in the *Goldstone Report*.[137] After all, the average pre-fragmentation number of pieces in each Claymore-type devices (~700)[138] falls far below the number of darts in each flechette round (5000-9000), let alone their effective range (180m for the latter while the average for the former stands on ~50m).

However, after concluding that, in the final analysis, flechette weapons should be considered a particularly "dubious" weapon under the customary principles of international law prohibiting the infliction of unnecessary suffering, it is time to investigate whether this unflattering title is justified in reference to the second customary principle regarding a weapon's legality, its indiscriminate use. We turn to this issue in the next chapter.

---

136 They remain outside the purview of the Convention only when they are equipped to function through command detonation. In such cases, they are not "designed ... to be exploded by the presence, proximity or contact of a person or a vehicle" and do not therefore fall within the mine definition as set out in Art. 2 to the Convention. In practice, some States Parties (e.g., Norway, Switzerland and the U.K.) have indeed reconfigured them by removing or destroying their fuses, while leaving the devices themselves intact. *See*, Stuart Maslen, *Commentaries on Arms Control Treaties: Convention on the Prohibition of the Use, Stockpiling, Production and Transfer of Anti-Personnel Mines and on Their Destruction*, Vol. I., 2nd Ed. (Oxford: Oxford University Press, 2005), 108, at paras. 1.83, 1.21, para. 2.31.

137 *Human Rights in Palestine and Other Occupied Arab Territories*, Rep. of the UN Fact-Finding Mission on the Gaza Conflict (A/HRC/12/48), (Sept. 25, 2009), at 425, para. 1971(d).

138 Other than the MON-90, the Russian copy of the U.S. M18A1 Claymore mine, which contained 2000 steel balls (*see* <http://ordatamines.maic.jmu.edu/displaydata.aspx?OrDataId=7447>), and the Swedish FFV 013 (with 1200 steel balls), most of these devices contain far less than 700 steel balls. *See* <http://ordatamines.maic. jmu.edu/displaydata.aspx?OrDataId=7544>. Note: *Ordata Online* is a U.S. government database of landmines and other unexploded ordnance, hosted on the Mine Action Information Center website of James Madison University.

# Chapter 8

## Flechette and the Principle of Distinction

In the 1997 edition of his famous book on international law, Professor Shaw opened his discussion over methods of warfare with the following statement: "A principle of long standing, if not always honoured in practice, is the requirement to protect civilians against the effects of hostilities".[1] This principle was clearly in the minds of the UNGA when it formulated its invitation to the UN Secretary-General (SG) to study, *inter alia*, "the prohibition and limitation of the use of certain methods and means of warfare":

> ... it is prohibited to launch attacks against the civilian population as such; that distinction must be made at all times between persons taking part in the hostilities and members of the civilian population to the effect that the latter be spared as much as possible.[2]

This chapter is devoted to the application of this principle, *mutatis mutandis*, to the Law of Weaponry in reference to flechette weapons. As the ICJ stated in its *Nuclear Weapons Advisory Opinion*, the application of this principle demands that States never "use weapons that are incapable of distinguishing between civilian and military targets".[3] We therefore begin with an overview of the principle's formulation (i.e., the prohibition on the use of indiscriminate weapons) prior to the cited ICJ *Advisory Opinion*. We then examine the feasibility of defining any type of weapon as inherently discriminatory, especially in the wake of the *Opinion*. Next, we conduct a detailed discussion regarding the extent to which flechette weapons – in the ter-

1 Malcolm Shaw, *International Law* 4th ed., (Cambridge: Cambridge University Press, 1997), at 812. This principle had already appeared in the Lieber Code, see *Instructions for the Government of Armies of the United States in the Field*, 24 April 1863, Art 22 [hereinafter *Lieber Code*], Dietrich Schindler and Jiri Toman (eds.), *The Laws of Armed Conflicts* (Leiden: Martinus Nijhoff, 2004), 3, at 6 [hereinafter, *Laws of Armed Conflicts*].

2 *See*, UNGA Res. 2444 (XXIII), *Respect for Human Rights in Armed Conflict*, (Dec. 19, 1968) (adopted by unanimous vote).

3 *See*, Legality of the Threat or Use of Nuclear Weapons, Advisory Opinion (International Court of Justice, July 8, 1996), *International Law Materials* 35 (1996), 809 & 1343 [hereinafter, *Nuclear Weapons Advisory Opinion*], para. 78. In our context (anti-personnel weapons), the distinction applies exclusively between combatants and non-combatants.

minology used by the proponents of the 2008 Cluster Munition Convention – are prone to cause *unacceptable civilian harm*.[4] A brief summary concludes this chapter.

## A. The Prohibition on Use of Indiscriminate Weapons – A Legal History

As the former chapter detailed, the principle prohibiting the infliction of unnecessary suffering was enshrined in an early attempt to issue codes for the conduct in war; with the 1868 St. Petersburg Declaration being the first modern multilateral treaty addressing weaponry use.[5] In contrast, the second principle mentioned, the prohibition on the use of indiscriminate weapons, has not yet been explicitly incorporated into IHL. As Judge Rosalyn Higgins – the only judge who has delved deeply into this issue in the *1996 Nuclear Weapons Advisory Opinion*[6] – mentioned in her *Dissenting Opinion* , this requirement "is not a general principle of humanitarian law specified in the 1899, 1907 or 1949 law, but flows from the basic rule that civilians may not be the target of attack".[7] The ICRC Study Report, which specifically notes that "(s)tates practice establishes this rule as a norm of customary international law applicable in both international and non-international armed conflicts",[8] refrained from pointing out any explicit prohibition in treaty law that might act as a source for such a stipulation. Alternatively, it did mention that "the prohibition of such weapons is also supported by the general prohibition of indiscriminate attacks".[9]

Therefore, long before the 1996 ICJ *Advisory Opinion* (where the Court simply deduced this prohibition from the general distinction between combatants and non-combatants), proponents of such a prohibition pointed to the preamble of the 1868 St. Petersburg Declaration as the sole (yet indirect) source for such a requirement[10]

4 For a discussion of the expression *"unacceptable harm" see,* John Borrie, *Unacceptable Harm: A History of How the Treaty to Ban Cluster Munitions Was Won*. (New York: United Nations, 2009), 333-38.

5 Declaration Renouncing the Use, in Time of War, of Explosive Projectiles under 400 Grammes Weight, (1868), *Laws of Armed Conflicts*, 91. *See,* also Art. 16 of the *very* first military manual prepared for the U.S. Army (the Lieber Code, 1863), *Laws of Armed Conflicts*, 3, at 6.

6 Other Judges have sufficed, at best, with general statements. Judge Gilbert Guillaume, for instance, stated that "Customary humanitarian law thus contains only one absolute prohibition: the prohibition of so-called 'blind' weapons which are incapable of distinguishing between civilian targets and military targets", Individual Opinion of Judge Gilbert Guillaume, The *Nuclear Weapons Advisory Opinion*, at para. 5. Note: all declarations, separate and dissenting opinions of the Judges are available at <http://www.icj-cij.org/docket/index.php?p1=3&p2= 4&k=e1&p3=4&case=95>.

7 Dissenting Opinion of Judge Higgins, the *Nuclear Weapons Advisory Opinion*, at para. 24.

8 Jean-Marie Henckaerts and Louise Doswald-Beck, *Customary International Humanitarian Law*, Vol. I., (Cambridge: Cambridge University Press, 2005) at 237 [hereinafter, *ICRC Study Report*].

9 *Id*, at 244.

10 *See*, for instance, Hans Blix, "Current Efforts to Prohibit the Use of Certain Conventional Weapons" *Instant Research on Peace and Violence* 4 (1) (1974) 21, at 24; ICRC, Weapons that may Cause Unnecessary Suffering or Have Indiscriminate Effects

("... the only legitimate object which States should endeavour to accomplish during war is *to weaken the military forces of the enemy*"[11]).

In consequence of this lack of formalization, a legal debate continued into the 1970s as to whether such a prohibition "had already acquired the status of a rule of positive international law".[12] After all, until the 1977 conclusion of Additional Protocol I to the 1949 Geneva Conventions [hereinafter AP-I],[13] the main instrument in which this prohibition was incorporated was the 1956 ICRC Draft Rules.[14] This draft, coming on the heels of Europe's WWII experiences, was designed to increase civilian protection during war. It included an article entitled *Prohibited Methods of Warfare*:

> ... the use is prohibited of weapons whose harmful effects – resulting in particular from the dissemination of incendiary, chemical, bacteriological, radioactive or other agents – could spread to an unforeseen degree or escape, either in space or in time, from the control of those who employ them, thus endangering the civilian population.[15]

This article, with its explicit reference to nuclear weapons, was an important reason for the entire draft's failure.[16] The absence of an explicit prohibition on the use of weapons which may considered inherently indiscriminate would finally be addressed to the satisfaction of many legal experts only in 1977 (AP-I). As Article 51(4) – characterized by the *ICRC's Commentary* as "one of the most important articles in the Protocol"[17] – reads:

> Indiscriminate attacks are prohibited. Indiscriminate attacks are... (b) those which employ a method or *means* of combat which cannot be directed at a specific military objective;[18] or (c) those which employ a method or *means* of combat the effects of

(Report of the Work of Expert) Geneva, (1973), (hereinafter, *1973 Geneva Report*), at 13, para. 24.

11 Emphasis added. Declaration of St. Petersburg, Preamble, para. 2. *Laws of Armed Conflicts*, 91, at 92.

12 *See,* Conference of Government Experts on the Use of Certain Conventional Weapons (Lucerne, 24.9-18.10.1974) ICRC Report (1975) [hereinafter, *1975 Lucerne Report*], at 10, para. 29.

13 Protocol Additional to the Geneva Conventions of 12 August 1949, and relating to the Protection of Victims of International Armed Conflicts, opened for signature Dec. 12, 1977, (Protocol I), [hereinafter, AP-I], *Laws of Armed Conflicts*, 711.

14 ICRC, Draft Rules for the Limitation of the Dangers Incurred by the Civilian Population in Time of War, (1956), [hereinafter, *ICRC Draft Rules*], *Laws of Armed Conflicts*, 339.

15 *ICRC Draft Rules*, Chap, IV, Art. 14. *Id.*, at 342.

16 *See, supra* chapter 2, A.

17 *See,* C. Pilloud *et al.*, Commentary on the Additional Protocols of 8 June 1977 to the Geneva Conventions of 12 August 1949 (Geneva: ICRC 1987) (hereinafter, *ICRC Commentary*), at 615, para. 1923.

18 Emphasis added. As the *ICRC Commentary* stressed, the term 'means of combat' "generally refers to the weapons being used, while the expression 'methods of combat' gen-

> which cannot be limited as required by this Protocol; and consequently, in each such case, are of a nature to strike military objectives and civilians or civilian objects without distinction.[19]

Yet, as the following section details, several states made sure to stress at the time of the Protocol's drafting that no such explicit prohibition could be deduced from this text.[20] Indeed, as the *Ad Hoc Committee's* Rapporteur, Prof. Kalshoven, noted: "Viewed thus, the rules of Article 51, paragraph 4, while reinforcing the general principle of protection of the civilian population, do not themselves amount to prohibitions on [the] use of specific weapons".[21] In fact, besides military manuals[22] and various UNGA resolutions passed as of 1972[23] only a few international legal instruments, notably the 1994 *San Remo Manual,*[24] clearly stated this prohibition prior to the 1996 ICJ *Advisory Opinion.*[25] Even the preamble to the 1980 CCW, while explicitly reaffirmed both the prohibition on inflicting unnecessary suffering and the relatively new prohibition regarding environmental protection,[26] refrained from mentioning any prohibition on the use of indiscriminate weapons.[27]

---

erally refers to the way in which such weapons are used". *The ICRC Commentary*, 621. para. 1957.

19 *Laws of Armed Conflicts*, 711, at 736, sub-paragraph (a) according to which: "Indiscriminate attacks are: (a) those which are not directed at a specific military objective", clearly pertains to ad hoc decisions on targets and as such is not related the law of weaponry.

20 *See*, e.g., U.K. explanations for the vote (then Art. 46). *Official Record of the Diplomatic Conference on the Reaffirmation and Development of International Humanitarian Law Applicable in Armed Conflict*, Geneva (1974-1977), [hereinafter, *CDDH O.R.*.] Vol. VI, SR. 41, (May 26, 1977), at 164.

21 Frits Kalshoven, "Arms, Armaments and International Law", *Recueil Des Cours* 19, (1985[II]) 181, at 247.

22 *ICRC Study Report*, Vol. II, at 1556-1560.

23 For a list of UNGA resolutions mentioning this prohibition *see, id.*, at 1575-78. Note that the first 1961 resolution mentioned (UNGA Res. 1653 (XVI), Nov. 24, 1961) referred solely to nuclear weapons and should be viewed within this context.

24 Article 42, in which the text of Article 51 of AP-I was reiterated in slightly different wording, reads: "In addition to any specific prohibitions...it is forbidden to employ methods or means of warfare which...(b) are indiscriminate, in that: (i) they are not, or cannot be, directed against a specific military objective; or (ii) their effects cannot be limited as required by international law as reflected in this document." *See, San Remo Manual on International Law Applicable to Armed Conflicts at Sea,* 12 June 1994, *Laws of Armed Conflicts*, 1153, at 1161.

25 Even the *ICRC Study Report*, while detailing various activities of the ICRC concerning the prohibition on unnecessary suffering, pointed to this issue as including only a single activity prior to 1995 (a 1991 *ICRC Press Release*). *ICRC Study Report*, Vol. II, at 1581 and compare to id., at 1551-1552.

26 The text reads: "Also recalling 'that it is prohibited to employ methods or means of warfare which are intended, or may be expected, to cause widespread, long-term and severe damage to the natural environment'."

27 Instead, it recalled "the general principle of the protection of the civilian population against the effects of hostilities." CCW Preamble, *Laws of Armed Conflicts*, 181, at 184.

However, whether the prohibition's source has been recognized already in the Preamble of the *1868 St. Petersburg Declaration,* or in Article 51(4) of the 1977 AP-I, after the 1996 ICJ *Advisory Opinion*, the prohibition's existence is no longer questioned. Since then it has been restated in various legal instruments, such as the 1998 *Rome Statute.*[28] The prohibition's implementation, however, has turned out to be a different story.

### *B. Defining Inherently Indiscriminate Weapons – A Mission Impossible?*

As the comprehensive *ICRC Study Report* found in reference to the task of defining indiscriminate weapons, "(s)everal military manuals and official statements mention weapons that 'have indiscriminate effects', 'strike military objectives and civilians indiscriminately' or 'cannot distinguish between military objective and civilians', *without further detail*".[29]

The lack of detail is far from coincidental. As found in the experts' meetings during the 1970s, defining inherently indiscriminate weapons was an almost impossible mission once it was agreed that this principle "has not been taken to mean a ban upon weapons… which, though directed against military targets, entail the risk *of incidental civilian casualties* or damage to civilian objects in the vicinity of the targets".[30] In any case, they agreed that in principle, the prohibition on indiscriminate warfare relates more often to methods of warfare or to the way that weapons are used than to the weapons themselves.

More than two decades later, the issue was again raised before the ICJ in the *Nuclear Weapons Advisory Opinion* after many states argued that any use of nuclear weapons should be considered indiscriminate given their "necessarily… cataclysmic and indiscriminate effects".[31] In response, states advocating their legality "under certain circumstances"[32] pointed to the cited distinction between a weapon's legality and the collateral damage it is likely to inflict.[33] As a prominent legal scholar noted, the fact that weapons can be employed in a manner breaching this prohibition (by being either instruments of direct or indiscriminate attack against civilians), "… does not

---

28 *See, Rome Statute of the International Criminal Court*, art 8(b)(xx), *Laws of Armed Conflicts*, 1309, at 1319. It should be noted, however, that the prohibition is conditional on the fact that "such weapons… are the subject of a comprehensive prohibition and are included in an annex to this Statute."

29 Emphasis added. *ICRC Study Report*, Vol. I, at 247.

30 Emphasis added. *1973 Geneva Report*, at 13, para. 25.

31 *See, India Written Statement* 3 (June 1995). *See,* also, e.g., *Egypt Written Statement* 12 (June 1995). All Written Statements are available at <http://www.icj-cij.org/docket/index.php?p1=3&p2=4&k=e1&case= 95&code =unan&p3=1>.

32 As they were referred to by the Court. *See*, e.g., *Nuclear Weapons Advisory Opinion*, at para. 94.

33 As the U.S. clarified: "The Law of Armed Conflict precludes making civilians the object of attack as such. This, of course, does not mean that attacks on military objectives are prohibited simply because they may cause collateral civilian injury or damage– as is often the case in armed conflict". *See U.S. Written Statement* 22 (June 1995). *See,* also the U.K. Written Statement 52-3 (June 1995).

stain the weapons themselves with an indelible mark of illegitimacy, since in other operations the same weapons may be used within the framework of LOIAC."[34]

Yet, experts participating in the 1973 experts meeting agreed that under certain circumstances, means of warfare could be indiscriminate by their very nature; in other contexts, it is their "normal or typical use" that has indiscriminate effects.[35] This difficulty in defining means of warfare as indiscriminate *per se* was reiterated in the 1974 Lucerne Conference when partial agreement was reached with respect to the fact that "all conventional weapons could be used in circumstances where the risk of hitting civilians was virtually non-existent".[36] However, some experts continued to argue that "a prohibition of the use of indiscriminate weapons *would be exceptionally hard to apply.*"[37] Notwithstanding the troublesome features of the issue, the crucial question raised at the 1973 experts meeting was whether "weapons which *by their nature* are incapable of being directed with any certainty at specific military targets" violated the prohibition, or whether the case in question concerned those weapons that, in "*typical or normal use* are not delivered with any certainty to such targets".[38] However, while the report on the 1973 experts meeting concluded that *both criteria* (i.e., "their nature or... the way in which they are commonly used") should be applied in prohibiting various *types of weapons*,[39] a different view was offered in the 1974 Lucerne Conference. For instance, a British opinion paper, distributed to the participants, rejected the "*typical or normal use*" criterion and significantly restricted the prohibition by suggesting that a generic prohibition would apply only to "the use of any weapon which cannot be accurately directed against military targets".[40]

In light of the debate, a few experts adopted what the Conference's principal Rapporteur (responsible for legal issues), Professor Kalshoven, chose to call a "middle course":[41] indiscriminate weapons were those that with "regard to their effects in time or space, *cannot be employed with sufficient or with predictable accuracy against the chosen target*".[42] This position, he noted a decade later, "is probably as close to the truth as one can get".[43] Given the ICRC's failure to settle this issue in the subsequent conference

34 Yoram Dinstein, *The Conduct of Hostilities under the Law of International Armed Conflict,* (Cambridge: Cambridge University Press, 2004), 55.

35 ICRC *1973 Geneva Report*, at 14, para. 27.

36 *1975 Lucerne Report*, at 10, para. 30.

37 Emphasis added. *Id.*

38 Emphasis added. *1973 Geneva Report*, at 13, para. 25.

39 *Id.*, at 71, para. 244.

40 *1975 Lucerne Report*, at 10, para. 31. This "excellent paper", according to Kalshoven (*Reflections on the Law of War: Collected Essays.* [Leiden: Martinus Nijhoff 2007], 158), entitled "Legal Criteria for the Prohibition or Restriction of Use of Categories of Conventional Weapons", was submitted by Sir David Hughes-Morgan. Kalshoven also noted that the large number and unequal quality of the documents introduced at the Conference convinced its bureau to decide against annexing any of them to the report. For numerous segments of Hughes-Morgan's paper, however, *see*, Kalshoven, *Id*, at 384-85. For a list of the documents distributed see *1975 Lucerne Report*, at 103.

41 Kalshoven, "Arms, Armaments..." *supra* note 20, at 236.

42 Emphasis added. *1975 Lucerne Report*, at 10-11, para. 31.

43 Kalshoven, "Arms, Armaments..." *supra* note 20, at 237.

(Lugano, 1976),[44] as Judge Higgins rightly noted 30 years later in the *Nuclear Weapons Advisory Opinion*, this thorny problem has remained unresolved.[45]

This "middle course" was eventually verbalized in AP-I (Art. 51(4)(b)), as cited above, while sub-paragraph (51(4)(c)) laid down the second criterion: Weapons having effects that cannot be limited as required by the Protocol are prohibited. In other words, once the very employment of a weapon violates either the rules regarding protection of the natural environment (Arts. 35(3) and 55) or protection of works and installations containing dangerous forces (Art. 56), or – mainly – the proportionality requirement (Arts. 51(5) and 57(2)), use of the weapon is prohibited.[46] Hence, as the *ICRC Study Report* summarized:

> The two criteria [for defining indiscriminate weapons-E.B.] that are most frequently referred to are whether the weapon is capable of being targeted at a military objective and whether the effects of the weapon can be limited as required by international law ....[47]

Once criteria were agreed-upon and made available, the next step required explicit parameters to determine whether a particular type of weapon failed to meet at least one of them. After all, in the absence of parameters, the criteria would be devoid of any value – as was soon discovered.

The first criterion: A weapon's capability of being directed at a specific military objective. As the *ICRC Study Report* noted, this criterion "is referred to in several military manuals, official statements and reported practice".[48] New Zealand's Military Manual, for instance, states that "weapons which cannot be directed at military objectives ... are prohibited."[49] However, review of the two Commentaries to the AP-I as well as the *ICRC Study Report* itself quickly reveals the immense difficul-

44 In light of the "great diversity" of opinion over the legal criteria applicable when assessing conventional weapons, it was suggested at the Lucerne Conference that a working group of legal experts be set up. However, no progress was made in this respect. *See, 1975 Lucerne Report*, at 13, para. 42.

45 Judge Higgins, *Nuclear Weapons Advisory Opinion*, at para. 24.

46 Michael Bothe, K. J. Partsch and W. A. Solf (eds.) *New Rules for Victims of Armed Conflicts: Commentary on the Two 1977 Protocols Additional to the Geneva Conventions of 1949*. (Leiden: Martinus Nijhoff, 1982), at 305-6, para. 2.5 .2.3 [hereinafter, *New Rules*]. These are the three explicit limitations on the effects of attacks noted in the AP-I.

47 *ICRC Study Report*, Vol. I, at 247. As its authors noted, "(t)hese criteria are part of the definition of indiscriminate attacks under customary international law". This state of affairs (i.e., existence of the two criteria) has remained unchanged. *See*, e.g., Bill Boothby, "The Law of Weaponry: Is It Adequate?" in: Michael Schmitt and Jelena Pejic, (eds.) *International Law and Armed Conflict: Exploring the Faultline: Essays in Honour of Yoram Dinstein* (Leiden: Martinus Nijhoff, 2007), 297, at 306.

48 *Id.*

49 New Zealand, *Military Manual* (1992) para. 509(4) as cited in *ICRC Study Report*, Vol. II, at 1558.

ties associated with the criterion's application – as some experts noted in the 1974 Lucerne Conference.[50] While the 1982 Commentary stated that it "prohibits 'blind' weapons which cannot, with any reasonable assurance, be directed against a military objective",[51] the *ICRC Commentary* goes on to explain that the relevant means of warfare "are primarily long-range missiles which cannot be aimed exactly at the objective"; German's V2 rockets in WWII provide an example of such means.[52] As the authors of the 2006 draft *Manual on the Law of Non-International Armed Conflict* explained in regard to this classic example: "Their guidance systems were so rudimentary that the rocket could not be effectively aimed".[53] Prominent legal scholars have likewise pointed to the 1991 use of Scud missiles by Iraq, "given the low chances to strike solely military targets with such missiles from such a long range".[54] As a proof for their "inaccuracy", one writer even pointed out that while targeting Israel, "on at least two occasions, missiles fell in the Israeli-occupied territories".[55] These two examples have since found their way into military manuals.[56]

Yet, as the *ICRC Commentary* itself noted more than twenty years ago (1987),[57] the trend in "most armies" – albeit for purely military reasons – is to increase the precision of weaponry. Therefore, "the margin of error of missiles is gradually reduced."[58] Indeed, modern long-range missiles, especially cruise missiles, are generally very pre-

50 *See, 1975 Lucerne Report,* at 10, para. 30.

51 *New Rules* at 305, para. 2.5 .2.2.

52 *ICRC Commentary*, at 621, para. 1958. This historical instance is widely known amongst legal scholars, probably due to the "storm of protest concerning the illegality of the weapon" during WWII. Yet, this protest was made under Article 25 of the Hague Regulation (i.e., the prohibition against attacking or bombarding undefended towns, villages, dwellings or buildings by any means). *See,* Von G. Glahn, *Law Among Nations: an Introduction to Public International Law.* 7th ed., (Boston: Allyn & Bacon, 1996) 632-3.

53 *See, The Manual on the Law of Non-International Armed Conflict (with Commentary), International Institute of Humanitarian Law,* San Remo, 2006, (Drafting Committee: Michael N. Schmitt, Charles H.B. Garraway, Yoram Dinstein), Rule 2.2.1.2. Reprinted in Yoram Dinstein & Fania Domb (eds.) *Israel Yearbook of Human Rights* 36 (2006) Supp., 333, at 30.

54 *See,* Orna Ben-Naftali and Yuval Shany, *International Law between War and Peace* (Tel Aviv: Tel Aviv University Press, 2006) (in Heb.), 152. Launchings of these missiles (including those against Saudi Arabia) were criticized as indiscriminate use of a weapon by the US Department of Defense as well as by the U.K. Foreign and Commonwealth Office. *See, ICRC Study Report,* Vol. II, at 1573 (U.K.), 1574 (U.S.).

55 *See,* Françoise J., Hampson "Means and Methods of Warfare in the Conflict in the Gulf," in Rowe, Peter (ed.), *The Gulf War 1990-91 in International and English Law,* (New-York: Routledge, 1993), 89, at 103.

56 For the German V2 rockets example (as well as the V1 type that, for unclear reasons, is omitted by the ICRC Commentary though most writers make sure to include it) *see* Ecuador's Naval Manual (1989), para. 1 as cited in *ICRC Study Report*, Vol. II., at 1557. For the Iraqi Scuds as an example *see*, e.g., Canada's LOAC Manual (1999) at 5-2.

57 English language edition; the French edition had been published a year earlier (1986).

58 *ICRC Commentary*, at 621, para. 1958.

cise.[59] The *Commentary's* statement was, in essence, based on the declining validity of the argument that contemporary deployed (as opposed to incomplete or experimental) "conventional" weapons, are considered – rightly or not – to be inherently indiscriminate. After all, missile accuracy heavily depends on a single component – the guidance system – which can be replaced or improved while leaving the missile intact. As one expert warned in the debate during the 1974 Lucerne Conference: "Accuracy was a relative concept in all cases and could never provide a clear guideline".[60]

The second criterion: weapons whose effects cannot be limited

As expected, disagreements over this criterion (Art.51(4)(c) of AP-I) arose during the drafting in the 1974-77 Diplomatic Conference. No solution could be found other than to point out that this it was "the vaguest and most problematical" of all conditions.[61] Several delegations firmly held the position that the "means of combat of which the use would constitute an indiscriminate attack in all circumstances" are virtually non-existent.[62] As explicitly suggested in the *ICRC Commentary*,[63] that view mainly – if not entirely – focused on a would-be scenario in which such a provision might be invoked in support of a prohibition against the use of nuclear weapons.[64] Hence, the report adopted by the Diplomatic Conference, Committee III, stressed:

> The main problem was that of defining the term 'indiscriminate attacks' ... Many but not all of those who commented were of the view that the definition was not

59 The LGM-118A ICBM, known as "the Peacekeeper" when it was still in service, or by its pre-deployment designation "MX" [Missile-Experimental), for instance, had a Circular Error Probable (CEP) of 100m. *See*, <http://nuclearweaponarchive.org/Usa/Weapons/Mx.html>. As for cruise missiles, for instance, Raytheon Company, the Tomahawk's manufacturer, boasted of its "precision strike" capability. *See*, <http://www.raytheon.com/ capabilities/rtnwcm/groups/public/documents/content/rtn_rms_products_t omahawk_data.pdf>.

60 *1975 Lucerne Report*, at 10-11, para. 31.

61 *New Rules*, at 305, para. 2.5 .2.3 . This provision (46 at the time of drafting) was not adopted unanimously, but by 77 votes (including the U.S.S.R., U.S. and U.K.) with one vote against (France) and 16 abstentions. See *CDDH O. R.*., Vol. VI, (1978) SR. 41, at 163.

62 *ICRC Commentary*, 623, para. 1964.

63 *Id.*, at para. 1966.

64 A clear indication of this view is the nuclear powers' insistence, as expressed during the pertinent discussion (*see*, CDDH O.R. Vol. VI, SR. 41, at 168-172,) that their views would be confirmed in their "explanations of vote". *See, ICRC Commentary*, 623, para. 1964. In fact, only after the ICJ rendered its *Nuclear Weapons Advisory Opinion* in 1996, the U.K. and France joined the Protocols in 1998 and 2001, respectively. At that time, the majority of Third World states, many NATO States, and the majority of former Warsaw Pact members (including Russia), were already bound by the Protocols. The U.S. currently remains the last declared nuclear state to join the Protocols, which it has signed but not yet ratified. One can, therefore, only speculate as to the role played by these states' concern over the legality of nuclear weapons given the provisions of AP-I chapter II and III (Art. 50-6) in general, and this provision in particular, in their former (and the current U.S.) refusal to join the Protocols.

> intended to mean that there are means or methods of combat whose use would involve an indiscriminate attack in all circumstances.[65]

The *ICRC Commentary* nevertheless stubbornly insisted that "there are some means of warfare of which the effects cannot be limited in any circumstances",[66] biological weapons being "an obvious illustration of this point" regarding the "means" whereas "poisoning sources of drinking water" provided an example of the "method".[67] There is little doubt that in most circumstances, both such means and methods would indeed strike combatants and non-combatants without distinction; both had also been explicitly prohibited when the AP-I was concluded (1977).[68] Not surprisingly, biological weapons were the example most frequently cited by legal experts intent on demonstrating the prohibition on "indiscriminate weapons".[69]

Given the intensity of the debate over whether the use of nuclear weapons (the so-called "ultimate weapon") could ever be compatible with the Law of Weaponry in general and with the prohibition against indiscriminate use in particular, the 1996 ICJ *Nuclear Weapons Advisory Opinion* centered in good part on how any such use could cope with this prohibition. The ICJ cited the stance taken by nuclear weapons opponents, according to which "[i]n the event of their use, nuclear weapons would

---

65 *See, CDDH O.R.., Vol.* XV, at 274, para. 55. As the Committee chairmen were two U.S. representatives (Richard Baxter and George H. Aldrich), much weight was probably given to opposing voices. For a list of the various Committee Chairmen see *ICRC Commentary*, XXXIII. However, Blix, who apparently shared the *ICRC Commentary's* conjectures as to the motives behind the nuclear states' position in this debate, noted that in this case, "the positions would seem to have been taken *ex abulldanra cautela* since the drafts submitted by the ICRC were not intended to introduce rules governing atomic, bacteriological and chemical weapons". *See,* Hans Blix. "Means and Methods of Combat" in: *International Dimensions of Humanitarian Law* (Geneva: UNESCO, 1988), 135, at 144.

66 *ICRC Commentary*, at 623, para. 1963.

67 *Id.*, at para. 1965. This was a common example during the formulation of the additional Protocols. As Blix stated during discussions in the Ad Hoc Committee on Conventional Weapons: "All weapons could be used indiscriminately but some were incapable of being directed at military objectives alone. One example was bacteriological weapons: germs could not distinguish between soldiers and civilians ... Some of the incendiary weapons had turned out to be quite indiscriminate." Sweden, Statement at the CDDH, *O. R.* Vol. XVI, CDDH/IV/SR.I, (Mar. 13, 1974), at 12, para. 20-21.

68 For the prohibition on poisoning wells *see, e.g., The 1907 Hague Regulation*, Article 23(a), *Laws of Armed Conflicts*, 55, at 73. For the prohibition on the use of biological weapons *see, The Geneva Protocol for the Prohibition of the Use in War of Asphyxiating, Poisonous or Other Gases and of Bacteriological Methods of Warfare*, 17 June 1925, *Laws of Armed Conflicts*, at 105. *See,* also the 1972 Biological Weapons Convention (BWC), which required States Parties to eliminate "all agents, toxins, weapons, equipment and means of delivery" within nine months. *See Convention on the Prohibition of the Development, Production and Stockpiling of Bacteriological (Biological) and Toxin Weapons and on their Destruction* (entry into force Mar. 26, 1975), Art. II, *Laws of Armed Conflicts*, 135, at 137.

69 *See, e.g., 1975 Lucerne report*, at 10, para. 30; Michael N. Schmitt, "Future War and the Principle of Discrimination" *Israel Yearbook on Human Rights* 28 (1999) 51, at 55; Ben-Natfali & Shany, *between War and Peace, supra* note 52, at 153; Dinstein, *The Conduct Of Hostilities*, *supra* note 33, at 56.

in all circumstances be unable to draw any distinction between the civilian population and combatants".[70] Yet, by the same token, the Court cited the U.K.'s Written Statement, which referred to a scenario involving the employment of low-yield nuclear weapons against "warships on the High Seas or troops in sparsely populated areas". According to this scenario, "[t]he reality ... is that nuclear weapons might be used in a wide variety of circumstances with very different results in terms of likely civilian casualties".[71] While noting that in neither instance it could adequately be predicted "whether such limited use would escalate into the all-out use of high-yield nuclear weapons",[72] the Court evaded making any decision in this matter, stating that "the Court does not consider that it has a sufficient basis for a determination on the validity of this view".[73]

As Professor Greenwood noted when referring to this ICJ *Advisory Opinion*, the Court rightly avoided exploring circumstances in which nuclear weapons could be used without violating general principles (i.e., the prohibition on the use of indiscriminate weapons, weapons which caused unnecessary suffering and the principle of neutrality[74]). By applying principles of such generality to the use of weapons in an indefinite variety of circumstances, he stressed that "the Court could not have determined that as a matter of *law* a nuclear weapon could not be used without violating one or more of those principles'".[75]

It is therefore no wonder that with perhaps the exclusion of biological weapons,[76] none of the means of warfare perceived as indiscriminate *per se* by any members of the international community have ever been the targets of a sweeping ban *solely* on this basis. As to the 1997 *Anti-Personnel Mine Ban Convention* (Ottawa Convention)[77] and

---

70 *Nuclear Weapons Advisory Opinion*, para. 92.

71 United Kingdom, *Written Statement*, 53 (June 1995); *see*, also *Nuclear Weapons Advisory Opinion*, para. 91. The court also referred to the U.S. Oral Statement, see *Id.*, para. 91.

72 *Nuclear Weapons Advisory Opinion*, para. 94.

73 *Id.*

74 Christopher Greenwood, *Essays on War in International Law*, (London: Cameron May 2006), 263.

75 *Id.*, at 265. For an institutional reason for which the Court "needed to avoid deciding it [the case] definitively", *see*, Michael N. Schmitt, "The International Court of Justice and the Use of Nuclear Weapons", *U.S. Air Force Annual Journal of Legal Studies* 7 (1996-1997), 57, at 81.

76 As the Polish delegate, who had proposed adding this weapon to the emerging Geneva Protocol on chemical weapons during the 1925 Geneva Conference, stated: "It is impossible to limit the field of action of bacteriological factors once introduced into warlike operations. The consequences of bacteriological warfare will thus be felt equally by the armed forces of the belligerents and the whole civilian population, even against the desire of the belligerents, who would be unable to restrict the action of the bacteriological weapons to an area decided upon beforehand". *See, League of Nations, Geneva Conference Proceedings of the conference for the Supervision of the International Trade in Arms and Amunition and in Implements of War, at 340.*

77 Convention on the Prohibition of the Use, Stockpiling, Production and Transfer of Anti-Personnel Mines and on their Destruction, 18 September 1997, *Laws of Armed Conflicts*, 285.

the 2008 *Cluster Munitions Convention* (CCM),[78] although considerations of unnecessary suffering played a significant role in gaining support for a ban, mainly with respect to landmines,[79] the fact that the intense efforts to ban both weapons failed in the CCW Review Conferences (where a consensus was required) confirmed how difficult it was to reach such decisions.[80]

Unlike those few weapons still perceived as inflicting unnecessary suffering, weapons considered as having indiscriminate effects have, instead, been subjected merely to restrictions over their use. Historically, sweeping prohibitions were accepted only when such arguments were conflated with those regarding unnecessary suffering.[81]

---

78 *Convention on Cluster Munitions (CCM),* (adopted: May 30, 2008), *available at* <http://treaties.un.org/doc/ Publication/CTC/26-6.pdf>.

79 As the ICRC President (1987-1997), Dr. Cornelio Sommaruga, clarified in 1996 in response to the question of whether blast anti-personnel mines inflict unnecessary suffering: "If a person steps on a buried anti-personnel mine, his or her foot or leg is blown off. The force of the blasts drives earth, grass, the vaporised mine case and portions of the victim's shoe and foot upwards into the tissues of the other leg, buttocks, genitals, arms and sometimes the eyes. ... If the wounded person gets to a hospital with the necessary facilities and expertise (both of which are rare in mine-affected countries) he or she will require several operations. Awaiting the survivor is permanent and severe disability with all the social, psychological and economic implications of being an amputee. Mines are designed to produce these effects." Cited in Jakkie Cilliers, "South Africa and the International Campaign to Ban Anti-Personnel Landmines," *African Security Review* 6(1), (1997), on-line publication, <http://www.iss.co.za/pubs/ASR/6No1/Cilliers.html>. It is also worth noting that the principle of superfluous injury or unnecessary suffering is mentioned, albeit in the context of booby traps, in the 1980 CCW Protocol II, which was removed in its 1996 amended version. *See*, CCW, Protocol on Prohibitions or Restrictions on the Use of Mines, Booby-Traps and Other Devices (Protocol II). Geneva, 10 October 1980, Art. 6(2), *Laws of Armed Conflicts*, 191, at 193. As to the argument that cluster munitions cause superfluous injury or unnecessary suffering and should be banned as inhumane *see, e.g.*, Karen Hulme, "Of Questionable Legality: The Military Use of Cluster Bombs in Iraq in 2003" *Canadian Yearbook of International Law* 42, 143, at 162-69; W. Hays Parks, "Conventional Weapons and Weapons Reviews", *Yearbook of International Humanitarian Law 8* (2005) 55, at 102, fn.169.

80 On these failures *see, supra* chapter 2, E(II) [for the landmines issue and the 1995 CCW First Review Conference] and E(IV) [for the issue of Cluster Munitions and the 2006 CCW Third Review Conference].

81 Indeed, to a large extent, arguments pertaining to the principle of unnecessary suffering were responsible for the first prohibition on chemical weapons as early as the 1899 Hague International Peace Conference [*Declaration (IV,2) concerning Asphyxiating Gases. The Hague, 29 July 1899, Law of Armed Conflicts*, 95] to be reiterated during the 1907 Second Peace Conference), where they played a *crucial* role in achieving consensus regarding the 1925 Geneva Protocol and its prohibition of the use of chemical and biological weapons. *See*, e.g., R.R. Baxter & Thomas Buergenthal, "Legal Aspects of the Geneva Protocols of 1925" *The American Journal of International Law* 63 (1970) 853, at 857-866. The military manuals of both Australia and France, for instance, mention chemical and biological weapons as weapons that inflict unnecessary suffering while the Australian manual mentions the effect on the "civilian population" as well. *See, ICRC Study Report* Vol. II, 1509-10 (Australia), 1513 (France). *See*, also Greenwood, *Essays on War, supra* note 74, at 246. The prohibition was also motivated by the so-called "chemical weapons taboo" (*see*, Richard Price, *The Chemical Weapons Taboo.* [New-York:

A worthy example of the improbability of approving a sweeping prohibition once unnecessary suffering ceases to be a consideration is the *1907 Convention (VIII) relative to the Laying of Automatic Submarine Contact Mines,*[82] which has never been controverted.[83] The U.S. Secretary of State Elihu Root, in replying to a 1906 French query as to whether the U.S. would suggest discussing the issue of submarine mines at the 1907 Hague Peace Conference in light of the problems caused on the open sea by mines planted during the 1905 Russo-Japanese War, responded as follows:

> While the use of submarine mines in war is as legitimate as that of torpedoes and other dirigible explosive devices, the mutual rights of the belligerents and their obligation toward the whole neutral world forbid their employment except under conditions of reasonable control... The belligerent's remedy and duty is *obviously to improve his contact mine* and prevent its becoming a terrible and uncontrollable peril to innocent mankind.[84]

In addition, a statement from the Chinese government detailing the injuries caused to its citizens by the use of these naval mines in the 1905 Russo-Japanese War – "in spite of every precaution being taken" – was submitted to the pertinent Conference Commission.[85] As no total ban was requested, only selected torpedoes (those "which do not become harmless when they have missed their mark") were banned. Unanchored mines were permitted if they became harmless within an hour after they were beyond human control.[86] As already noted,[87] this approach has been sustained

---

Cornell University Press, 1997], chaps. 2-4); For a concise version of Price's attempt to explore the origins and meanings of the taboo, *see,* Richard Price, "A Genealogy of the Chemical Weapons Taboo", *International Organization* 49 (1995), 73-103.

82 Convention (VIII) relative to the Laying of Automatic Submarine Contact Mines, The Hague, 18 Oct. 1907, *Law of Armed Conflicts,* 1071. As Prof. Cassese noted "... the desire 'to safeguard the life and interests of neutrals and non-combatants' lay behind some basic provisions of the VIII Hague Convention of 1907...". *See,* Antonio Cassese, "Means of Warfare: The Traditional and the New Law" in: Antonio Cassese (ed.), *The New Humanitarian Law of Armed Conflict,* (Napoli: Editoriale Scientifica, 1979), 161, at 169.

83 Yet, only twenty-seven States have ratified or acceded to the Convention. Russia, China, India and Italy, among the major naval powers, are not parties. With the exception of two cases of declared succession (Fiji in 1973 and South Africa in 1978), there have been no accessions since 1935. *See, Id,* at 1076. In any case, as O'Connell has noted, the Convention's applicability in practice, and in the light of technological developments, has been questioned. *See,* D.P. O'Connell (I.A. Shearer, ed.), *The International Law of the Sea,* Vol. II, (Oxford: Clarendon Press, 1984), 1103.

84 Emphasis added. *See,* S. Howard Levie, *Mine Warfare at Sea,* (Dordrecht: Martinus Nijhoff, 1992), 23-4.

85 The statement reads: "[I]t is calculated that from five to six hundred of our countrymen in the pursuit of their peaceful occupations have met a cruel death through these dangerous engines". James Brown Scott (ed.) *The Reports of the Hague Conferences of 1899 and 1907,* (London: Clarendon Press, 1909), 657.

86 *See, Law of Armed Conflicts,* at 1071. For a detailed overview of the negotiation behind each article see Levie, *Mine Warfare, supra* note 84, at 30-63.

87 *Supra,* chapter 2.

by the CCW as well: While non-detectable fragmentation weapons or permanent blinding laser weapons were simply banned, other weapons, such as anti-personnel mines, mentioned in the *ICRC Commentary* discussion on Article 51(b) of AP-I[88] – were only subjected to restrictions over the methods of their use as long as the "civil society" (i.e., public opinion) remained dormant (see the 1997 Ottawa Convention).

In concluding this section, it is worth recalling some reflections on this debate made at that time by Prof. Kalshoven:

> The criterion of indiscriminateness… was recognized by the experts… Yet, this did not imply an equal readiness to accept the existence of a rule prohibiting "indiscriminate weapons". To many, especially among the Western experts, the latter question really was an academic one, as they did not accept the existence, *within the category of conventional weapons*, of types of weapon which are intrinsically indiscriminate. In their opinion, *all conventional weapons can be used with or without discrimination….*[89]

More than twenty years later, the ICJ has effectively added nuclear weapons, albeit inadvertently, to "all conventional weapon which can be used with or without discrimination". Leaving aside biological weapons, which are unquestionably prohibited,[90] the question of "indiscriminate weapons" *per se* has now become academic. Therefore, as the CCW imposes only restrictions on weapons (e.g., anti-personnel mines) when it comes to protection of civilians,[91] scholars have noted that the express references to "indiscriminate effects" (as well as "excessive injury"), added to the Convention's title "(a)t the very last minute"[92] (i.e., Convention… Which May be Deemed to be Excessively Injurious or to Have Indiscriminate Effects), "is more political rhetoric than an accurate picture of the views of the majority of governments".[93]

---

88 *ICRC Commentary*, at 621, para. 1959.

89 Emphasis added. *See*, Kalshoven, *Reflections of War*, *supra* note 41, at 160. [Originally published in *NYIL* 6 (1975), 77-102].

90 As noted, biological weapons ("bacteria") had been singled out as examples of inherently indiscriminate weapons. *See*, e.g. *1975 Lucerne Report*, at 10, para. 30.

91 Indeed, as Parks noted in reference to the CCW protocols: "No protocol concludes that the weapon in question is indiscriminate as such". See Parks, "Conventional Weapons …" *supra* note 79, at 79.

92 *See*, Frits Kalshoven, "The Conventional Weapons Convention: Underlying Legal Principles" *International Review of the Red Cross* 279 (1990) 510, at 515.

93 *See*, Hays Parks, "Means and Methods of Warfare" *The George Washington International Law Review* 38 (2006) 511, at 518. As Parks explained, "had the weapons in question been deemed by governments to be 'excessively injurious or to have indiscriminate effects' *per se*, as the resolution title suggests, their use already would be prohibited, making new treaty provisions unnecessary". *Id. See*, also Kalshoven, "The Conventional Weapons Convention..." *supra* note 92, at 515.

### C. Are Flechette Weapons Prone to Inflicting Unacceptable Harm to Civilians?

Given the preceding discussion, it is obvious that, *a priori*, flechette weapons cannot be considered inherently indiscriminate weapons. Yet, an assessment of the extent to which flechette weapons are prone to cause *unacceptable harm to civilians* or – using the wording of the "Swedish Group" – to have indiscriminate effects, is still warranted. In doing so, it seems reasonable to follow Judge Higgins' choice[94] of the "middle course", meaning that flechette weapons – like all types of weapons – should be considered as having indiscriminate effects if they "*cannot be employed with sufficient or with predictable accuracy against the chosen target*".[95]

The question of benchmarks for assessing a weapon's indiscriminate effects nonetheless arises. A careful analysis of the 1970s experts' discussions indicates that only one "dominant factor" was agreed upon, the so-called 'area of effectiveness'.[96] This criterion was defined as the area within which an unprotected person, standing in the open, has at least a 50% chance of becoming a casualty after the weapon is employed.[97] Yet, before deepening our discussion of this criterion, a preliminary stipulation should be mentioned, one that dramatically reduces the types of weapons to be subjected to its provisions: the definition of a weapon as an "Area Weapon" or a "Point Weapon".

According to this filter, means of warfare that are effective over a broad area, *even if they are used against a pointed target*, are labeled "area weapons", as distinct from "point weapons"; such as rifles or even hand grenades, which have a limited area of effectiveness. This distinction, as an expert argued at the 1974 Lucerne Conference, represents the first step in determining whether a particular weapon is necessarily indiscriminate.[98] In other words, "point weapons" should be automatically excluded from application of this principle. Instead, discussion should focus on whether a weapon that is defined as an "area weapon" (e.g., fragmentation weapons, as described in the 1973 Geneva Experts Meeting report) veritably "ha[s] an obvious and uncontrollable tendency towards indiscrimination".[99]

Comparing a weapon's area of effectiveness

A comparison of the areas of effectiveness attributed to various "representative" types of anti-personnel weapons reveals the high position of fragmentation weapons on the list. While a high-explosive 105 mm shell found itself on the lowest rung (among six), together with a hand grenade (i.e., areas of effectiveness ranging between 10m to 50m), a 155mm artillery flechette shell was placed on the fourth rung (i.e., a threshold

94 Dissenting Opinion of Judge Higgins, *Nuclear Weapons Advisory Opinion*, at para. 24.

95 Emphasis added. *1975 Lucerne Report*, at 10-11, para. 31.

96 *1973 Geneva Report*, at 24, para. 57.

97 *Id.*, at 26, fn. a.

98 *1975 Lucerne Report*, at 11, para. 32.

99 *1973 Geneva Report*, at 48, para. 150.

below 1000m).[100] Unsurprisingly, one of the final remarks made in the 1973 *Experts Meeting Report* warned that the trend towards developing fragmentation weapons with smaller fragments for the purpose of increasing their number, combined with their wide area of effectiveness, "increase[d] … the possibility that civilians will be affected."[101]

As expected, similar to the case of cluster munitions,[102] flechettes, as pre-controlled fragmentation weapons *designed* to cover a wide area, were deemed indiscriminate by its opponents during the 1970s.[103] In fact, the following correlation (originally made with reference to cluster munitions) between a specific weapon's area of effectiveness and its inherent tendency toward indiscriminate effects is doubtlessly applicable to flechette weapons:

> Cluster bombs, however, may be effective even if they miss the targeted spot. 'The advantage of this type of warhead is that it *gives a wide area of coverage, which allows for a greater margin of error in delivery*.' The advantage for the military is a distinct disadvantage, however, for any civilians who happen to be within about a football field or more of the point target.[104]

As an additional rationale supporting his proposal to ban anti-personnel fragmentation weapons in general (CDD/IV/201) and flechette weapons in particular (CDDH/DT/2), Blix, at the February 1975 Second Session of the Diplomatic Conference's Ad Hoc Committee on Conventional Weapons, pointed out that the delivery of fragmentation weapons was usually planned so as to give the targeted combatants almost no advance warning.[105] In effect, the victims hear the darts' buzz or see the dye marker's yellow cloud – if at all – only after the rounds have already split over their heads.[106] Hence, any civilians who happened to be present in the area attacked have little chance to take cover.[107] That being the case, Blix warned the participants that "those weapons offered a temptation by their effectiveness but they

100 *Id.*, at 26. The experts emphasized, however, that regarding "the tendency towards indiscriminateness…much, but not all, depends upon the manner and circumstances in which the weapons are used." *Id.*, at 48, para. 150.

101 *1973 Geneva Report*, at 71, para. 246.

102 Cluster munitions are classified as a classic area weapon given their large 'footprint'. *See*, e.g., Robert Cryer, "The Fine Art of Friendship: Jus in Bello in Afghanistan" *Journal of Conflict and Security Law* 7 (2002) 37, at 61.

103 *See*, e.g., Blix's statement, *CDDH O. R.*., Vol. XVI, at 47.

104 *See*, Thomas Michael McDonnell, "Cluster Bombs over Kosovo: A Violation of International Law" *Arizona Law Review* 44 (2002) 31, at 97 [italics in original].

105 This is unsurprising given weapon's design aim.

106 Tragic evidence for the veracity of this description is video footage that was filmed by Reuters TV cameraman, Fadel Shana, a few seconds before he was killed by the darts; *see, supra*, chapter 6, (i). The yellow cloud (i.e., the round's dye markers), as caught by his camera, can be viewed at <http://www.reuters.com/news/video?videoId =80475>.

107 *See, CDDH O. R.*, Vol. XVI, at 122.

were likely to cause a greater number of civilian casualties at any rate, when the attack occurred in populated areas."[108]

As might be anticipated, the issue of whether fragmentation weapons do have an inherent tendency for indiscriminateness became a bone of contention at the Lugano Conference, with the "Swedish Group" proposing a prohibition on these weapons in general, and flechettes in particular (CDDH/IV/201), their opponents were primarily experts from the U.S.-led camp. While the former clarified that their proposal was limited to those *anti-personnel* fragmentation weapons having "wide area coverage",[109] the latter advocated the weapons' legality on the following grounds: (a) The military's need for anti-personnel weapons having a wide area of coverage for defensive purposes; (b) the indiscriminate effects of these means were similar to those of the weapons they could replace; (c) most importantly, indiscriminateness-depended on the method of use rather than the weapon's nature;[110] and (d) the ban's acceptance would induce increased use of "general use" munitions (i.e., blast weapons), which would lead to more severe wounds and greater damage over a wider area. Therefore – as one expert concluded – "to limit the use of controlled fragmentation weapons would not contribute to humanizing war, but would have exactly the opposite effect".[111] Irrespective of these arguments' veracity, it is clear that there was a little dispute regarding a flechette weapon's wide area of effectiveness".

## D. Closing Remarks

We have tried to demonstrate why defining any weapon as inherently indiscriminate soon became a mission impossible. Not surprisingly, Judge Higgins, the only contributor to the *Nuclear Weapons Advisory Opinion* to seriously delve into the subject, arrived at an impasse after concluding that for this principle to acquire "a separate existence", distinct from that of collateral harm, it must be shown that "a weapon will be unlawful *per se* if it is incapable of being targeted at a military objective only, even if collateral harm occurs".[112] As Higgins presciently recognized, such a stipulation would be so broad that it would not disqualify the use of nuclear weapons, as "that very term [nuclear weapons] covers a variety of weapons [i.e., high yield and tactical low-yield nuclear weapons alike - E.B.]".[113]

After all, if a weapon is to be developed so as to be incapable of being directed at a military target, such a characteristic would *ipso facto* empty it of its presumed military effectiveness, meaning that its development would be inherently futile and thus unlikely. It is no surprise, therefore, that the "spectacular example of an inherently indiscriminate weapon",[114] raised by Prof. Greenwood (specifically, the WWII

108 *See, Id.*, at 121. For Blix's remark, made in February 1975, *see, Id.*, at 122.

109 *See, 1975 Lugarno Report*, at 17, para. 44.

110 *Id.*, at 72, para. 6.

111 *Id.*, at 71-2, para. 6.

112 Dissenting Opinion of Judge Higgins, the *Nuclear Weapons Advisory Opinion*, at 24.

113 *Id.*, at para. 24.

114 Christopher Greenwood, "Current Issues in the Law of Armed Conflict: Weapons, Targets and International Criminal Liability" *Singapore Journal of International and*

U.S. Navy "bat bomb") was – as he rightly noted – "never used".[115] In fact, other than Scud missiles (missiles presumed to have a primitive guidance system), the only known examples of "blind weapons" available in the literature that went beyond failure, incompletion (e.g., the bat bomb), or experimental status (e.g., the infamous "balloon bomb": balloons with attached bombs, launched by Japan in WWII[116]), were "[l]and mines, laid without customary precautions, and which are unrecorded, unmarked, or which are not designed to destroy themselves within a reasonable time…".[117] Again, this example captures the complexity of differentiating between the means of warfare and the way they are used. As the 1996 CCW Amended Protocol II[118] endorsed the continuing practice of laying landmines once the above conditions (as well as some technical issues) were addressed, it is clear that for those Protocol's members who refused to join the 1997 *Ottawa Convention*, landmines were legitimate weapons.

To conclude this chapter, there is little doubt that flechette weapons are area weapons *par excellence*. In fact, these weapons leave such a relatively "large footprint" that the U.S. Army came to prohibit their use "over an area with exposed friendly troops, even in combat emergencies".[119] Yet, given the numerous circumstances in

*Comparative Law* 1 (1997) 441, at 443, fn.9. Greenwood also mentioned the commonly cited examples of the German V1 and V2 missiles during WWII and the Iraqi Scud missiles launched during the 1990-1991 Gulf War. Id

115 The plan was to release bats, with a small timed incendiary bomb attached to each, over Japanese cities. The bats were expected to roost in eaves and attics and set fire to otherwise inaccessible places. Although the program was officially canceled due to its slow progress, the fact is that the project encountered far too many complications. For example, in one 1943 test, a U.S. Army base was set alight when incendiary-armed bats were accidentally released. The fact that no military interest has pursued the project's revival despite the $2 million already invested, at least as far as is known publicly, speaks for itself. For a comprehensive overview of the project *see*, Jack Couffer, *Bat Bomb: World War II's Other Secret Weapon* (Austin, TX: University of Texas Press, 1992); *see*, also C. V. Glines, *The Bat Bombers*, *Air Force Magazine*, 73(10) (Oct. 1990) available online at <http://web.archive.org/web/20080531082803/ http://www.afa.org/magazine/1990/1090bat.html>.

116 Over 9000 of these experimental weapons (named Fu-Go) were launched near the end of WWII (Nov. 1944-Apr. 1945); they were designed to wreak havoc over Canadian and American cities while relying on the jet stream over the Pacific Ocean for their propulsion. Despite initial forecasts, they inflicted no more than six casualties and a small amount of damage. For a comprehensive overview of the project, beginning with the weapon's production in Japan and extending to the US response *see*, Robert C. Mikesh, *Japan's World War II Balloon Bomb Attacks on North America*, (Washington D.C.: Smithsonian Institution Scholarly Press, 1990). On the U.S. response see also Cornelius W. Conley, "The Great Japanese Balloon Offensive", *Air University Review*, (Jan.-Feb. 1968), available at: <http://www.airpower.maxwell.af.mil/airchronicles/aureview/1968/jan-feb/conley. html>.

117 *New Rules*, at 305, para. 2.5.2.2.

118 CCW Protocol on Prohibitions or Restrictions on the Use of Mines, Booby-Traps and Other Devices, Amended Protocol II (entered into force Dec. 3, 1998), *Laws of Armed Conflicts*, 196.

119 *See*, *The Federation of American Scientists* (*FAS*), available at <http://www.fas.org/man/dod-101/sys/land/m546.htm>.

which no proximity between combatants and non-combatants can be anticipated (e.g., underwater submarine warfare or isolated military bases), it is patently clear that an exceptionally large area of effectiveness, in and of itself, cannot determine whether a given means of warfare is inherently indiscriminate. Therefore, no alternative criteria for determining whether a specific weapon has indiscriminate effects, or the degree of those effects, have been identified to date. Given the ICJ *Nuclear Weapons Advisory Opinion,* no change in this state of affairs can be envisaged for the foreseeable future.[120] However, as one writer has already noted in light of the new CCM's success, "the notion of 'unacceptable harm' to civilians could be used to promote a higher standard for the precautions to be taken in attacks in or near urban or densely populated areas".[121] Time will tell if, and to what extent, the criterion of 'unacceptable harm' earns a place of its own in IHL, to the point of serving as a substitute for the ill-fated distinction criterion. The fact that this phrase is already embedded in one IHL instrument (i.e., the CCM's preamble) and implied in another (i.e., the 1997 Ottawa Convention's preamble) strongly suggests that this may occur. After all, as John Borrie concluded in his detailed study on the history of the "Oslo process", the success of international efforts to address the humanitarian impacts of cluster munitions (i.e., the CCM), "is not only the end of the beginning, it could mark a significant milestone in how we can identify ways to protect civilians in armed conflict more broadly".[122]

120 On the quagmire in which the ICJ found itself and the reasons why it needed to avoid deciding on the matter *see, e.g.,* Michael N. Schmitt, "The International Court of Justice and the Use of Nuclear Weapons", *USAF Annual Journal of Legal Studies* 7 (1996-1997), 57, at 80-82.

121 Tommaso Di Ruzza, "The Convention on Cluster Munitions: Towards a Balance Between Humanitarian and Military Considerations?", *Military Law and the Law of War Review* [Revue de droit militaire et de droit de la guerre], 47 (2008) 405, at 441.

122 Borrie, *Unacceptable Harm…*, *supra* note 4, at 338.

# Summary and Conclusions

As we have noted, the literature offers various criteria (the three relevant for our case are treaty obligations and the two *generic rules* regulating weaponry – the prohibitions against inflicting unnecessary suffering and the weapon's capability of distinguishing between targets) to assist AP-I States Parties in their determination of the legality of "new" weapons as required by Article 36.[1] These criteria were carefully explored within the limited context of the legality of the flechette weapon, with lessons and insights discussed in each chapter's concluding remarks. Any thorough review of a specific type of weapon under the rubric of international law contributes to our understanding of the intricacy entailed in determining the legality of weapons as a comprehensive category. In the context of conventional weapons, such a review may address, albeit indirectly, the question of whether the standard mechanism for generating *specific rules* to regulate such weapons (i.e., the CCW), together with the two relevant *generic rules,* adequately respond to the humanitarian concerns arising in connection with the use of certain types of (conventional) weapons.

Hence, in the context of flechette weapons, the survey (Chapter 2) of the specific rules subsumed under weapon-related treaty law (e.g., the exclusion of WMD-related instruments) indicates that in spite of impressive efforts to the contrary, no rule targeting flechette weapons has been formulated. Furthermore, insufficient public awareness and successful U.S. attempts to undermine arguments for the weapon's illegality[2] (i.e., the severity and extent of damage caused) have ensured that the issue of flechette weapons – despite the proposed ban on their use – has so far been brushed aside.

However, once we delineated the political context of the respective efforts and the reasons for their failure, we found that the *absence* of an explicit rule regarding a specific weapon within the CCW framework is far from ensuring the weapon's compliance with any aspect of the two relevant *generic rules* under the law of war. This shortcoming is primarily the product of a voting procedure, that requires a consensus for a rule's passage. Indeed, shortly before the first session of the UN Preparatory

1 *See*, for instance, Timothy McCormack *et al.*, *Report on States Parties' Responses to the Questionnaire: International Humanitarian Law and Explosive Remnants of War*, (Melbourne, Asia Pacific Centre for Military Law, 2006), 35-36; Isabelle Daoust *et al.*, "New Wars, New Weapons? The Obligation of States to Assess the Legality of Means, and Methods of Warfare", *International Review of the Red Cross* 846 (*IRRC*) (2002), 345, at 347-350; Justin McClelland, "The Review of Weapons in Accordance with Article 36 of Additional Protocol I", *IRRC* 850 (2003), 397, at 406-410.

2 *See*, *supra* chapter 3(e).

Conference (Aug.–Sept. 1978), when the "thorny question" of a consensus versus a majority vote was still being deliberated, Prof. Kalshoven reported as follows:

> Another group of States, convinced of the desirability of the agreements they seek to achieve, feel that to *accept a consensus rule would be tantamount to bowing to the will of an erring minority.* Time will show which side wins the argument and, subsequently, what comes of the 1979 United Nations Conference.[3]

In retrospect, those states rightly predicted the current gloomy situation. As our survey has detailed, the CCW was and remains almost useless when it comes to prohibiting or, at least, heavily restricting the use of certain weapons when a few states – or even a single powerful state – prefers military considerations over humanitarian concerns.[4]

It is therefore not surprising that the CCW has yet to produce a single meaningful *prohibition* on any existing weapon. In fact, given the very few and limited scope of weapon prohibitions in treaty law, the specific rule criterion for reviewing a weapon has so far been proven useless. First, it is difficult to conceive of many cases in which the weapon under review can be found in conflict with existing treaty law.[5] Second, it is quite plausible that the state that acquired or developed the weapon under review is not party to the pertinent specific rule. In the present context, one can only speculate about the outcome of Israel's HCJ decision had the Swedish proposal to ban flechette weapons been accepted and if Israel, which acceded to the CCW only in 1995, had not yet become a State Party. After all, when the 1995 First Review Conference was opened, the CCW had *only* 49 State Parties;[6] as of July 2010,

---

3 Frits Kalshoven, *Reflections on the Law of War: Collected Essays* (Leiden: Martinus Nijhoff 2007), 258 [hereinafter, *Reflections of War*] (first published in *NYIL* 9 (1978) 107-171). For the expression "thorny question" *see, Id.*, at 257.

4 Almost 30 year later, complaints about this paralyzing requirement were voiced by a delegate to the CCW Meeting of States Parties, who speculated: "Does this fact in itself justify abandoning the consensus approach?" *See,* Bill Boothby, "The Law of Weaponry: Is It Adequate?" in: Michael Schmitt and Jelena Pejic (eds.), *International Law and Armed Conflict: Exploring the Faultline: Essays in Honour of Yoram Dinstein*, (Leiden: Martinus Nijhoff, 2007), 297, at 312. The writer was a member of the U.K. delegation to the 2005 meeting, whose failure to agree on a new protocol on Anti-Vehicle Mines (AVM) or even to adopt a mandate to negotiate such a protocol in 2006 led him to complain that "progress on matters already identified as the source of humanitarian concerns can be blocked by one or more states". *Id.*

5 A rare case is detailed in reference to the CCW Protocol I. As Parks recalled in 1998, when he conducted a legal review of the Modular Crowd Control Munition (a less-lethal version of the M18 Claymore mine), which contains hard rubber balls in lieu of steel projectiles. As by that time the U.S. had finally ratified the CCW and Protocols I and II (March 1995), "the rubber projectiles were implanted with five percent Barium Sulfate to make them detectable by X-ray and compliant with Protocol I, CCW". W. Hays Parks, "Conventional Weapons and Weapons Reviews" *Yearbook of International Humanitarian Law 8* (2005) 55, at 76-77, fn. 82.

6 *See,* speech by the Chinese representative, Sha Zukang, in the Second Review Conference. Rep. of the Second Review Conference of the States Parties to the Convention on Prohibitions or Restrictions on the Use of Certain Conventional

some 30 years later, it still has only 112 State Parties (with 5 states yet to ratify their signatures).[7] Furthermore, because joining the CCW is permissible after adopting only two protocols,[8] even if the proposed Swedish prohibition had been accepted, Israel could have refrained from accepting the relevant protocol and still remained a State Party. The purpose of non-acceptance of the protocol would, after all, have allowed it to retain the option to use flechettes (consider the case of Protocol III on incendiary weapons, which Israel and four other states have yet to ratify). This state of affairs could have been overcome had Israel's HCJ viewed such a Protocol as customary law. However, given the ICJ's statement on the *North Sea Continental Shelf Cases* regarding the relationship between treaty law and customary law,[9] it is difficult to imagine Israel's HCJ doing so.[10]

Yet, a review of treaty law indicates that other than the various efforts and direct proposals to ban the weapon, there is an additional good reason to examine its legality under international customary law, and perhaps even better ones for limiting – if not prohibiting – its use. As jurists have noted, the CCW Protocols "fall into one of two normative baskets ... either to protect combatants from 'excessively injurious' weapons or to protect civilians from weapons deemed to 'have indiscriminate effects'".[11] The arguments raised against the use of flechettes addressed both "norma-

---

Weapons which May Be Deemed to Be Excessively Injurious or to Have Indiscriminate Effects, *Summary Record of the 2nd Plenary Meeting* (Dec. 11, 2001), Conf. II/2, at 74, para. 43.

7 Current data are available at <http://www.icrc.org/ihl.nsf/WebSign?ReadForm&id=500&ps=P>.

8 *See,* CCW, Art. 4(3), Dietrich Schindler and Jiri Toman, (eds.) *The Laws of Armed Conflicts*, (Leiden: Martinus Nijhoff, 2004), 181, at 186 [hereinafter, *Laws of Armed Conflicts*]

9 As stated by the ICJ, "in order to achieve this result [that actions by the Convention's non-parties would constitute the *opinio juris* -E.B.], two conditions must be fulfilled. Not only must the acts concerned amount to a settled practice, but they must also be such, or be carried out in such a way, as to be evidence of a belief that this practice is rendered obligatory by the existence of a rule of law requiring it...The States concerned must therefore feel that they are conforming to what amounts to a legal obligation". *North Sea Continental Shelf Cases* (Federal Republic of Germany v Denmark and the Netherlands), (Judgment) 1969 ICJ Rep. 3 (Feb. 20, 1969), at 44.

10 On the ways in which multilateral treaties may develop into custom when referring to weaponry treaties *see,* David Turns. "Weapons in the ICRC Study on Customary International Humanitarian Law" *Journal of Conflict and Security Law* (J. OF C&S LAW) 11 (2006) 201-237. For a discussion of this issue within the context of the *Nuclear Weapons Advisory Opinion see,* Roger S. Clark, "Treaty and Custom" in: Laurence Chazournes de Boisson and Philippe Sands (eds.). *International Law, the International Court of Justice and Nuclear Weapons*, (Cambridge: Cambridge University Press, 1999) 171-180.

11 *See,* David Kaye & Steven A. Solomon, "The Second Review Conference of the 1980 Convention on Certain Conventional Weapons" *The American Journal of International Law* 96 (*AJIL*) (2002) 922, at 923. Protocol I on Non-Detectable Fragments and Protocol IV on Blinding Laser Weapons were presented as generally falling into the first basket while Protocol II on Mines, Booby-traps and Other Devices, Protocol III on Incendiary Weapons and Protocol V on Explosive Remnants of War, fall into the second basket.

tive baskets",[12] although – as the 1979 Second Preparatory Session's final report noted – "time did not allow for their discussion in more detail".[13]

Under the principle underlining the first basket – the prohibition on inflicting superfluous injury or unnecessary suffering – the flechette weapon was found to be a good candidate for the imposition of a prohibition or, at least, restrictions on its use. *First*, there is the issue of the multiplicity of wounds, which many human right activists perceive as a weighty reason *in and of itself* to prohibit a specific weapon. As noted, flechette weapons, *by their very nature*, inflict multiple wounds, they may *potentially* be the weapon inflicting the greatest number of multiple wounds among those currently available in world military arsenals. In fact, as opposed to various accusations raised by belligerents on both sides on the use of dum dum bullets in the early stages of WWI,[14] the circumstances surrounding the September 1918 German protest over the use of shotguns by U.S. soldiers (mainly the timing) appear to have made the argument genuine.[15] Hence, leaving aside the protracted debate about the shotgun's legality in warfare, the 1918 German protest strongly demonstrates the problematic issues associated with any anti-personnel weapon that inflicts multiple wounds. *Second*, flechette weapons are highly problematic given the "quantitative aspect"[16] of suffering or injury. From the onset, flechettes were clearly designed to inflict mass casualties. *Third*, as to the qualitative aspect, a three-pronged argument can be made against this weapon:

(a) If multiple wounds were insufficient to determine a weapon's violation of this principle, multiple wounds are said to cause a high level of pain and raise the mortality rate. Even though disputed by U.S. experts, the fact that many other experts did agree that "(a)s a very general rule, the more tissue that is damaged the more painful will be the wound"[17] is significant. By the same token, it is clear

12 In fact, throughout the years, far after the diplomatic negotiations on the CCW were over, arguments of both types were used concurrently. In the specific context of Israel's use of flechettes in South Lebanon, for instance, while HRW focused on the flechette weapon's alleged indiscriminate effects, others chose to focus primarily on the unnecessary suffering it may inflict. For the first claim *see*, e.g., HRW, *Civilian Pawns: Laws of War Violations and the Use of Weapons on the Israel-Lebanon Border*, HWR, May 1996,at 44. For the second one *see*, e.g., David Turns, "Some Reflections on the Conflict in Southern Lebanon: The 'Qana Incident' and International Humanitarian Law" J. of C&S LAW 5 (2000) 177, at 205.

13 *Rep. of the Preparatory Conferences for the UN Conference on the Use of Certain Weapons*, UN Doc. A/Conf.95/3 (May 25, 1979), at 13, para. 37.

14 For such accusations, and the propaganda considerations motivating them, *see*, Neil Pilford "Controversy over the Use of Dum Dum Bullets in the First World War", *Imperial War Museum Review* 5 (1990) 98-109.

15 In other words, unrelated to the propaganda war in the War's early stages. Indeed, while the protest was issued in September 1918 as Pilford found, the last accusation on the use of Dum-Dum bullets was raised in 1916. After that, he noted, "(u)se of the dum dum might still be an atrocity, but after Europe's first experience of total war it could no longer be accounted the nadir of frightfulness." *Id.*, at 108.

16 McClelland , "The Review of Weapons…" *supra* note 1, at 407.

17 ICRC, *Weapons that may Cause Unnecessary Suffering or Have Indiscriminate Effects* (Report of the Work of Expert) Geneva, (1973), at 27, para. 67. [hereinafter: *1973 Geneva Report*] We should recall, however, that the amount of tissue damaged by a single dart

that permanent disability is a possible outcome for any flechette weapon victim as the darts may spread throughout the body. A review of the autopsy reports made available by Israel's National Center of Forensic Medicine confirms the plausibility of this argument.

(b) As the clear outcome of the multiple wounds problem and the quantitative aspect, there is genuine concern over the adequacy of the medical treatment provided in the field for most flechette-injured persons, especially in developing states.

(c) Though yet to be substantiated, the argument of deformation on impact cannot be overlooked, especially as flechette-filled weapons can strike at such a high velocity that they may *tumble* rapidly inside the victims' bodies, thereby causing extensive damage to tissues, adjacent blood vessels, nerves, and organs.

In sum, if not prohibited, flechette weapons should certainly be deemed dubious under the first basket, which served as the central type or argument for prohibiting their use in the 1970s.[18] As one expert noted during the 1974 Lucerne Conference, the use of flechettes induces "an extremely high degree of suffering, long and difficult treatment, and a high mortality rate".[19] However, even if consent over the arguments' validity were to be reached, a similar consent on the weapon's presumed military necessity remained pending.[20]

As to the second basket, while the existence of its underlying principle (i.e., the third criterion: the weapon's capability to distinguish between civilian and military targets) was no longer questioned after the 1996 ICJ discussion on the legality of nuclear weapons, the feasibility of defining any type of conventional weapon as inherently discriminatory, ironically, no longer exists in the wake of this *Advisory Opinion.*

It should be mentioned that even prior to the ICJ *Advisory Opinion*, Hans Blix, in his role as a legal expert, stated the following:

> The question as to whether there can be means – weapons – the effect of which cannot in any circumstances be limited as required by the rules of the Protocol [AP-I], notably the rule of proportionality, *is most likely to be answered in the negative*... If the means – weapons – are at all capable of being directed, one can always envisage

---

would not be very great, relative to other munitions, if there is no cavitation or break-up.

18 Based on his personal knowledge, as a U.S. Delegation representative, Hays Parks claimed that at least during the 1978-1979 debate held in the preparatory sessions, this was the case for all weapons discussed. See Parks, "Conventional Weapons and..." *supra* note 5, at 75.

19 *See, Conference of Government Experts on the Use of Certain Conventional Weapons (Lucerne, 24.9-18.10.1974) ICRC Report* (1975) at 61, para. 214.

20 After all, as the U.S. negotiator for several protocols, Hays Parks, noted "(t)hese negotiations in many respects are like litigation...". See W. Hays Parks, "Means and Methods of Warfare", *George Washington Law Review (GWILR) 38* (2006), 511, at 535. However, in our view, this state of affairs stresses the importance of returning the ICRC SIrUS project to the negotiation table.

> the location of military targets so isolated that an infringement of the rule of proportionality is inconceivable.[21]

Yet under the umbrella of the recent successful international campaign to ban cluster munitions, worded as the extent to which a specific weapon is prone to cause unacceptable harm to civilians), flechette weapons, as classic area weapons, seem destined for prohibition. As one jurist stressed in regard to this third review criterion, the judgment as to whether any weapon can be directed exclusively against military objectives "clearly centres in the first place on its accuracy".[22] As an Israeli military correspondent noted after observing the first known instance of collateral damage caused by the use of flechettes in the Gaza Strip: "The dispersal of darts turns the flechette into an almost statistical weapon … The standard round, on the other hand, generally ensures a precise injury."[23]

It is no wonder, therefore, that Blix's 1975 statement, in which he referred to flechette use as containing "the risk of indiscriminately wounding combatants and civilians might be very great",[24] was echoed 30 years later by Israel's HCJ when it noted, albeit as an *obiter dictum*, that "[r]espondents do not dispute that the use of flechettes… entails increased danger towards civilians that inadvertently enter the zone of fire".[25] Six years later, the 2009 Gaza Fact-Finding Mission reached a similar conclusion:

> Discharged from tank shells and aircraft or UAV-launched missiles, they [flechettes] are fired in salvo and are therefore an area anti-personnel weapon. *They are, therefore, by their nature lacking in discrimination.*[26]

21 *See,* Hans Blix. "Means and Methods of Combat" in: *International Dimensions of Humanitarian Law,* (Geneva: UNESCO, 1988), 135, at 145. A classic example for "location of military targets so isolated" was given in 1996 by the ICJ Vice-President, Stephen M. Schwebel, in his *Dissenting Opinion* to the *Nuclear Weapons Advisory Opinion.* Schwebel pointed to the use of nuclear depth-charges to destroy submarines, stating that "(b)y the circumstance of its use, the nuclear depth-charge would not give rise to immediate civilian casualties." *See, Nuclear Weapons Advisory Opinion,* Dissenting Opinion of Vice-President Schwebel, at 98, available at <http://www.icj-cij.org/docket/files/95/7515.pdf?PHPSESSID=7c86212of720348a257c717b13fb6707>.

22 McClelland, "The review of weapons …" *supra* note 1, at 408.

23 "It is possible – he added – that we should examine whether the import of "[South] Lebanese" norms [regarding rules of engagement] indeed justifies the dangers inherent in this weapon." *See,* Amos Harel, "Weapons Effective in South Lebanon Become Problematic in the Gaza Strip", *Haaretz,* (June 11, 2001) at A1-2 (in Heb.)

24 *Official Record of the Diplomatic Conference on the Reaffirmation and Development of International Humanitarian Law Applicable in Armed Conflict, Geneva (1974-1977),* Vol. XVI, at 47.

25 HCJ 8990/02, *Physicians for Human Rights v. O.C. Southern Command*, P. D. 57(4) 193, at 194d.

26 Emphasis added. *Human Rights in Palestine and other Occupied Arab Territories,* Report of the United Nations Fact-Finding Mission on the Gaza Conflict (A/HRC/12/48) Sept. 25, 2009, at 193, para. 880.

However, our brief discussion on the principle of proportionality and the use of flechette weapons in the Gaza Strip concluded that any unequivocal determination that this use *in and of itself* violated this rule is premature. Our conclusion was reached either according to the common approach of "case-by-case assessment of the specific circumstances of the particular situation"[27] or in accordance with the "progressive statement of the applicable law with regard to the obligation to protect civilians", as devised in the Kupreskic Judgment.[28] The latter, "an *overall* assessment of the totality of civilian victims as against the goals of the military campaign"[29] – as the NATO Review Committee understood it – required essential data (i.e., the number of times in which flechette rounds were safely employed), which are missing in the South Lebanon and the Gaza Strip cases. Yet, we have taken care to point out that the IDF has developed its own policy for circumscribing the possible scenarios for flechette use. After all, it did "experience" the weapon's indiscriminate effects along the years in both South Lebanon and the Gaza Strip. In a May 27, 2004 incident, which surprisingly found its way to the media, a flechette round was fired toward suspicious figures in the Gaza Strip. Those who authorized the volley failed, however, to notify the "friendly" forces preparing an ambush in the target zone. As expected, those forces fell victim to this "statistical weapon".[30] As a senior IDF officer in the Gaza Strip headquarters explained the situation, "it was a serious incident that could have had catastrophic results," while his colleague – suddenly adopting the Petitioners' claims – added that "... injuries from flechette fragments are considered exceptionally fatal, and the fragments' dispersal is wide range".[31]

This statement, undoubtedly an impartial view on flechettes, reflects–to say the least–the associated problematics. The gap between the weapons' controversial aspects and the inability to draw conclusions regarding their legality under any of the two generic rules brings us back to the opening question about the capacity of these generic rules to adequately address humanitarian concerns once raised in association with certain conventional weapons.

As previously noted, both generic rules do not "provide concrete mathematical restrictions on what is legitimate".[32] That is, in practice, generic rules have a lim-

27 Michael J. Matheson, "The Opinions of the International Court of Justice on the Threat or Use of Nuclear Weapons" *AJIL* 91 (1997) 417, at 428.

28 *See,* Kupreskic *et al.* Judgment, (2000), Case No. IT-95-16-T (ICTY; Trial Chamber), para. 526. For the quotation *see,* Final Rep. to the Prosecutor by the Committee Established to Review the NATO Bombing Campaign Against the Federal Republic of Yugoslavia, *International Legal Materials* 39 (*ILM*) (2000), 1257, at 1271-2, para. 52 [hereinafter, *NATO Bombing Report*].

29 *NATO Bombing Report,* 1271, at para. 52.

30 Harel "Weapons Effective...", *supra* note 23.

31 For the incident in which an IDF officer was badly injured, and a solider lightly injured, *see,* Arieh O'Sullivan, "Paratrooper Captain Killed on Nablus Mission", *The Jerusalem Post,* (May 30, 2004), at A1; Haaretz staff and Arnon Regular, "Two Reservists Hurt by Friendly Fire", *Haaretz,* (May 28, 2004), at A14 (in Heb.) [The English version, Haaretz Staff, "Two Reservists Hurt by Friendly Fire", *Haaretz,* (May 28, 2004), is a partial one without the quotations noted]; Itzik Saban, "Mistaken Identity: Tank Fires Shell, Injures Officer and Soldier", *Yediot Ahronot,* (May 28, 2004), at A13.

32 Boothby, "The Law of Weaponry..." *supra* note 4, at 301.

ited effect given states' interests and tendencies to prefer military over humanitarian considerations whenever they perceive a threat to their national security. The U.S., for instance, firmly holds the position – as it clarified in its written statement to the ICJ in the nuclear weapons case – that the respective "prohibition was intended to preclude weapons *designed to* increase the injury or suffering of the persons attacked beyond that necessary to accomplish the military objective".[33] Neither the fact that the subjective element (i.e., *designed to*) has no mention in the authentic French text of the 1899 and the 1907 Hague Regulations (23b)[34] (i.e., *of a nature to*) nor the subjective element's clear elimination in AP-I (Art 35(2)), in which the principle is reaffirmed) have changed the U.S. position,[35] irrespective of the fact that 169 states have already joined this Protocol.[36]

In fact, in order to substantiate his argument made during the 1978-1980 CCW formulation period that "(g)overnment delegations distinguished between suffering inflicted by a weapon or projectile *designed* to enhance wounding effect and occasional effects as a result of its normal employment",[37] Hays Parks chose rejection of the proposal to ban flechette weapons as an example of this distinction. After citing arguments according to which a flechette strike would result in "painful and severe effects",[38] he stressed that "(t)he effects-based argument was regarded as too speculative to warrant regulation or prohibition".[39] While the validity of this contention is in doubt given the final wording of AP-I (Art 35(2)), this example points to the acute sensitivity of flechette weapons to this principle's application. Furthermore, inflicting pain for its own sake is not an overt military objective, let alone among modern armies. Hence, the fact that any decision under the CCW must be consensual implies that the U.S., or any other State Party sharing its view,[40] can obstruct this principle transformation into practical restrictions.[41]

---

33 Emphasis added. *U.S. Written Statement*, 28-29 (June 1995). All Written Statements are available at <http://www.icj-cij.org/docket/index.php?p1=3&p2=4&k=e1&case=95&code=unan&p3=1>.

34 *See, supra* chapter 5 (B).

35 As Parks has written in defense of such a position, "The latter phrase ('calculated to cause unnecessary suffering'-E.B.) is a clearer expression of the intent of governments to focus on design and intended purpose rather than every remote possibility of weapon injury. Art. 35(2) AP I returned to 'of a nature to cause', prompting the novel but historically incorrect ICRC argument". Parks, "Conventional Weapons and…" *supra* note 5, at 86-7, fn. 123.

36 For an updated list of State Parties to AP-I *see,* <http://www.icrc.org/ihl.nsf/WebSign?ReadForm&id =470&ps=P>.

37 Emphasis added. Parks, "Conventional Weapons and…" *supra* note 5, at 82.

38 *Id.*, at 81.

39 *Id.*, at 82.

40 The second prominent military power, Russia, argued before the ICJ that by virtue of this principle, an *explicit prohibition was required* to outlaw the use of a specific weapon. *See, Russian Written Statement* 12 (June 1995). *See* also the similar French position in this respect, *France Written Statement* 40-41 (June 1995).

41 It is difficult to conceive of any weapon that the U.S. would ban *solely* according to this principle. In fact, as Professor Wells noted in referring to this position in the context of U.S. Manual (FM 27-10 The Law of Land Warfare, Art. 34), "…the interpretation of

Therefore, and regrettably, it seems that Prof. Cassese's conclusion of 30 years ago regarding the prohibition of unnecessary suffering has not lost its validity:

> ... no common consent has ever evolved among States as to the actual normative value of the principle. Each State has interpreted the principle in its own way and international disagreement over whether a given weapon fell under the prohibition of the principle has never resulted in the reaching of a common view. It is therefore my opinion that Article 23e [of the Huge Convention] *as it stands now plays in practice a normative role only in extreme cases ...*[42]

A decade later, on the occasion of the CCW tenth anniversary (1990), Prof. Kalshoven addressed the question of whether these principles [i.e., the two generic rules] should be applied as either yardsticks or guidelines:

> If invoked as yardsticks, they should be able to support a fairly straightforward determination of the legal issue. As guidelines, they need provide little more than a set of basic, perhaps mutually conflicting, considerations that must be weighed in arriving at such a decision.[43]

As expected, he concluded that these two principles "are not particularly suited to serve as yardsticks as defined above".[44] However, before concluding that generic rules have little practical meaning, especially in this case, it seems that their role, as predicted some 30 years ago, did materialize after the mid 1990s, and is now more relevant than ever. As Prof. Cassese's noted in reference to the prohibition on inflicting unnecessary suffering, a comment that can be as readily applied to the criterion of indiscriminancy:

> It stands to reason that Article 23*e* can also play a role as a moral and political standard by which world public opinion assesses how belligerent States behave or misbehave.... This meta-legal value of the principle under consideration should not be underestimated; it *could turn out to be more important than the merely legal value, for the impact that public opinion can have, through [the] mass media, on governments.*[45]

---

what this means to the United States is simply that it is forbidden to use weapons which are not in the arsenal. The presumption is that careful scrutiny was given as each new weapon was invented to establish that it did not cause unnecessary suffering. There is no evidence to support such an assumption". *See,* Donald A. Wells, *The Laws of Land Warfare: A Guide to the U.S. Army Manuals*, (Westport: Greenwood Press, 1992), 180-81.

42 Emphasis added. *See,* Antonio, Cassese, "Means of Warfare: The Traditional and the New Law" in: Antonio Cassese (ed.), *The New Humanitarian Law of Armed Conflict*, (Napoli: Editoriale Scientifica, 1979), 161, at 163.

43 Frits Kalshoven, "The Conventional Weapons Convention: Underlying Legal Principles" *International Review of the Red Cross* 279 (1990) 510, at 517.

44 *Id.*

45 Emphasis added. Cassese, "Means of Warfare...", *supra* note 42, at 163.

The disintegration of the Soviet Union and the end of the Cold War resulted in increased NGO interest and efforts to regulate or prohibit conventional weapons – efforts that have enshrined these two generic rules.[46] However, given that the voting procedure induced paralysis of the CCW and the mentioned weaknesses of the two generic rules (i.e., the non-translation into practical restrictions on specific weapons), NGOs and like-minded States had no choice but to change the rules of the game.[47] In other words, they were forced to circumvent the CCW's framework for drafting new *specific rules* on conventional weapons. The 1997 Ottawa Convention[48] and the 2008 Convention on Cluster Munitions (CCM)[49] are the outcomes of their successful campaigns.[50] It is therefore not surprising that both Conventions contain meaningful prohibitions that the CCW has yet to introduce.

The question still remains regarding the practical effects of this alternative and its outcomes (i.e., the two Conventions). After all, the states that demanded voting by consensus in the CCW during the late 1970s had a quite convincing argument, to wit:

> The idea of 'broadest possible agreement' implies that consensus should be required in each case: only thus...can a prohibition or restriction on the use of a specific conventional weapon be expected to achieve its purpose. Conversely, *an agreement with no more than a majority vote behind it risks remaining without effect, especially if important States are among those which did not cast an affirmative vote.*[51]

Given the fact that many military powers have strongly opposed the new CCM (the U.S. and Russia, for instance, are not State Parties even to the Ottawa Convention), this question remains highly relevant.[52] However, if we turn to the literature on the

46 *See, e.g.*, Parks, "Conventional Weapons and ...", *supra* note 5, at 82.

47 It seems that even delegates to the CCW may sympathize with the approach of those NGOs and like-minded states. As Boothby noted regarding the failure of its members to adopt a new CCW protocol on AVR: "Such apparent failure may cause certain states and NGOs to be tempted to go outside the consensus-based CCW process in search of a 'quicker fix' elsewhere". Boothby, "The Law of Weaponry..." *Supra* note 4, at 312.

48 *See*, the Convention on the Prohibition of the Use, Stockpiling, Production and Transfer of Anti-Personnel Mines and on their Destruction, Art. 2, *Laws of Armed Conflicts*, 285, at 287.

49 The Convention's text in all official UN languages is available at <http://treaties.un.org/doc/ Publication/CTC/26-6.pdf>.

50 On the process that led to the Ottawa Convention *see, e.g.*, Richard Price, "Reversing the Gun Sights: Transnational Civil Society Targets Land Mines" *International Organization* 52 (1998) 613-644; Maxwell A. Cameron *et al.*, (eds.) *To Walk without Fear: The Global Movement to Ban Landmines*, (Oxford: Oxford University Press, 1999). For the most comprehensive overview on the process that led to 2008 CCM see John Borrie, *Unacceptable Harm: A History of How the Treaty to Ban Cluster Munitions Was Won*, (New-York: United Nations, 2009).

51 Emphasis added. *Reflections of War*, at 257-8.

52 For an updated list of State Parties and State Signatories to this Convention *see* <http://www.icrc.org/ihl.nsf/ WebSign?ReadForm&id=580&ps=P>.

so-called "tradition of non-use" regarding nuclear weapons[53] and, to a lesser extent, chemical weapons,[54] it appears that certain weapons can be so stigmatized that states will refrain from their use even in situations where they may conceivably be highly effective. Furthermore, the successful campaign to generate a convention banning anti-personnel land mines, born in the early 1990s, signified the crossing of a further threshold in this regard. Stigmatization was no longer limited to weapons perceived as WMD. The emerging prohibition indicated that the legitimacy attached to so-called "conventional weapons" was no longer an unassailable barrier to their limitation.[55]

The case of anti-personnel land mines confirms that the very existence of an instrument denouncing the use of a certain weapon may strongly reinforce – or generate – a new international norm prohibiting such weapons. As research on weapon stigmatization has shown, "[s]ingling out certain weapons or classes of weapon from the rest" is the first step in "differentiating them from other means of force that result in death and injury."[56] Thanks to a weapon's de-legitimization, states, including powerful ones, may, we can hope, reconsider their resort to – at least large-scale – use of such weapons, especially against targets in populated areas.

If the CCM does succeed in delegitimizing the use of cluster munitions (CMs) – as the Convention's supporters hope[57] – it will provide a viable and practicable alterative to the failure of the CCW and generic rules to adequately address humanitarian concerns. After all, unlike anti-personnel mines,[58] CMs are still perceived by modern militaries as effective weapons. As the U.S. Secretary of Defence Robert

53 For an extended discussion of this point *see*, Nina Tannenwald, *The Nuclear Taboo: The United States and the Non-Use of Nuclear Weapons Since 1945*. (Cambridge: Cambridge University Press, 2007); T.V. Paul, *The Tradition of Non-use of Nuclear Weapons*. (Princeton: Princeton University Press, 2009).

54 *See*, Richard Price, *The Chemical Weapons Taboo*, (New-York: Cornell University Press, 1997); Kenneth Adelman, "Chemical Weapons: Restoring the Taboo" *Orbis* 30 (1986) 443-455. For a succinct analysis of the two mentioned cases (nuclear weapons and chemical weapons) *see*, Richard Price and Nina Tannenwald, 'Norms and Deterrence: The Nuclear and Chemical Weapons Taboos", in: Peter J. Katzenstein, (ed.), *The Culture of national Security: Norms and Identity in World Politics*, (New York: Columbia University Press, 1996) 126-142.

55 As Prof. Dinstein argued in 2004: "There is every reason to believe that the prohibition of anti- personnel mines will gradually be endorsed by customary International law". *See*, Yoram Dinstein, *The Conduct of Hostilities under the Law of International Armed Conflict*, (Cambridge: Cambridge University Press, 2004), 69.

56 *See*, Brian Rappert, *A Convention Beyond the Convention: Stigma, Humanitarian Standards and The Oslo Process*, (Landmine Action Rep., May 2008), at 18.

57 *See*, for instance, the speech made by Ireland's Foreign Minister Michael Martin at the Dublin Diplomatic Conference's closing ceremony on May 30, 2008: "...even though we all know that there are important states not present, I am also convinced that together we will have succeeded in stigmatizing any future use of cluster munitions", available at <http://www.clustermunitionsdublin.ie/pdf/Ireland.pdf>.

58 On the limited military utility attributed to AP mines *see*, *e.g.*, Robert G. Gard, Jr., "The Military Utility of Anti-Personnel Mines", in: Maxwell A. Cameron *et al.*, (eds.), *To Walk without Fear, supra* note 51, at 136-157.

Gates stated in June 2008: "(t)hey ... provide distinct advantages against a range of targets".[59]

In such circumstances, the question naturally arises as to the future of flechette weapons: Could the *already* heavy involvement of prominent human rights NGOs (e.g., Amnesty international, HRW) in the effort to stop flechette use in urban settings, coupled with the September 2009 call by the Goldstone Report for urgent UNGA discussions on its legality, generate a process similar to that which took place following Israel's extensive use of CMs in the 2006 Second Lebanon War?

We may recall that accusations claiming Israeli use of illegal weapons (various types of incendiary weapons, mainly CMs) in Lebanon appeared in the international media amidst the intense fighting of summer 2006.[60] Yet, the world's outcry against the use of CMs was heard mainly in the war's aftermath.[61] In addition to numerous legal studies of the CMs' legality,[62] the tremendous harm to South Lebanon's civilians, together with clearance and risk awareness activities, attracted considerable interest on the part of human rights NGOs.[63] Their reports have helped keep the negative implications of CM use before the public. Thanks to their activities, the Israeli use became a pivotal factor in reinforcing the campaign to ban CMs

59 As stated by U.S. Secretary of Defence Robert Gates upon announcing a new U.S. policy on CMs in June 2008, available at <http://www.defenselink.mil/news/d20080709cmpolicy.pdf>. In principle, this policy (June 19, 2008) sets the goal of using CMs with less than one percent HDR within the next ten years.

60 Reports were issued mainly by the NGOs participating in the Cluster Munition Coalition (CMC). *See*, for instance, *Israeli Cluster Munitions Hit Civilians in Lebanon*, Human Rights Watch Press Release (July 24, 2006), available at <hrw.org/english/docs/2006/07/24/isrlpa13798_txt.htm>; "Lebanon President Says Israel Uses Phosphorous Arms" *Reuters*, (July 24, 2006), available at <www.alertnet.org/thenews/newsdesk/L24911888.htm>.

61 The sitting UNSG, Kofi Annan, condemned Israel while his aide, the Humanitarian Affairs and Emergency Relief Coordinator, Under-Secretary-General Jan Egeland, focused his criticism on the "shocking and...completely immoral" fact that "90 percent of the cluster bomb strikes occurred in the last 72 hours of the conflict, when we knew there would be a resolution". Yoav Stern & AP, "Annan Condemns IDF Use of Cluster Bombs," *Haaretz*, (Sep. 1, 2006), at 3; Todd Pitman, "Unexploded Israeli Bombs Menace Lebanese", *Washington Post*, (Aug. 31, 2006), at A3.

62 *See*, for instance, Bonnie Docherty, "The Time is Now: A Historical Argument for a Cluster Munitions Convention" *Harvard Human Rights Journal* 20 (2007) 53-87; Alexander Breitegger, "Preventing Human Suffering During and After Conflict? The Complementary Case for a Specific Convention on Cluster Munitions" *Austrian Review of International & European Law* 10 (2005) 3-40.

63 The most notable reports are: Thomas Nash, *Foreseeable Harm: the Use And Impact of Cluster Munitions In Lebanon*, (Landmine Action Rep., Oct. 2006); HRW, *Flooding South Lebanon: Israel's Use of Cluster Munitions in Lebanon in July and August 2006*, (HRW Rep., Feb. 2008); Greg Crowther, *Counting the Cost: The Economic Impact of Cluster Munition Contamination in Lebanon* (Landmine Action Rep., May 2008). *See*, also UNIDIR, *The Humanitarian Impact of Cluster Munitions* 29-47 (Geneva: UNIDIR, 2008).

(known as the "Oslo Process" as of February 2007),[64] and enabling the December 2008 CCM's successful conclusion.[65]

Interestingly, in its initial response to the world outcry in the 2006 war's aftermath, Israel's MAG hurriedly issued a legal opinion arguing that "the IDF may use cluster, phosphorus, and *flechette rounds*".[66] The fact that a legal opinion was issued for these three types of weapons clearly speaks for their special status: All have been categorized as dubious weapons. However, as Judge Rosalyn Higgins has stated, repugnancy *in and of itself* does not suffice to outlaw a certain type of weapon, while "especially repugnant means of conducting hostilities", such as incendiary weapons, remain legal.[67]

The first type of weapon among the three has since been banned by a targeted treaty (i.e., the CCM), the second type is partially regulated by the CCW Protocol III on Incendiary Weapons; the outcry over its use in the December 2008-January 2009 *Operation Cast Lead* ensured it will be used carefully – if at all – in future fighting in urban settings.[68] The last type of weapon – the flechette – thus far remains orphaned, waiting for NGOs and like-minded states to take action.

---

64 For an updated overview of that process *see*, Borrie, *Unacceptable Harm*, *supra* note 51.

65 As Jonas Gahr Støre, Norway's Foreign Minister and a prominent figure behind this convention clearly stated in October 2006, one month prior to calling a conference of like-minded governments seeking to ban CMs: "[t]he case of Lebanon clearly demonstrates that there is a real need to strengthen humanitarian law in this area …. This is why Norway will take the lead … to put in place an international prohibition against cluster munitions". Jonas Gahr Støre, Norwegian Ministry of Foreign Affairs, Reply to Olav Akselsen's (Labour Party) Question Regarding the War in Lebanon (*sic*) and the Use of Cluster Munitions, Written Question No. 61 (2006-2007), ( Oct. 24, 2006), available at Norway Foreign Ministry website. Experts and CMC's activists who refused to perceive Israel's 2006 use of CMs as a *sine qua non* for the CCM's emergence, admit that it was crucial for formulating the strong convention, which entered into force on Aug. 1, 2010. As prominent CMC activists Richard Moyes and Steve Goose [HRW's Arms Division Director] summed it up, Lebanon was "necessary, but not sufficient", emphasizing that multilateral processes were already underway. Email correspondence with Richard Moyes, Dec. 12, 2008, and with John Borrie, Dec. 15, 2008 (on file with the author).

66 Emphasis added. *See*, Uri Yablonka, "In Response to the Criticism – the Military Advocate's Office, 'The IDF is Authorized to Shoot Cluster, Phosphorus and Flechette Bombs'", *Maariv*, (Sept. 9, 2006), at 5A.

67 Dissenting Opinion of Judge Higgins, the *Nuclear Weapons Advisory Opinion*, at para. 16. available at <http://www.icj-cij.org/docket/index.php?p1=3&p2=4&k=e1&p3=4&case=95>.

68 *See*, for example, *Rain of Fire: Israel's Unlawful Use of White Phosphorus in Gaza* (HRW Rep., Mar. 25, 2009).

# Sources and Works Cited

## Documents

* U.S. Government
Congress Record,
Department of Defense
U.S. Department of State
US Marine Corps Field Manual
Ballistics Research Laboratory (BRL) Documents
Picatinny Technical Reports
Congressional Research Service

* Government of Israel
National Center of Forensic Medicine
* UN Documents
* ICRC Documents

## Interviews and Correspondences

Elazar Barak, Brig.-Gen., IDF's Chief Ordnance Officer (1973-78); Oct. 26; Nov. 9, 2006.

Robin M. Coupland, Medical Advisor, ICRC Chief Medical Officer; May 12, 2003.

Ken Kohut, Team Leader, Magellan Aerospace Marketing; Sept 14, 18, 2008.

Ehud Lahav, Management's Member, Reshef Technologies LTD; Oct. 27, 2009.

Noam Lubell; International Law Adviser and Project Director, Prisoners and Detainees, PHR-I; Dec. 21, 2006.

Richard Moyes, Cluster Munition Coalition (CMC), Co-Chair, Dec. 12, 2008.

Eric Prokosch; Friend World Committee's representative [1975 Lucerne Conference]; Oct. 8, 2008; Nov. 10, 2009; Jan. 9, 2010.

Yossi Zur; Flechette Petitioners' advocate; May 5, 12, 2003; Dec.13, 2006.

## Archives

Physicians for Human Rights–Israel (Tel-Aviv).

*B'Tselem* – The Israeli Information Center for Human Rights in the Occupied Territories (Jerusalem)

US Defense Documentation Center

## Books

### *1 English*

*A Brief Chronology of Events and Achievments 1962-1972*. United States: United States Army Materiel Command, 1972.

Ader, Clement. *Military Aviation*. [first published as L'Aviation militaire, 1909] [Lee Kennett Editor & Translator], Montgomery, AL: Air University Press, 2003.

Bailey, Jonathan B. A. *Field Artillery and Firepower*. 2nd ed, Annapolis, MD: Naval Institute Press 1993.

Barker, Ralph. *A Brief History of the Royal Flying Corps in World War I*, [first published as *The Royal Flying Corps in France* 2 Vols., 1995] rep. London: Robinson 2002.

Bassiouni, Cherif. *The Status of the International Criminal Court: A Documentary History*, Ardsley, N.Y.: Transnational Publishers, 1998.

Beate Hamizrachi, *The Emergence of the South Lebanon Security Belt*. New-York: Praeger 1988.

Beller, S. William, *The American Arsenal*. New-York: G. P. Putnam's Sons 1968.

Bernstein, Jonathan. *US Army AH-1 Cobra Units in Vietnam*. Oxford: Osprey Publishing, 2003

Bishop Chris and Laurier, Jim. *Apache AH-64 Boeing (McDonnell Douglas) 1976-2005*. Oxford: Osprey Publishing, 2005

Borrie, John. Unacceptable Harm: *A History of How the Treaty to Ban Cluster Munitions Was Won*. Geneva: UNIDIR, 2009.

Bothe, Michael, Karl J. Partsch and Waldemar A. Solf (eds.) *New Rules for Victims of Armed Conflicts: Commentary on the Two 1977 Protocols Additional to the Geneva Conventions of 1949*. Leiden: Martinus Nijhoff 1982.

Bowden, Mark. *Black Hawk Down: a Story of Modern War*. New-York: Atlantic Monthly Press 1999.

Bowen, Ezra. *Knights of the Air*. New-York: Time-Life Books 1980.

Braybrook, Roy. *Aircraft Weaponry of Today*. Sparkford: Haynes Publishing Group 1988.

Brown, Fredrich. *Chemical Warfare: a Study in Restraints*. Princeton: Princeton University Press 1968.

Brown, J. B. *The Hague Peace Conference of 1899 and 1907* [Documents] London: Clarendon Press 1909.

Carlson R., Neil, *Physiology of Behavior* 9th ed. Boston: Pearson Allyn & Bacon 2007.

Cassese, Antonio. *International Law*. New York: Oxford University Press 2005.

Chickering Roger and Stig Forster (eds.) *Great War, Total War: Combat and Mobilization on the Western Front, 1914-1918*. Cambridge: Cambridge University Press 2000.

Clayton, Anthony. *Paths of Glory: the French Army*, 1914-1918. London: Cassell 2003.

Couffer, Jack. *Bat Bomb: World War II's Other Secret Weapon*. Austin, TX: University of Texas Press 1992.

Coupland, Robin M. (ed.) *The SIrUS Project: Towards a Determination of which Weapons Cause Superfluous Injury or Unnecessary Suffering*. Geneva: International Committee of the Red Cross 1997

Coupland, Robin M. *The Red Cross Wound Classification*. Geneva: International Committee of the Red Cross 1997

Crane, Conrad C. *American Airpower Strategy in Korea 1950-1953*, Kansas City: University Press of Kansas 2000.

Davilla, James J. & Soltan, Arthur M. *French Aircraft of the First World War*. Boulder, CO: Flying Machines Press. 2002.

Delissen astrid and Tanja Gerard (eds.), *Humanitarian Law of Armed Conflict, Challenges Ahead, Essays In Honour Of Fritz Kalshoven*. Leiden: Martinus Nijhoff 1991.

Detter, Ingrid. *The Law of War*, Cambridge: Cambridge University Press, 1st ed.1987, 2nd ed. 2000.

Dickson, Paul. *War Slang: American Fighting Words and Phrases since the Civil War*. London: Brassey's 2003.

Dinstein, Yoram. *The Conduct of Hostilities* under *the* Law *of International Armed Conflict*. Cambridge: Cambridge University Press 2004.

Dockery, Kevin. *Future Weapons*. Berkley: Berkley Books, 2007.

Doleman, Edgar C. et al., *Tools of War (Vietnam Experience)*. Boston: Boston Publishing 1985.

Dommett, William Erskine. *Aeroplanes and Airships*. New-York: Whittaker & Co 1915.

Doswald-Beck, Louise. (ed.), *Blinding Weapons: Reports of the Meetings of Experts Convened by the International Committee of the Red Cross on Battlefield Laser Weapons, 1989-1991*. Geneva: ICRC 1993.

Duffett, John, (ed.), *Against the Crime of Silence: Proceedings of The International War Crimes Tribunal*. New-York: Simon & Schuster 1970.

Durant Michal J. & Hartov, Steven. *The Night Stalkers*. New-York: G. P. Putnam's Sons 2006.

Eshel, David. *Chariots of the Desert: Story of the Israeli Armoured Corps*. London: Brasseys 2004.

Fall, Bernard & Fall, Bernard B. *Hell in a Very Small Place: the Siege of Dien Bien Phu*. Cambridge: Da Capo Press 2002.

Falls, Cyril B. *The First World War*. London: Longmans 1960.

Fisk, Robert. *Pity the Nation*. 4th edition, London: Nation Books 2002.

Fleischer, Wolfgang German *Air-Dropped Weapons to 1945*. [translator: Ted Oliver] London: Midland Publishing, 2003.

Fox, Martin. *Corporals All: with the Special Brigade R.E. 1915-1919*. New-York: C.H. Harman 1967.

Gardam, Judith Gail. *Necessity, Proportionality and the Use of Force by States*. Cambridge: Cambridge University Press 2004.

Glahn, Von G. *Law Among Nations: an Introduction to Public International Law*. 7th ed. Boston: Allyn & Bacon 1996.

Goldblat, Jozef. *Arms Control: the New Guide to Negotiations and Agreements*. 2nd ed. Washington D.C.: Sage Publications Ltd 2002.

Green, Michael & Stewart, Greg *M1 Abrams at War*. New-York: Zenith Imprint 2005.

Greenwood, Christopher. *Essays on War in International Law*. London: Cameron May 2006.

Gunston, Bill & Spick Mike, *Modern Fighting Helicopters*, Little Rock, AR: Tiger Books 1991.

Hawthorne-Tagg, Lori S. *On the Front Line of R&D: Wright-Patterson Air Force Base in the Korean War, 1950-1953*. Wright-Patterson AFB, OH: Wright-Patterson AFB 2001.

Henckaerts, Jean-Marie and Doswald-Beck , Louise (eds.). *Customary International Humanitarian Law*. Cambridge: Cambridge University Press 2005.

Higgins, Pearce. The Hague Peace Conferences and Other International Conferences Concerning the Laws and Usages of War. Cambridge: Cambridge University Press 1909.

Hogg, Ian V. & Gander, Terry. *Jane's Ammunition Handbook 1993-94*. London: Jane's Information Group 1992.

------------- & Weeks, John S. *Military Small Arms of the 20th Century*. 7th Ed. Iola, WIS: Krause Publications 2000.

------------ *The Illustrated Encyclopedia of Ammunition*. Berkeley: Quantum Books 2003.

ICRC, *International Red Cross Handbook* 11th ed. Geneva: ICRC 1971.

Jaber, Hala. *Hezbolla: Born with a Vengeance.* New York: Columbia University Press 1997.

Joenniemi, Pertti & Rosas, Allan. *International Law and the Use of Conventional Weapons.* Tampere: Tampere Peace Research Institute 1975.

Kam Ephraim & Shapir, Yiftah (eds.), *The Middle East Strategic Balance 2002–2003.* Tel Aviv: Jaffee Center for Strategic Studies 2003.

Kalshoven, Fritz. *Reflections on the Law of War: Collected Essays.* Leiden: Martinus Nijhoff 2007.

Katz, Sam. *Merkava Main Battle Tank MKs I, II & III.* Oxford: Osprey Publishing 1997.

Kneubuehl Beat P. and Sellier, Karl G. *Wound Ballistics and the Scientific Background,* Amsterdam: Elsevier 1994.

Kretzmer, David. *The Occupation of Justice: the Supreme Court of Israel and the Occupied Territories.* Albany NY: State University of New York Press 2002.

Lauterpachat Elihu and Greenwood Christopher, *International Law Reports* (Vol. 82) (Cambridge: Cambridge University Press. 1990.

------------------------, *International Law Reports* (Vol. 88) (Cambridge: Cambridge University Press. 1992.

------------------------, *International Law Reports* (Vol. 99) (Cambridge: Cambridge University Press. 1994.

Lawand, Kathleen (principle author). *A Guide to the Legal Review of New Weapons, Means and Methods of Warfare: Measures to Implement Article 36 of Additional Protocol I of 1977.* Geneva: ICRC, revised version Nov. 2006.

Lee, Kennett. *A History of Strategic Bombing.* New-York: C. Scribner's Sons 1982.

--------------. *The First Air War: 1914–1918.* New-York: Free Press 1999.

Levie, S. Howard. *Mine Warfare at Sea.* Dordrecht: Martinus Nijhoff 1992.

Lewis, John W. & X Litai, *China Builds the Bomb.* Stanford: Stanford University Press 1988.

Martel, Rene. French *Strategic and Tactical Bombardment Forces of World War I.* [first published as L'Aviation Francaise de Bombardement (Des Origines au 11 Novembre 1918), 1939] (translated by Allen Suddaby; edited by Steven Suddaby). Lanham: Scarecrow Press 2006.

McCormack, Timothy, *et al.*, *Report on States Parties' Responses to the Questionnaire: International Humanitarian Law & Explosive Remnants of War,* Melbourne: Asia Pacific Center For Military Law 2006 [reprinted also as UN Doc. CCW/GGE/XIII/WG.1/WP.12].

McCudden, James Thomas Byford. *Flying Fury: Five Years in the Royal Flying Corps.* [First Published in 1919 as *Five Years in the Royal Flying Corps*], Barnsly: Greenhill Books, Reprinted edition 2000.

Melzer, Nils. *Targeted Killing in International Law.* Oxford: Oxford University Press 2008.

Mikesh, Robert C. *Japan's World War II Balloon Bomb Attacks on North America.* Washington D.C.: Smithsonian Institution Scholarly Press 1990.

Morrow, John H. *The Great War in the Air: Military Aviation from 1909 to 1921.* Washington D.C.: Smithsonian Institution Press 1993.

Mountcastle, Jhon Wyndham. *Flame On: U.S. Incendiary Weapons, 1918–1945.* Shippensburg, PA: White Mane Books 1999.

Norton, Augustus Richard. *Hezbolla: A Short History.* Princeton: Princeton University press 2007.

Nossov, Konstantin. *Ancient and Medieval Siege Weapons: A Fully Illustrated Guide to Siege Weapons and Tactics.* Guilford, CT: The Lyons Press 2005.

O'Connell, D. P. *The Influence of Law on Sea Power.* Annapolis, MD: Naval Institute Press 1975.

---------------. *The International Law of the Sea*. Oxford: Clarendon Press 1984.

Office of United States Chief of Counsel for the Prosecution of Axis Criminals, *The Nazi Conspiracy and Aggression: Opinion and Judgment*. Washington D.C: GPO 1947.

Ott, David Ewing. *Vietnam Studies: Field Artillery*, 1954-73. Washington D.C.: Department of the Army 1975.

Orgill, Douglas. *The Tank*. London: William Heinemann Ltd 1970.

Parker, Sybil P. *McGraw-Hill Encyclopedia of Science & Technology*. 8th ed. New-York: McGraw-Hill 1997.

Paul, T.V., *The Tradition of Non-use of Nuclear Weapons*. Princeton: Princeton University Press 2009.

Payne-Gallwey, Ralph. *The Book of the Crossbow*. New-York: Dover Publications, Inc. 1995, first published: London: Longmans Green, and Co. 1903.

Prentiss, Augustin M. *Chemicals in War; A Treatise on Chemical Warfare*. New-York: McGraw-Hill Book Company 1937.

Preston, Richard & Wise, *Sydney. Men in Arms: A History of Warfare and it's Interrelationships with Western Societies*. New-York: Praeger 1979.

Price, Richard. *The Chemical Weapons Taboo*. New-York: Cornell University Press 1997.

Prokosch, Eric. *The Simple Art of Murder: Anti-personnel Weapons and Their Developers*. Philadelphia: NARMIC 1972.

--------------. *The Technology of Killing: A Military and Political History of Anti-personnel Weapons*. London: Zed Books Ltd 1995.

Qassem, Naim. *Hizbullah: The Story From Within*. [Dalia Khalil trans.] London: SAQI 2005.

Rogers, A. P. V, *et al.*, *Fight It Right* (Model Manual on the Law of Armed Conflict for Armed Forces). Geneva: ICRC, 1999.

----------. *Law on the battlefield*. 2nd ed. Manchester : Manchester University Press 2004.

Pilloud, Claude et al. *Commentary on the Additional Protocols of 8 June 1977 to the Geneva Conventions of 12 August 1949*. Geneva: ICRC 1987.

Shaw, Malcolm. *International Law* 4th ed. Cambridge: Cambridge University Press 1997.

Schindler, Dietrich & Toman, Jiri (eds.) *The Laws of Armed Conflicts*, Leiden: Martinus Nijhoff 2004.

Simpson, A. & Weine, E. S. C. (eds.), *The Oxford English Dictionary* 2nd ed. Oxford: Clarendon Press 1989.

Singh, Nagendra and McWhinney, Edward, *Nuclear Weapons and Contemporary International Law* 2nd ed. Leiden: Martinus Nijhoff, 1989.

Stockholm International Peace Research Institute (SIPRI), *Incendiary Weapons*. Stockholm: Almqvist & Wiksell 1975.

--------, The Problem of Chemical and Biological Warfare: CBW and the Law of War Stockholm: Almqvist & Wiksell 1973.

--------, *The Law of War and Dubious Weapons*. Stockholm: Almqvist & Wiksell l976

--------, *Anti-Personnel Weapons*. New-York: Taylor and Francis, 1978.

Swedish Working Group Study, *Conventional Weapons, Their Deployment and Effects from Humanitarian Aspects, Recommendations for the Modernization of International Law*, Stockholm: Almqvist & Wiksell 1973.

Swearengen, Thomas F. *The World's Fighting Shotguns*. New-York: T.B.N. Enterprises 1978.

Tannenwald, Nina. *The Nuclear Taboo: The United States and the Non-Use of Nuclear Weapons Since 1945*. Cambridge: Cambridge University Press 2007.

Taylor, Thomas. *Lightning in the Storm – The 101st Air Assault Division in the Gulf War*. New-York: Hippocrene Books 1994.

Terry, Treadwell C. *America's First Air War: The United States Army, Naval and Marine Air Services in the First World War*. New-York: Airlife, 2000.

Tuchman, Barbara W. *The Proud Tower: A Portrait of the World Before the War, 1890-1914*. New-York: Bantam Books, 1967.

Turner, M.S. *The Not So Diabolical Crossbow: A Re-Examination of Innocent II's Supposed Ban of the Crossbow at the Second Lateran Council*. London: Lulu 2004.

UNIDIR. *The Humanitarian Impact of Cluster Munitions*. Geneva: UNIDIR 2008.

United Nations. *The United Nations and Disarmament 19450-1970*. New-York: UN 1970.

Weir, William. *50 Weapons That Changed Warfare*. New-Jersey: New Page Books 2005.

Wells Donald A., *The Laws of Land Warfare: A Guide to the U.S. Army Manuals*. Westport: Greenwood Press1992.

Westmoreland, William C. *A Soldier Reports* New-York: Doubleday 1976.

Whiteman, Marjorie M. *Digest of International law*, Vol. VI. Washington, D.C.: U.S. Government Printing Office 1943.

Williams, Anthony G. & Ness, Leland, (eds). *Jane's Ammunition Handbook 2005-2006* 14th ed. London: Jane's Information Group 2005.

Windrow, Martin. *The Last Valley: Dien Bien Phu and the French Defeat In Vietnam*. Cambridge: Da Capo Press 2005.

Winter Jay and Blaine Bagget, *The Great War and the Shaping of the 20th Century*. New-York: Penguin Studio 1996.

Wise, Harold Lee. *Inside the Danger Zone – the U.S. Military in the Persian Gulf 1987-1988*. Annapolis, MD: Naval Institute Press 2007.

Zaloga, Steve and Sarson, Peter, *M1 Abrams Main Battle Tank 1982-1992*. Oxford: Osprey 1993.

Foreign Language Books

Barak Aharon, *Judicial Discretion* [Hebrew] Tel Aviv: Papyrus 1987.

Ben-Naftali, Orna and Shany, Yuva. *International Law between War and Peace*. [Hebrew] Tel Aviv: Tel Aviv University Press, 2006.

HalamishYuval. *Israel's Security Concept in Southern Lebanon 1985-1998* [Hebrew] Ramat-HaSharon: National Security College 1998.

Harel Amos & Avi Isacharoff. *The Seventh War*. [Hebrew] Tel-Aviv: Miskal 2004.

Leroy, Jean-Pierre. *Les Fléchettes D'avion De La Grande Guerre À Nos Jours* [The Aerial Flechettes of the Great War of Our Time] Paris: Emotion Primitive 2007.

Rubinstein Eliakim, *Judges of the Country*. [Hebrew] Tel Aviv: Shocken 1981.

## Unpublished Materials

Avni, Lior. *Israel-Hezbolla Relations from "Operation Accountability" to the "Second Lebanon War" (July 1993 – July 2006) Limited Security Regime or Violent Dialogue?* (2007) (Hebrew).

Prokosach, Eric. Notes on the Development of Canister and Beehive Rounds for Artillery, July 1974.

'U.S. War Technique and Genocide in Vietnam', (16mm, 1971, 35 min.), Film Transcript.

## Journal Articles, Essays, and Book Chapters

### *English*

Adelman, Kenneth, "Chemical Weapons: Restoring the Taboo" *Orbis* 30 (1986) 443-455.

"Aeroplane Darts and Fire Darts", 2042 *Scientific American Supplement* (Feb. 20, 1915).

Anderberg, Bengt et al. "Blinding Laser Weapons and International Humanitarian Law", *Journal of Peace Research* 29 (1992) 287-297.

Ayton, Andrew & Sir Preston, Phillip. "Topography and Archary: Further Reflection on the Battle of Crécy" in: *The Battle of Crécy*, 1346, Andrew Ayton et al. (eds.) 351-378 Rochester, NY: Boydell Press 2007.

Barak, Aharon. "The Role of a Supreme Court in a Democracy, and the Fight against Terrorism" *University of Miami Law Review* 58 (2003) 125-141.

---------------. "Human Rights in Israel" *Israel Law Review* 39(2) (2006) 12-34.

Barak, Eitan. "Doomed to be Violated ? The U.S.-Israeli Clandestine End-User Agreement and the Second Lebanon War: Lessons for the Convention on Cluster Munitions", *Denver Journal of International Law and Policy* 38 (2009) 1-40.

-----------. "None to be Trusted: Israel's Use of Cluster Munitions in the Second Lebanon War and the Case for the Convention on Cluster Munitions" *American University International Law Review* 25 (2010) 423-483.

Barak-Erez, Daphne. "Israel: The Security Barrier – Between International Law, Constitutional Law, and Domestic Judicial Review" *International Journal of Constitutional Law* 4(3) (2006) 540-552.

Barazili Gad, "Political Institutions and Conflict Resolution – The Israeli Supreme Court and the Peace Process" *The Middle East Peace Process – Interdisciplinary Perspectives.* Peleg Ilan (ed.), 87-105 Albany NY: State University of New-York Press, 1988.

Baxter, R.R. & Buergenthal Thomas. „Legal Aspects of the Geneva Protocols of 1925" *The American Journal of International Law* 63 (1970) 853-879.

Baxter, R. R. "Criteria of the Prohibition of Weapons in International Law" in: *Festschrift fur Ulrich Scheuner – Zum 70. Geburtstag*, Horst Ehmke *et al.*, (eds.) 41-52 Berlin: Duncker & Humblot, 1973.

-------------. "Humanitarian Law or Humanitarian Politics? The 1974 Diplomatic Conference on Humanitarian Law", *Harvard International Law Journal* 16 (1975) 1-26.

-------------. "Conventional Weapons under Legal Prohibitions" *International Security* 1(3) (Winter 1977) 42-61.

Beckett, Jason A. "Interim Legality: A Mistaken Assumption? An Analysis of Depleted Uranium Munitions Under Contemporary International Humanitarian Law" *Chinese Journal of International Law* 3 (2004) 43-86.

Bello, Judith Hippler & Bekker, Peter H. F. "Legality of the Threat or Use of Nuclear Weapons" *The American Journal of International Law* 91(1) (1997) 126-133.

Bendor, Ariel L. Bendor "Justiciability of The Israeli Fight Against Terrorism" *The George Washington International Law Review* 39 (2007) 149-163.

Benvenisti, Eyal. "Human Dignity in Combat: The Duty to Spare Enemy Civilians" *Israel Law Review* 39 (2006) 81-109.

Benvenuti, Paolo "The ICTY Prosecutor And The Review Of The NATO Bombing Campaign Against The Federal Republic of Yugoslavia" *European Journal of International Law* 12 (2001) 503-530.

Ben-Naftali, Orna & Michaeli. Keren R."We Must Not Make a Scarecrow of the Law: A Legal Analysis of the Israeli Policy of Targeted Killings" *Cornell International Law Journal* 36 (2003) 233-292.

--------------------------------------- "Justice-Ability: A Critique of the Alleged Non-Justiciability of Israel's Policy of Targeted Killings" *Journal of International Criminal Justice* 1 (2003) 368-405.

Blix, Hans. "Current Efforts to Prohibit the Use of Certain Conventional Weapons" *Instant Research on Peace and Violence* 4 (1) (1974) 21-30.

----------- "Means and Methods of Combat" in: *International Dimensions of Humanitarian Law* 135-151 Geneva: UNESCO 1988.

Boothby, Bill. "The Law of Weaponry: Is It Adequate?" in: *International Law and Armed Conflict: Exploring the Faultline: Essays in Honour of Yoram Dinstein*, Michael Schmitt and Jelena Pejic (eds.) 297-316 Leiden: Martinus Nijhoff 2007.

Borrie, John. "The 'Long Year': Emerging International Efforts to Address the Humanitarian Impacts of Cluster Munitions" 10th *Yearbook of International Humanitarian Law* (2007) 251-275.

Breitegger, Alexander. "Preventing Human Suffering During and After Conflict? the Complementary Case for a Specific Convention on Cluster Munitions" *Austrian Review of International & European Law* 10 (2005) 3-40.

Bring, Ove. "Regulating Conventional Weapons in the Future- Humanitarian Law or Arms Control?" *Journal of Peace Research* 24 (1987), 275-286.

Bugnion François, "The International Committee of the Red Cross and Nuclear Weapons: From Hiroshima to the Dawn of the 21st Century" *International Review of the Red Cross* 859 (2005) 511-524.

Bundt S.,Thomas. "An Unconventional War: The Philippine Insurrection 1899" *Military Review* (May-June 1984) 9-10.

Calvo-Goller, Karin. "Legal Background and Analysis of the 1997 Convention Banning Anti-Personnel Mines" *Israel Law Review* 32 (1998) 645-680.

Cannon, Garland & Mendez Egle, Beatrice. "New Borrowings in English" *American Speech* 54 (1979) 23-37.

Carnahan, Burrus M. "The Law of Land Mine Warfare: Protocol II to the United Nations Convention on Certain Conventional Weapons" *Military Law Review* 105 (Summer 1984) 73-95.

------------------ & Robertson, Marjorie. "The Protocol on 'Blinding Laser Weapons': A New Direction for International Humanitarian Law" *The American Journal of International Law* 90(3) (1996) 484-490.

-------------------- "Lincoln, Lieber and the Laws of War: The Origins and Limits of the Principle of Military Necessity", *The American Journal of International Law* 92(2) (1998), 213-231.

Cassese Antonio, "Means of Warfare: The Traditional and the New Law" in: *The New Humanitarian Law of Armed Conflict*. Antonio Cassese (ed.) 161-198 Napoli: Editoriale Scientifica 1979.

-----------------."The Statute of the International Criminal Court: Some Preliminary Reflections" *European Journal of International Law* 10 (1999) 144–171.

----------------- "The Martens Clause: Half A Loaf Or Simply Pie In The Sky?" *European Journal of International Law* 11 (2000) 187-216.

Chabasse, Philippe. "The French Campaign" in: Maxwell A. Cameron et al (Eds.), *To Walk Without Fear: The Global Movement to Ban Landmines*. Maxwell A. Cameron et al (Eds.) 60-67 Oxford: Oxford University Press, 1999.

Clark, Roger S. "Human Rights Strategies of the 1960s within the United Nations: A Tribute to the Late Kamleshwar Das" *Human Rights Quarterly* 21 (2) (1999) 308-341

--------------. "Treaty and Custom" in: *International Law, the International Court of Justice and Nuclear Weapons*. Chazournes Boisson Laurence & Philippe Sands (eds.) 171-180 Cambridge: Cambridge University Press, 1999.

Conley, Cornelius W., "The Great Japanese Balloon Offensive", *Air University Review*, (Jan.-Feb. 1968).

Coupland, Robin. "Abhorrent Weapons and 'Superfluous Injury and Unnecessary Suffering': from Field Surgery to Law" *British Medical Journal* 315 (1997) 1450

----------------& Herby, Peter. "The SirUS Project and Reviewing the Legality of New Weapons" *International Review of the Red Cross* 835 (1999) 583-592.

Cryer, Robert. "The Fine Art of Friendship: Jus in Bello in Afghanistan" *Journal of Conflict and Security Law* 7 (2002) 37-83.

Danaher, Kevin. "Israel's Use of Cluster Bombs in Lebanon" *Journal of Palestine Studies* 11/12(4/1) (1982) 48-57.

Daoust, Isabelle. "ICRC Expert Meeting on Legal Reviews of Weapons and the SIrUS Project" *International Review of the Red Cross* 842 (June 2001) 539-542.

------------ *et al.*, "New Wars, New Weapons? The Obligation of States to Assess the Legality of Means, and Methods of Warfare", *International Review of the Red Cross* 846 (June 2002) 345-363.

Di Ruzza, Tommaso. "The Convention on Cluster Munitions: Towards a Balance Between Humanitarian and Military Considerations?", *Military Law and the Law of War Review* [Revue de droit militaire et de droit de la guerre], 47 (2008) 405–448.

Dinstein, Yoram. "Collateral Damage and the Principle of Proportionality" in: *New Wars, New Laws? Applying the Laws of War in 21st Century Conflicts.* David Wippman and Matthew Evangelista (eds.) 211-224 Ardsley NY: Transnational Publishers 2005.

----------------. "The ICRC Customary International Humanitarian Law Study" *Israel Yearbook on Human Rights* 36 (2006) Dinstein, Yoram and Domb, Fania (eds.). 1-15

Docherty, Bonnie. "The Time Is Now: A Historical Argument for a Cluster Munitions Convention" *Harvard Human Rights Journal* 20 (2007) 53-87.

Domb, Fania. "Judgments of the Supreme Court of Israel Relating to the Administered Territories" *Israel Yearbook on Human Rights 34* (2004) 305-347.

Erickson, John R. "Protocol I: A Merging of the Hague and Geneva Law of Armed Conflict" *Virginia Journal of International Law* 19(3) (1979) 557-592.

Falk, Richard A. "The Shimoda Case: A Legal Appraisal of the Atomic Attacks upon Hiroshima and Nagasaki" *The American Journal of International Law* 59(4) (1965) 759-793.

----------------. "Nuclear Weapons International Law and the World Court: A Historic Encounter" *The American Journal of International Law* 91(1) (1997) 64-75.

Fenrick William J. "The Rule of Proportionality and Protocol in Conventional Warfare" *Military Law Review* 98 (Fall 1982) 91-127.

Gard, Robert G. Jr., "The Military Utility of Anti-Personnel Mines" in: *To Walk Without Fear: The Global Movement To Ban Landmines.* Maxwell A. Cameron *et al.*, (eds.) 136-157 Oxford: Oxford University Press 1999.

Gardam, Judith Gail. "Proportionality and Force in International Law" *The American Journal of International Law* 87(3). (1993) 391-413.

Garver, Milton. "French Army Slang" *Modern Language Notes* 32 (1917) 151-154.

Granat, Mark. G. "Modern Small-Arms Ammunition in International Law" *Netherlands International Law Review* 40 (1993) 149-169.

Greenwood Christopher, "Current Issues in the Law of Armed Conflict: Weapons, Targets and International Criminal Liability" *Singapore Journal of International and Comparative Law* 1 (1997) 441-467.

-------------------, "The Law of Weaponry at the Start of the New Millennium" in: *The Law of Armed Conflict: Into the Next Millennium.* Michael N. Schmitt & Leslie C. Green (eds.) 185-232 Newport RI: Naval War College International Law Studies 1998.

Gross, Aeyal M. "Human Proportions: Are Human Rights the Emperor's New Clothes of the International Law of Occupation?" *The European Journal of international Law* 18(1) (2007) 1-35.

Gross, Michael L. "Fighting by Other Means in the Mideast: A Critical Analysis of Israel's Assassination Policy" 51 *Political Studies* (2003) 350–368.

Hampson, Françoise J. "Means and Methods of Warfare in the Conflict in the Gulf" in: *The Gulf War 1990-91 in International and English Law*. Rowe, Peter (ed.) 89-110 New York: Routledge 1993 .

Herby, Peter & Nuiten, Anna. "Explosive Remnants of War: Protecting Civilians Through an Additional Protocol to the 1980 Convention on Certain Conventional Weapons" *International Review of the Red Cross* 841 (2001) 195-205.

Herthel, Thomas J. "On the Chopping Block: Cluster Munitions and the Law of War – Unexploded Submunitions from Cluster Bombs" *Air Force Law Review* 51 (2001) 229-269.

Hulme,Karen, "Of Questionable Legality: The Military Use of Cluster Bombs in Iraq in 2003" *Canadian Yearbook of International Law* 42 (2004), 143-169.

Jochnick, af Chris & Normand, Roger. "The Legitimation of Violence: A Critical History of the Laws of War" *Harvard International Law Journal* 35 (1994) 49-95

Kalb, Marvin & Saivetz,Carol. "The Israeli-Hezbollah War of 2006: The Media as a Weapon in Asymmetrical Conflict" *The International Journal of Press/Politics* 12(3) (2007) 43-66.

Kalivretakis, Elaina. "Are Nuclear Weapons above the Law–A Look at the International Criminal Court and the Prohibited Weapons Category" *Emory International Law Review* 15 (2001) 683-732.

Kalshoven, Frits. "The Soldier and his Golf Clubs" in: *Studies and Essays in International Humanitarian Law and Red Cross Principles in Honour of Jean Pictat*. Christophe Swinarski, (ed.) 369-385 Leiden: Martinus Nijhoff 1984.

---------------. "Arms, Armaments and International Law", *Recueil Des Cours* 19 (1985)

----------------. "The Conventional Weapons Convention: Underlying Legal Principles" *International Review of the Red Cross* 279 (1990) 510-520.

Kaye, David & Solomon, Steven A. "The Second Review Conference of the 1980 Convention on Certain Conventional Weapons" *The American Journal of International Law* 96(4) (2002) 922-936.

Kirsch, Philippe & Holmes, John T. "The Rome Conference on an International Criminal Court: The Negotiating Process" *The American Journal of International Law* 93(1) (1999) 2-12.

Krepon, Michael. "Weapons Potentially Inhumane: The Case of Cluster Bombs" *Foreign Affairs* 52 (1974) 595-611.

----------------. "Blanket Coverage: Two Case Studies of Area Weapons in Indochina" in:, *The World Military Order: The Impact of Military Technology on the Third World* Mary Kaldor and Asbjorn Eide (eds.) 49-63 London: Macmillan Press 1979.

Kretzmer, David. "Targeted Killings of Suspected Terrorists: Judicial Executions or Legitimate Means of Defence" *European Journal of International Law* 16(2) (2005) 172-212.

Lapidoth, Ruth. "International Law within the Israel Legal System" *Israel Law Review* 24 (1990) 451-484.

Lawrence, Juda, "Negotiating a Treaty on Environmental Modification Warfare: The Convention on Environmental Warfare and its Impact upon Arms Control Negotiations", *International Organization* 32(4) (1978), 975-991.

Leitenberg, Milton. "America in Vietnam-Statistics of a War" *Survival*, 14(6) (1972) 268 – 274.

Levey, Zach. "Anatomy of an Airlift: United States Military Assistance to Israel During the 1973 War" *Cold War History* 8 (2008) 481-501.

Levie, Howard S. "Prohibitions or Restrictions on the Use of Conventional Weapons" *St. John's Law Review* 68 (1994).643-672.

Levie, Howard S. "The Law of War" *The American Journal of International Law* 82 (1989) (book review) 194-198.

Luft, Gal. "Israel's Security Zone in Lebanon – A 'Tragedy'?" *Middle East Quarterly* 7(3) (Sept. 2000) 13-20.

Lumsden, Malvern. "The Prohibition of Inhumane and Indiscriminate Weapons" *in: SIPRI, SIPRI Yearbook 1979: World Armaments and Disarmament* (New-York: Taylor & Francis 1979) 453-462. Mallison, T. W. "The Laws of War and the Juridical Control of Weapons of Mass Destruction in General and Limited Wars" *George Washington Law Review* 36 (1967) 308-346.

Maresca, Louis. "A New Protocol on Explosive Remnants of War: The History and Negotiation of Protocol V to the 1980 Convention on Certain Conventional Weapons" *International Review of the Red Cross* 856 (2004) 815–835.

Matheson, Michael J. "The Opinions of the International Court of Justice on the Threat or Use of Nuclear Weapons" *The American Journal of International Law* 91(3) (1997) 417-435.

Mathews, Robert J. "The 1980 Convention on Certain Conventional Weapons: A Useful Framework Despite Earlier Disappointments" *International Review of the Red Cross* 83(844) (2001) 991-1012.

McClelland, Justin. "The Review of Weapons in Accordance with Article 36 of Additional Protocol I" *International Review of the Red Cross* 85(850) (2003) 397-415.

-------------------."Conventional Weapons: A Cluster of Developments" *International Law and Comparative Law Quarterly* 54 (2005) 755-767.

McCormack Timothy & Sue Robertson. "Jurisdictional Aspects of the Rome Statute for the New International Criminal Court", *Melbourne University Law Review* 23 (1999) 635-667.

McDonnell, Thomas Michael "Cluster Bombs over Kosovo: A Violation of International Law" *Arizona Law Review* 44 (2002) 31-130.

McNab, Ian R. "Early Electric Gun Research" *IEEE Transactions on Magnetics* 35(1) (1999) 250-261

Meron, Theodor. "The Martens Clause, Principles of Humanity, and Dictates of Public Conscience" *The American Journal of International Law*, 94(1) (2000) 78-89

Meyrowitz, Henri. "The Principle of Superfluous Injury or Unnecessary Suffering" *International Review of the Red Cross* 299 (1994) 98-122

Millet, Allen. R. 1986. "Over Where ? The AEF and the American Strategy for Victory, 1917-1918" in: *Against All Enemies: Interpretations of American Military History from Colonial Times To The Present.* Kenneth J. Hagan & William R. Roberts (eds.) 223-273 Westport, CT: Greenwood Press 1986.

Milton, Garver. "French Army Slang" *Modern Language Notes*, 32 (1917) 151.

Moxley Charles, "The Unlawfulness of the United Kingdom's Policy of Nuclear Deterrence: The Invalidity of the Scottish High Court's Decision in *Zelter", 58 Disarmament Diplomacy* (June 2001).

Nash, Marian (Leich). "Contemporary Practice of the United States Relating to International Law" *The American Journal of International Law* 91 (1997) 697-717.

Nell, Edward & Nell, Onora. "War Words" *College Engl*ish 28 (1967) 603-606.

Nelson, Steven. C. "Contemporary Practice of the United States Relating to International Law" *American Journal on International Law* 66 (1972) .378-395.

Oeter, Stefan. "Methods and Means of Combat" in: *The Handbook of Humanitarian Law in Armed Conflicts.* Dieter Fleck (ed.) 105-207 Oxford: Oxford University Press, 1995.

Otto, Roland. "Neighbours as Human Shields? The Israel Defense Forces' "Early Warning Procedure" and International Humanitarian Law" *International Review of the Red Cross* 86(856) (2004) 771-787.

Parks, Hays. "Memorandum of Law: Review of Weapons in the Advanced Combat Rifle Program" reprinted in: *The Army Lawyer*, DA PM 27-50-211 (July 1990) 16-22.

----------- "Air War and the Law of War", Air Force Law Review 32 (1990) 1-225.

----------. "The Protocol on Incendiary Weapons" *International Review of the Red Cross* 279 (1990) 535-550.

------------."Joint Service Combat Shotgun Program" *The Army Lawyer* DA-PAM 27-50-299. (Oct. 1997) 16-24.

----------. "Making Law Of War Treaties: Lessons From Submarine Warfare Regulation" in: *International Law across the Conflict Spectrum: Essays in Honor of Professor L.C. Green", M.N. Schmitt*, (ed.) 229-385 Newport RI: United States Naval War College 2001.

-----------. "Conventional Weapons and Weapons Reviews" *Yearbook of International Humanitarian Law 8* (2005) 55-142.

-----------. "The ICRC Customary Law Study: A Preliminary Assessment" *American Society of International Law Proceedings* 99 (2005) 208-212.

-----------. "Means and Methods of Warfare" *The George Washington International Law Review* 38 (2006) 511-539.

Pilford, Neil, "Controversy over the use of dum dum bullets in the First World War", *Imperial War Museum Review* 5 (1990) 98-109.

Price, Richard. "A Genealogy of the Chemical Weapons Taboo", *International Organization* 49(1) (Winter 1995), 73-103.

-------------& Tannenwald, Nina, "Norms and Deterrence: The Nuclear and Chemical Weapons Taboos" in: *The Culture of national Security: Norms and Identity in World Politics.* Peter J. Katzenstein, (ed.) 114-152 New York: Columbia University Press 1996.

-------------- "Reversing the Gun Sights: Transnational Civil Society Targets Land Mines" *International Organization* 52(3) (1998) 613-644.

Prokosch, Eric. "Anti-personnel Weapon", *International Social Science Journal*, 28(2), (1976) 341-358.

-----------. "The Swiss Draft Protocol on Small-Calibre Weapon Systems" *International Review of the Red Cross* 307 (1995) 411-425.

------------- & Hogendoorn, Ernst Jan. „Anti-personnel Weapons" in: *War or Health? A Reader.* Ilkka Taipale *et al.*,(eds) 341-358 New-York: Zed Books 2002.

Roach, Ashley J. "Certain Conventional Weapons Convention: Arms Control or Humanitarian Law?" *Military Law Review* 105 (Summer 1984) 2-72.

Robblee Paul A., Jr. "The Legitimacy of Modern Conventional Weaponry" *Military Law Review* 71 (Winter 1976) 95-148.

Roberts, Adam. "The Laws of War in the 1990-91 Gulf Conflict" *International Security* 18(3) (1993-1994) 134 -148.

Robinson, Julian Perry. "Chemical Weapons Proliferation in the Middle East" in: Non-Conventional Weapons Proliferation in the Middle East: Tackling The Spread Of Nuclear, Chemical, and Biological Capabilities. Efraim Karsh et al., (eds) 68-97 London: Clarendon Press, 1993.

Schmitt M. N. "The International Court of Justice and the Use of Nuclear Weapons", *U.S. Air Force Annual Journal of Legal Studies* 7 (1996-1997), 57- 95.

---------------. "Future War and the Principle of Discrimination" *Israel Yearbook on Human Rights* 28 (1999) 51-90.

------------, Charles H.B. Garraway, Yoram Dinstein (Drafting Committee) *The Manual on the Law of Non-International Armed Conflict* (with Commentary), International Institute of Humanitarian Law, San Remo, 2006.), Reprinted in Yoram Dinstein & Fania Domb (eds.) *Israel Yearbook of Human Rights* 36 (2006) Supp. 333, 30.

----------------. "Human Shields in International Humanitarian Law" *Columbia Journal of Transnational Law* 47 (2009) 292-328.

Sela, Avraham. "Civil Society, the Military, and National Security: The Case of Israel's Security Zone in Lebanon" *Israel Studies* 12(1) (Spring 2007) 53-78.

Shamgar, Meir. "The Observance of International Law in the Administered Territories" *Israel Yearbook on Human Rights* 1 (1971) 262-266.

Shany, Yuval. "Israeli Counter-Terrorism Measures: Are They "Kosher" under International Law?", in *Terrorism And International Law: Challenges And Responses.*

Michael N. Schmitt & Gian L. Beruto (eds.) 96- 113 Ardsley, N.Y.: Transnational Publishers Inc, 2003.

---------------- "Faraway, So Close: The Legal Status of Gaza after Israel's Disengagement" *Yearbook of International Humanitarian Law* 8 (2005) 369-383.

Stone, Andrew M. "Conventional Weapons, Their Deployment and Effects from a Humanitarian Aspect: Recommendations for the Modernization of International Law" *The American Journal of International Law* 68(1) (1974) 149-150 (book review).

Støre, J. "Cluster Munitions: Special Comment" *Disarmament Forum* 4 (2006) 3-4.

Szasz, Paul. "The Conference on Excessively Injurious or Indiscriminate Weapons" *The American Journal of International Law* 74(1) (1980) 212-215.

Tirosh, Yair and Matan Goldblatt, "Judicial Review Over the IDF Response to Palestinian Violence since September 2000", *IDF Law Review* 1 (2003) 311-340.

Turns, David. "Some Reflections on the Conflict in Southern Lebanon: The 'Qana Incident' and International Humanitarian Law" *Journal of Conflict and Security Law* 5(2) (2000) 177-209.

---------------. "Weapons in the ICRC Study on Customary International Humanitarian Law" *Journal of Conflict and Security Law* 11(2) (2006) 201-237.

Vagts, Detlev F. "The Hague Conventions and Arms Control" *The American Journal of International Law* 94(1) (2000) 31-41.

Verchio, Donna Marie. "Just Say No! The SIrUS Project: Well-Intentioned, But Unnecessary and Superfluous" *Air Force Law Review* 51 (2001) 183-227

Walker, Jeffery K. "Introductory Note, Supreme Court: Physicians for Human Rights et al. v. Almoj and the State of Israel" *International Legal Material* 42 (2003) 880-882.

Weiler, Joseph *et al.*, (Editorial), "Gaza – From Warfare to Lawfare", *The European Journal of International Law* 20(2) (2009), 259-262.

Wiebe, Virgil. "Footprints of Death: Cluster Bombs as Indiscriminate Weapons Under International Humanitarian Law" *Michigan Journal of International Law* 22 (2000) 85-167.

Williams, Jody and Goose, Stephen. "The International Campaign to Ban Landmines" in: *To Walk Without Fear: The Global Movement To Ban Landmines.* Maxwell A. Cameron *et al* (eds.) 20-47 Oxford: Oxford University Press 1999.

## On Line Essays

Cilliers, Jakkie "South Africa and the International Campaign to Ban Anti-Personnel Landmines", *African Security Review* 6(1), (1997), [on-line publication], available at: http://www.iss.co.za/pubs/ASR/6No1/Cilliers.html.

Coupland, Robin M. "The Effect of Weapons: Defining Superfluous Injury and Unnecessary Suffering" *Medicine and Global Survival* 3 (1996), [online publication as of Jan.1996], available at: http://www.ippnw.org/MGS/V3Coupland.html.

Dunlap, Charles, "Law and Military Interventions: Preserving Humanitarian Values in 21st Conflicts,", *Working Paper* prepared for the Humanitarian Challenges in Military Intervention Conference, Washington D.C., (Nov. 29, 2001), available at: http://www.duke.edu/~pfeaver/dunlap.pdf.

Kay, David, "The Goldstone Report" *American Society of International Law Insight* 13(16) (Oct. 2009) available at: http://www.asil.org/insights091001.cfm

Lubell Noam, "Who's the Target? The Blurring of Distinction Between Civilians and Combatants During IDF Ground Operations in Urban Environments", *Working Papers, Ground Operations Workshop* (Oct. 17-18, 2002), Carr Center for human rights, Harvard University, available at: http://www.hks.harvard.edu/cchrp/programareas/conferences/october2002.php

Pillai, Priya. "Adoption of the Convention on Cluster Munitions" *American Society of International Law Insight* 12(20) (Oct. 2008) available at: http://www.asil.org/insights081001.cfm.

Samson, Elizabeth, "Warfare through Misuse of International Law" *Perspectives* 73 Papers on Current Affairs (Mach 2009) (Bar Ilan Univ. BESA Center), available at: http://www.biu.ac.il/SOC/besa/perspectives73.html

Stone, Troy R., "War Is Too Important to be Left to the Lawyers", (Naval War College Newport RI Joint Military Operations Dept. Report) (Oct.2008). available at: http://www.stormingmedia.us/06/0634/A063494.html

Watters, Daniel. A 5.56 X 45mm Timeline – Small Arms History. available at: http://www.thegunzone.com/556dw.html

-----------------. Special Purpose Individual Weapons: A Brief History of Fléchette and the SPIW Project. available at: http://www.thegunzone.com/spiw.html.

Wotherspoon, Jason. Introduction & Short History of Shuriken. available at: http://www.secrets-of-shuriken.com.au/history.htm#5.

Flechettes Shotgun Ammunitions – Historical Background, Sabot Designs LLC. available at: http://www.anti-personnel.net/sdllc/001.html.

### *Foreign Language Journal Articles, Essays, and Book Chapters*

Sheleff, Leon "The Green Line is the Border of Judicial Activism: Queries about Supreme Court Judgments in the Territories", *Iyunei Mishpat* (Tel Aviv University Law Review, Hebrew) 17(3) (1993) 757-809.

Tonnele, Jean. "Une Escadrille D'hydravions a Port-Said (1914-1916)" [A Seaplane Squadron at Port Said, 1914-16] *Revue Maritime* 228 (1966) 6-15.

## Professional Magazines

*Air University Review* (1947-1987)

*American Rifleman* (1885-present)

*Armor* (U.S. ARMY, 1888-present)

*Army Research and Development News Magazine* (U.S. ARMY, 1960-1978)

*Aviation Week* (1916-present)

*Field Artillery Journal* (U.S. ARMY, 1911-2007)

*Flight* (1909-2005)

*Flight International* (2005-present)

*Jane's Defence Weekly* (1984-present)

*Jane's International Defense Review* (1968-present)

*Infantry Magazine* (U.S. ARMY, 1957-present)

*Marine Corps Gazette* (1916-present)

*Strategic Assessment* (1998-present)

## Reports & Releases

Amnesty International, NATO/Federal Republic of Yugoslavia – "collateral damage" or unlawful killings? Violations of the Laws of War by NATO during Operation Allied Force, EUR 70/018/2000 (6 June 2000).

--------- Call for Investigation into Killing of Cameraman and Other Civilians, (Press Release, Apr. 18, 2008).

--------- Army's So-Called Inquiry Into Cameraman's Killing in Gaza a Scandal, (Press Release, Aug 15, 2008).

--------- Israeli Army Used Flechettes Against Gaza Civilians, (Press Release, Jan. 27, 2009).

--------- MDE. Israel-OPT – Fuelling Conflict: Foreign Arms Supplies to Israel/Gaza, (Feb. 2009).

--------- Israel/Gaza: Operation "Cast Lead": 22 Days of Death and Destruction (July 2, 2009).

B'Tselem – The Israeli Information Center for Human Rights in the Occupied Territories, Israeli Violations of Human Rights of Lebanese Civilians, ( Jan. 2000)

----------- Firing of Flechette Shell that killed Reuters Cameraman Violates Laws of war (Press Release, Apr. 17, 2008)

Human Rights Watch – Civilian Pawns: Laws of War Violations and the Use of Weapons on the Israel-Lebanon Border, (May 1996)

-------- Don't Use Fuel-air Weapons in Chechnya (News Release, Feb. 14, 2000)

------- Israel: Dart Shells Pose Civilian Threat, (Press Release, June 15, 2001)

------- A Global Overview of Explosive Submunitions Prepared for the Convention on Conventional Weapons (CCW) Group of Governmental Experts on the Explosive Remnants of War (ERW) (Memo to CCW delegates May 21-24, 2002)

------- Israel: Stop Using Flechettes in Gaza, (Press Release, Apr. 29, 2003)

------- Israeli Cluster Munitions Hit Civilians in Lebanon, (Press Release, July 24, 2006)

---------- Flooding South Lebanon: Israel's Use of Cluster Munitions in Lebanon in July and August 2006 (Feb. 2008).

-------- Israel Investigates Death of Gaza Civilians, (Press Release, Apr. 18, 2008)

------- Israel: Investigate Death of Gaza Civilians, (Press Release, Apr. 20, 2008)

-------- Rain of Fire: Israel's Unlawful Use of White Phosphorus in Gaza (Mar. 25, 2009)

Landmine Action – Foreseeable Harm: the Use and Impact of Cluster Munitions in Lebanon, (Nash, Thomas, Oct. 2006)

------ Anti-Vehicle Mines: Understanding the Impact and Managing the Risk, A Resource for Security Management, (Crowther, Greg & Moyes, Richard, Nov. 2007)

------ Counting The Cost: The Economic Impact of Cluster Munition Contamination in Lebanon (Crowther, Greg, May 2008)

-------- A Convention beyond the Convention: Stigma, Humanitarian Standards and the Oslo Process (Rappert,Brian, May 2008)

NGO Monitor – "Lawfare": Exploitation of Courts in the Israeli-Arab Conflict (Anne Herzberg , NGO Monitor Monograph Series, Sept. 2008)

------------------ – HRW's "Rain of Fire": Neither Thorough nor Impartial, (Apr. 2, 2009)

PCHR ref: 45/2003, (press release, Apr. 15, 2003).

-------- ref: 52/ 2003, (press release, Apr. 27, 2003)

## Newspapers

### *English language Newspapers*

*The Advertiser*
*British Illustrated War News*
Daily Mail (London)
Daily Telegraph (London)
Financial Times
The Gazette (Quebec)
Guardian
Haaretz (eng. Edition)
Independent (London)
Jerusalem Post
Newsday
Newark Advocate (Newark, Ohio)
New York Times
Los Angeles Times
Salt Lake Tribune
Scotsman
Sunday Express
The Sydney Morning Herald
Times
Washington Post
On-Line News Sources:
*Agence France-Presse (AFP)*
*BBC World News*
*CBS News*
*News24 (South African News Agency)*
*Palestinian Report Online*
*Reuters*
*XINHUA News Agency (China)*

### *Foreign Language Newspapers and online News Sources*

*Haaretz*
Ma'ariv
Yedioth Ahronoth
YNET

# Index of Persons*



* footnotes excluded

# Index of Subjects

# International Humanitarian Law Series

1 Michael J. Kelly, *Restoring and Maintaining Order in Complex Peace Operations: The Search for a Legal Framework*, 1999 ISBN 90 411 1179 4

2 Helen Durham and Timothy L.H. McCormack (eds.), *The Changing Face of Conflict and the Efficacy of International Humanitarian Law*, 1999 ISBN 90 411 1180 8

3 Richard May, David Tolbert, John Hocking, Ken Roberts, Bing Bing Jia, Daryl Mundis and Gabriël Oosthuizen (eds.), *Essays on ICTY Procedure and Evidence in Honour of Gabrielle Kirk McDonald*, 2001 ISBN 90 411 1482 3

4 Elizabeth Chadwick, *Traditional Neutrality Revisited: Law, Theory and Case Studies*, 2002 ISBN 90 411 1787 3

5 Lal Chand Vohrah, Fausto Pocar, Yvonne Featherstone, Olivier Fourmy, Christine Graham, John Hocking and Nicholas Robson (eds.), *Man's Inhumanity to Man: Essays on International Law in Honour of Antonio Cassese*, 2003 ISBN 90 411 1986 8

6 Gideon Boas and William A. Schabas (eds.), *International Criminal Law Developments in the Case Law of the ICTY*, 2003 ISBN 90 411 1987 6

7* Karen Hulme, *War Torn Environment: Interpreting the Legal Threshold*, 2004 ISBN 90 04 13848 X

8 Helen Durham and Tracey Gurd (eds.), *Listening to the Silences: Women and War*, 2005 ISBN 90 04 14365 3

9* ** Marten Zwanenburg, *Accountability of Peace Support Operations*, 2005 ISBN 90 04 14350 5

10 Hirad Abtahi and Gideon Boas (eds.), *The Dynamics of International Criminal Law*, 2006 ISBN 90 04 14587 7

11 Frits Kalshoven, *Belligerent Reprisals*, 2005 ISBN 90 04 14386 6

12 Pablo Antonio Fernández-Sánchez (ed.), *The New Challenges of Humanitarian Law in Armed Conflicts: In Honour of Professor Juan Antonio Carrillo-Salcedo*, 2005 ISBN 90 04 14830 2

13 Ustinia Dolgopol and Judith Gardam (eds.), *The Challenge of Conflict: International Law Responds*, 2006 ISBN 90 04 14599 0

14* Laura Perna, *The Formation of the Treaty Law of Non-International Armed Conflicts*, 2006 ISBN 90 04 14924 4

15 Michael Schmitt and Jelena Pejic (eds.), *International Law and Armed Conflict: Exploring the Faultlines, Essays in Honour of Yoram Dinstein*, 2007 ISBN 978 9004154 28 5

16 Ola Engdahl, *Protection of Personnel in Peace Operations: The Role of the 'Safety Convention' against the Background of General International Law*, 2007 ISBN 978 9004154 66 7

17 Frits Kalshoven, *Reflections on the Law of War: Collected Essays*, 2007 ISBN 978 90 04 15825 2

18 Héctor Olásolo, *Unlawful Attacks in Combat Situations: From the ICTY's Case Law to the Rome Statute*, 2007 ISBN 978 90 04 15466 7

19 José Doria, Hans-Peter Gasser and M. Cherif Bassiouni (eds.), *The Legal Regime of the International Criminal Court: Essays in Honour of Professor Igor Blishchenko*, 2008 ISBN 978 90 04 16308 9

20 David A. Blumenthal and Timothy L.H. McCormack (eds.), *The Legacy of Nuremberg: Civilising Influence or Institutionalised Vengeance?*, 2008 ISBN 978 90 04 15691 3

21 Omar Abdulle Alasow, *Violations of the Rules Applicable in Non-International Armed Conflicts and Their Possible Causes: The Case of Somalia*, 2008 ISBN 978 90 04 16475 8

---

* Winner of the ASIL Francis Lieber Prize.

** Winner of the 2006 Paul Reuter Prize.

22 Ola Engdahl and Pål Wrange (eds.), *Law at War: The Law as it Was and the Law as it Should Be – Liber Amicorum Ove Bring*, 2008 ISBN 978 90 04 17016 2

23 Pablo Antonio Fernández-Sánchez (ed.), *International Legal Dimension of Terrorism*, 2009 ISBN 978 90 04 17053 7

24 Rob McLaughlin, *United Nations Naval Peace Operations in the Territorial Sea*, 2009 ISBN 978 90 04 17479 5

25 Ian Henderson, *The Contemporary Law of Targeting*, 2009 ISBN 978 90 04 17480 1

26 Cecilia M. Bailliet (ed.), *Security: A Multidisciplinary Normative Approach*, 2009 ISBN 978 90 04 17296 8

27 Kristen Eichensehr and W. Michael Reisman (eds.), *Stopping Wars and Making Peace: Studies in International Intervention*, 2009 ISBN 978 90 04 17855 7

28 Noelle Higgins, *Regulating the Use of Force in Wars of National Liberation – The Need for a New Regime: A Study of the South Moluccas and Aceh*, 2009 ISBN 978 90 04 17287 6

29 Nout van Woudenberg and Liesbeth Lijnzaad (eds.), *Protecting Cultural Property in Armed Conflict: An Aside into the 1999 Second Protocol to the Hague Convention of 1954 for the Protection of Cultural Property in the Event of Armed Conflict*, 2010 ISBN 978 9004 18377 3

30 Yuki Tanaka, Tim McCormack and Gerry Simpson (eds.), *Beyond Victor's Justice?: The Tokyo War Crimes Trial Revisited*, 2011 ISBN 978 9004 20303 7

31 Marco Odello and Ryszard Piotrowicz (eds.), *International Military Missions and International Law*, 2011 ISBN 978 9004 17431 5

32 Eitan Barak, *Deadly Metal Rain: The Legality of Flechette Weapons in International Law: A Reappraisal Following Israel's Use in the Gaza Strip (2001–2009)*, 2011 ISBN 978 9004 16719 3

33 Arist von Hehn, *The Internal Implementation of Peace Agreements after Violent Intrastate Conflict: Guidance for Internal Actors Responsible for Implementation*, 2011 ISBN 978 9004 19587 5

34 Philipp Kastner, *International Criminal Justice in bello*, 2011 ISBN 978 9004 19312 3

35 Mary Ellen O'Connell, *What is War?*, 2011 ISBN 978 9004 17234 0

www.ingramcontent.com/pod-product-compliance
Lightning Source LLC
LaVergne TN
LVHW081515300926
849161LV00006B/1182
*9789004167193*